STUDY GUID

STANLEY C. HATFIELD
KENNETH G. PINZKE
Southwestern Illinois College

TARBUCK

LUTGENS

TASA

EARTH SCIENCE

TWELFTH EDITION

Upper Saddle River, NJ 07458

Editor-in-Chief, Science: Nicole Folchetti
Publisher, Geosciences: Dan Kaveney
Acquisitions Editor: Drusilla Peters
Assistant Editor: Sean Hale
Associate Managing Editor, Science: Gina M. Cheselka
Project Manager, Science: Wendy Perez
Supplement Cover Manager: Paul Gourhan
Supplement Cover Designer: Victoria Colotta
Senior Operations Supervisor: Alan Fischer
Operations Specialist: Amanda A. Smith
Marketing Manager: Amy Porubsky
Photo Credit: Climber repelling from a winter camp below Basin Mountain, Eastern Sierra, California (Galen Rowell/Mountain Light)

Pearson Prentice Hall
Pearson Education, Inc.
Upper Saddle River, NJ 07458

Printed in the United States of America

10 9 8 7 6 5 4 3 2 1

ISBN-13: 978-0-13-604920-3

ISBN-10: 0-13-604920-6

Pearson Education Ltd., *London*
Pearson Education Australia Pty. Ltd., *Sydney*
Pearson Education Singapore, Pte. Ltd.
Pearson Education North Asia Ltd., *Hong Kong*
Pearson Education Canada, Inc., *Toronto*
Pearson Educación de Mexico, S.A. de C.V.
Pearson Education—Japan, *Tokyo*
Pearson Education Malaysia, Pte. Ltd.

CONTENTS

PREFACE

Keys to Successful Learning

Learning involves much more than reading a textbook and attending classes. It is a process that requires continuous preparation, practice, and review in a dedicated attempt to understand a topic through study, instruction, and experience. Furthermore, most learning experts agree that successful learning is based upon association, repetition, and visualization.

Students often ask questions such as "What is the best way to learn Earth science?" or "How should I study?" Although there are no set answers to these questions, many successful students use similar study techniques and utilize many of the same study habits. For you to become a successful learner, you need to develop certain skills that include, among others, the following:

1. **Carefully observing the world around you.** Pay special attention to the relations between natural phenomena. For example, when studying weather, you might observe and note the direction that storms move in your area, or the relation between air temperature and wind direction. Or when studying minerals and rocks, you may want to list, locate, and identify those that are found in your area.

2. **Being curious and inquisitive.** Ask questions of your instructor, your classmates, and yourself, especially about concepts or ideas you do not fully understand.

3. **Developing critical thinking skills.** Carefully analyze all available information before forming a conclusion, and be prepared to modify your conclusions when necessary on the basis of new evidence.

4. **Visualizing.** Take time to create mental images of what you are studying, especially those concepts that involve change through time. If you can "picture" it, you probably understand it.

5. **Quantifying.** Accurate measurement is the foundation of scientific inquiry. To fully comprehend scientific literature, you must be familiar with the common systems and units of measurement, in particular the metric system. If you are not already comfortable with the metric system, take time to study and review it. Practice "thinking metric" by estimating everyday measurements such as the distance to the store or the volume of water in a can or pail using metric units.

6. **Being persistent.** Be prepared to repeat experiments, observations, and questions until you understand and are comfortable with the concepts in your own mind.

Remember that an important component of learning is identifying what you *should* know and understand and what you *do* know and understand about the material presented in the textbook and during class. With this in mind, the following steps may help you become an active, successful learner.

1. **Prepare.** Quickly scan the pages of the textbook chapter and accompanying section of this study guide. Try to visualize the organization of the chapter in simple, broad concepts. Read the chapter overview and learning objectives provided for each chapter in the Study Guide. These will help you focus on the most important concepts the authors present in the chapter.

2. **Read the assignment.** Read each assignment before attending class. Taking written notes as you read can be an effective means of helping you remember major points presented in the chapter. As you read the assignment, you may also find the following suggestions helpful.

- To help you visualize concepts, look at all the illustrations carefully and read the captions. Also try to develop mental images of the material you are reading.
- Pace yourself as you read. Often a single chapter contains more subject matter than can be assimilated in one long reading. Take frequent short breaks to help keep your mind fresh.
- If you find a section confusing, stop and read the passage again. Make sure you understand the material before proceeding.
- As you read, write down questions you may want to ask your instructor during class time.

3. **Attend and participate.** To get the most from any subject, you should attend all lecture classes, study sessions, review sessions, and (if required) laboratory sessions.

   - Bring any questions concerning the subject matter to these sessions.
   - Communicate with your instructor and the other students in the class.
   - Learn how to become a good listener and how to take useful class notes. Remember that taking a lot of notes may not be as effective as taking the right notes.
   - Come fully prepared for each session.

4. **Review.** Review your class notes and the chapter material by examining the section headings, illustrations, and chapter notes. Furthermore,

   - Identify the key facts, concepts, and principles in the chapter by highlighting or underlining them.
   - Examine the chapter summary statements presented in the Study Guide.
   - Write out the answers to the review questions presented at the end of the textbook chapter.

5. **Study.** Studying is one of the most important keys to successful learning. Very few individuals have the ability to retain vast amounts of detailed information after only one reading. In fact, most of us require seeing and thinking about a concept many times, perhaps over several days, before we fully comprehend it. Consider the following as you study and prepare for exams.

   - As you study, keep "putting the pieces together" by integrating the key concepts and facts as you go along.
   - As you review your textbook notes, lecture notes, and chapter highlights, keep in mind that most introductory science courses emphasize terms and definitions because scientists must have consistent, precise meanings for purposes of communication. To quiz your knowledge of each chapter's key terms, complete the vocabulary review section of the appropriate chapter in the Study Guide. Review your answers often prior to an examination.
   - Complete the comprehensive review section of the appropriate chapter in the Study Guide by writing out complete answers to the questions. By actually writing the answers, you will be able to retain the material with greater detail than simply answering the questions in your mind or copying the answers from the answer key.
   - Several days before an exam, test your overall retention of the chapter material by taking the practice test for the appropriate chapter(s) in the Study Guide. Thoroughly research and review any of the questions you miss.

Using the Study Guide Effectively

This Study Guide has been written to accompany the textbook *Earth Science*, Eleventh Edition, by Edward J. Tarbuck and Frederick K. Lutgens. For each chapter of the textbook there is a corresponding Study Guide chapter. For example, Chapter 1 of the Study Guide corresponds to Chapter 1 of the textbook. Each chapter in the Study Guide contains eight parts. They are

1. **Chapter Overview.** The overview provides a brief summary of the topics presented in the chapter. The overview should be read before reading the textbook chapter.

2. **Learning Objectives.** The learning objectives are a list of what you should know after you have read and studied the textbook chapter and completed the corresponding chapter in the Study Guide. You should examine the objectives before and after you read and study the chapter. Included with the objectives are brief summaries of the most important concepts presented in the chapter. The statements are useful for a quick review of the major chapter objectives as well as preparing for examinations.

3. **Key Terms.** The key terms list includes the new vocabulary terms introduced in the textbook chapter.

4. **Vocabulary Review.** This section consists of fill-in-the-blank statements that are intended to test your knowledge of the new terms presented in the textbook chapter. You should make sure you have a good understanding of the terms before completing the remaining sections of the Study Guide chapter. Answers to the vocabulary review are listed in the answer keys located at the back of the Study Guide.

5. **Comprehensive Review.** The comprehensive review contains several questions that test your knowledge of the basic facts and concepts presented in the chapter. In this section you will be asked to write out answers, label diagrams, and list key facts. Answers to the comprehensive review questions are listed in the answer keys located at the back of the Study Guide.

6. **Practice Test.** This section of each Study Guide chapter will test your total knowledge and understanding of the textbook chapter. Each practice test consists of multiple choice, true/false, word choice, and written questions. Taking the practice test is one way for you to determine how ready you are for an actual exam. You may find it helpful to write your answers to the practice test questions on a separate sheet of paper and leave the test in the Study Guide unmarked so you can review it several times while preparing for a unit exam, midterm exam, and/or final exam. Answers to the practice test questions are listed in the answer keys located at the back of the Study Guide.

7. **Answer Key.** Chapter answer keys are located at the back of the Study Guide.

To effectively use the Study Guide, it is recommended that you

- Quickly glance over the appropriate chapter in the Study Guide before reading the corresponding textbook chapter.

- When required, write out complete answers to the Study Guide questions.

- Check your answers with those provided at the back of the Study Guide.

- Research the answers you don't know by looking them up in your textbook and/or class notes. You will learn very little by simply copying the answers from the answer keys.

- Test yourself by taking the practice test provided at the end of each Study Guide chapter.
- Constantly review your answers to all the Study Guide questions.

Earth Science on the World Wide Web

Designed to give you an interactive means to further your studies and effectively utilize the Internet, an *Earth Science* companion website has been prepared by the authors. In addition to providing direct links to numerous websites that present interesting and relevant information that supplements and reinforces the topics presented in each chapter, the *Earth Science* site is one of the most outstanding educational Internet locations available today. Check out these interactive resources and interesting links for further thought via the Internet at **http://www.prenhall.com/tarbuck**.

ACKNOWLEDGMENTS

Our sincere thanks to each of the many individuals who assisted in the preparation of this edition of the *Study Guide* for *Earth Science*. Furthermore, we would like to acknowledge all of our students, past and present, whose comments and questions have helped us focus on the nature, meaning, and essentials of Earth science. A special debt of gratitude goes to Sean Hale, Geoscience Assistant Editor, and his Prentice Hall colleagues, who skillfully guided this manuscript from its inception to completion. They are true professionals with whom we feel fortunate to be associated.

Thanks,
Ken Pinzke
Stan Hatfield

I hear and I forget.

I see and I remember.

I do and I understand.

—Ancient Proverb

1

Introduction to Earth Science

Introduction to Earth Science sets the stage for the text by introducing many of the major themes and underlying principles of the Earth sciences. Following a brief discussion of the scope of Earth science and how it relates to the environment and resources, the nature of scientific inquiry is examined. To help provide a foundation for what will follow and to place the Earth in perspective, the current theories for the origin of the universe and solar system are also presented along with the general structure of the planet's surface and interior. Although discussed in greater detail in following chapters, an introduction to the theory of plate tectonics provides the impetus for viewing Earth as a system, a recurring theme throughout the remaining chapters.

Learning Objectives

After reading, studying, and discussing this chapter, you should be able to:

List the sciences traditionally included in Earth science.

Earth science is the name for all the sciences that collectively seek to understand Earth and its neighbors in space. It includes *geology, oceanography, meteorology*, and *astronomy*. Geology is traditionally divided into two broad areas—physical and historical.

Explain what is meant by the term "environment."

Environment refers to everything that surrounds and influences an organism. These influences can be biological, social, or physical. When applied to Earth science today, the term *environmental* is usually reserved for those aspects that focus on the relationships between people and the natural environment.

List the two types of resources.

Resources are an important environmental concern. The two broad categories of resources are (1) *renewable*, which means that they can be replenished over relatively short time spans, and (2) *nonrenewable*. As population grows, the demand for resources expands as well.

Describe some of the environmental problems faced by humans.

Environmental problems can be local, regional, or global. Human-induced problems include urban air pollution, acid rain, ozone depletion, and global warming. Natural hazards imposed by the physical environment include earthquakes, landslides, floods, and hurricanes. As population grows, pressures on the environment also increase.

Discuss the nature of scientific inquiry.

All science is based on the assumption that the natural world behaves in a consistent and predictable manner. The process through which scientists gather facts through observations and formulate scientific *hypotheses* and *theories* is called the *scientific method*. To determine what is occurring in the natural world, scientists often (1) collect facts, (2) develop a scientific hypothesis, (3) construct experiments to validate the hypothesis, and (4) accept, modify, or reject the hypothesis on the basis of extensive testing. Other discoveries represent purely theoretical ideas that have stood up to extensive examination. Still other scientific advancements have been made when a totally unexpected happening occurred during an experiment.

Describe the geologic time scale.

One of the challenges for those who study planet Earth is the great variety of space and time scales. The *geologic time scale* subdivides the 4.5 billion years of Earth history into various units.

Discuss the formation of the solar system.

The *nebular hypothesis* describes the formation of the solar system. The planets and Sun began forming about 5 billion years ago from a large cloud of dust and gases. As the cloud contracted, it began to rotate and assume a disk shape. Material that was gravitationally pulled toward the center became the *protosun*. Within the rotating disk, small centers, called *planetesimals*, swept up more and more of the cloud's debris. Because of their high temperatures and weak gravitational fields, the inner planets were unable to accumulate and retain many of the lighter components. Because of the very cold temperatures existing far from the Sun, the large outer planets consist of huge amounts of lighter materials. These gaseous substances account for the comparatively large sizes and low densities of the outer planets.

List and describe each of Earth's four "spheres."

Earth's physical environment is traditionally divided into three major parts: the solid Earth, or *geosphere*; the water portion of our planet, the *hydrosphere*; and Earth's gaseous envelope, the *atmosphere*. In addition, the *biosphere*, the totality of life on Earth, interacts with each of the three physical realms and is an equally integral part of Earth.

Describe Earth's internal structure.

Earth's internal structure is divided into layers based on differences in chemical composition and on the basis of changes in physical properties. Compositionally, Earth is divided into a thin outer *crust*, a solid rocky *mantle*, and a dense *core*. Based on physical properties, other layers of Earth include (1) the *lithosphere*—the cool, rigid outermost layer that averages about 100 kilometers thick, (2) the *asthenosphere*, a relatively weak layer located in the mantle beneath the lithosphere, (3) the more rigid *lower mantle*, where rocks are very hot and capable of very gradual flow, (4) the liquid *outer core*, where Earth's magnetic field is generated, and (5) the solid *inner core*.

Compare the two principal divisions of Earth's surface.

Two principal divisions of Earth's surface are the *continents* and *ocean basins*. A significant difference is their relative levels. The elevation differences between continents and ocean basins are primarily the result of differences in their respective densities and thicknesses.

List the subdivisions of continents and the ocean floor.

The largest features of the continents can be divided into two categories: *mountain belts* and the *stable interior*. The ocean floor is divided into three major topographic units: *continental margins, deep-ocean basins*, and *oceanic (mid-ocean) ridges*.

Discuss the Earth system and Earth system science.

Although each of Earth's four spheres can be studied separately, they are all related in a complex and continuously interacting whole that we call the *Earth system*. *Earth system science* uses an interdisciplinary approach to integrate the knowledge of several academic fields in the study of our planet and its global environmental problems.

Describe the differences between closed and open systems.

A *system* is a group of interacting parts that form a complex whole. *Closed systems* are those in which energy moves freely in and out but matter does not enter or leave the system. In an *open system*, both energy and matter flow into and out of the system.

List the sources of energy that power the Earth system.

The *two sources of energy that power the Earth system are* (1) *the Sun*, which drives the external processes that occur in the atmosphere, hydrosphere, and at Earth's surface; and (2) *heat from Earth's interior*, which powers the internal processes that produce volcanoes, earthquakes, and mountains.

Key Terms

abyssal plain	geologic time scale	oceanic (mid-ocean) ridge
asthenosphere	geology	oceanography
astronomy	geosphere	open system
atmosphere	hydrosphere	outer core
biosphere	hypothesis	paradigm
closed system	inner core	physical environment
continental margin	interface	plate tectonics
continental shelf	lithosphere	positive feedback mechanism
continental slope	lithospheric plate	renewable resource
core	lower mantle	seamount
crust	mantle	shield
deep-ocean basin	meteorology	stable platform
deep-ocean trench	model	system
Earth science	nebular hypothesis	theory
Earth system science	negative feedback mechanism	
environment	nonrenewable resource	

Vocabulary Review

Choosing from the list of key terms, furnish the most appropriate response for the following statements.

1. The solid Earth is divided into three principal units: the very dense ___________; the less dense ___________; and the ___________.
2. In scientific inquiry, a preliminary untested explanation is a scientific ___________.
3. The ___________ is a relatively steep drop-off that extends from the outer edge of the continental shelf to the floor of the deep ocean.
4. The science of ___________ is traditionally divided into two parts—physical and historical.
5. ___________ is the study of the atmosphere and the processes that produce weather and climate.
6. The ___________ subdivides Earth history into units and provides a meaningful time frame for the events of the geologic past.
7. A scientific ___________ is a well-tested and widely accepted view that scientists agree best explains certain observable facts.
8. The largest of Earth's four spheres is the ___________.
9. The dynamic water portion of Earth's physical environment is the ___________.
10. A ___________ is a relatively narrow, extremely deep depression in the ocean floor.
11. The science that involves the application of all sciences in a comprehensive and interrelated study of the oceans in all their aspects and relationships is called ___________.

12. A(n) ___________ is a resource that can be replenished over a relatively short time span.
13. The ___________ is a liquid zone in Earth's interior.
14. The "sphere" of Earth, called the ___________, includes all life.
15. That portion of the seafloor adjacent to a major landmass is referred to as the ___________.
16. A resource, such as oil, that cannot be replenished is classified as a(n) ___________.
17. ___________ is the study of the universe.
18. A ___________ is an expansive, flat region composed of deformed crystalline rock located in the stable interior of a continent.
19. The life-giving gaseous envelope surrounding Earth is the ___________.
20. The most prominent feature on the ocean floor is the ___________.
21. The ___________ suggests that the bodies of our solar system evolved from an enormous rotating cloud.
22. The ___________ refers to the rigid, outer layer of Earth, which includes the crust and part of the upper mantle.
23. Beneath Earth's lithosphere lies the soft and comparatively weak layer known as the ___________.
24. Loosely defined, a ___________ can be any size group of interacting parts that form a complex whole.
25. An incredibly flat feature, called an ___________, is located in the deep-ocean basin.
26. The term ___________ refers to anything that surrounds and influences an organism.
27. Mechanisms that enhance or drive change in a system are called ___________.
28. The theory, called ___________, provides geologists with a comprehensive model of Earth's internal workings.
29. A(n) ___________ is an extensively documented theory that is held with a very high degree of confidence.

Comprehensive Review

1. What are the sciences that are traditionally included in Earth science?

2. Describe the difference between renewable and nonrenewable resources. Give an example of each.

3. List and briefly describe Earth's four "spheres."

    1)

    2)

    3)

    4)

4. How would you explain to a friend what the science of Earth science involves?

5. What is the difference between a scientific hypothesis and a scientific theory?

6. Beginning with the Big Bang and ending with the formation of the solar system, summarize the sequence of major events that scientists believe have taken place in the universe.

7. What are the four steps that are often used in science to gain knowledge?

    1)

    2)

    3)

    4)

8. Label each of the four units of the solid Earth at the appropriate letter in Figure 1.1.

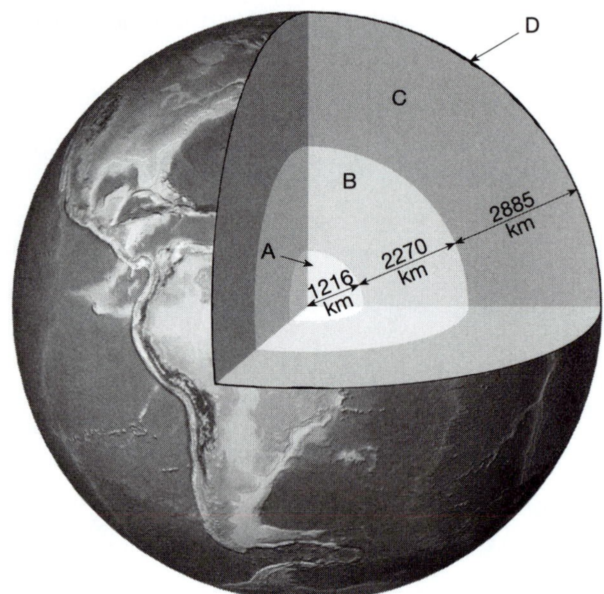

Figure 1.1

9. Describe the following two broad areas of geology:

   a) Physical geology:

   b) Historical geology:

10. Briefly describe how Earth acts as a system.

11. List and describe the five layers of Earth based on physical properties.

12. Write a brief paragraph describing the theory of plate tectonics. What drives plate motion?

13. List and describe the major features of the deep-ocean basin.

14. Briefly describe the most widely accepted view on the origin of our solar system.

Practice Test

Multiple choice. Choose the best answer for the following multiple-choice questions.

1. The _____ strongly influences the other three "spheres" because without life their makeup and nature would be much different.
   a) atmosphere
   b) hydrosphere
   c) geosphere
   d) biosphere

2. Earth science attempts to relate our planet to the larger universe by investigating the science of _____.
   a) geology
   b) oceanography
   c) meteorology
   d) astronomy
   e) biology

3. The science that studies the processes that produce weather and climate is _____.
   a) geology
   b) oceanography
   c) meteorology
   d) astronomy
   e) biology

4. The global ocean covers _____ percent of Earth's surface.
   a) 71
   b) 38
   c) 42
   d) 65
   e) 83

5. A scientific _____ is a preliminary untested explanation that tries to explain how or why things happen in the manner observed.
   a) estimate
   b) law
   c) fact
   d) hypothesis
   e) idea

6. The Earth "sphere," called the _____, includes the freshwater found in streams, lakes, glaciers, as well as that existing underground.
   a) atmosphere
   b) hydrosphere
   c) biosphere
   d) geosphere

7. The science that includes the study of the composition and movements of seawater, as well as coastal processes, seafloor topography, and marine life is _____.
   a) geology
   b) oceanography
   c) meteorology
   d) astronomy
   e) biology

8. Which one of the following is NOT an example of a nonrenewable resource?
   a) solar energy
   b) natural gas
   c) copper
   d) coal
   e) aluminum

9. An understanding of Earth is essential for _____.
   a) location and recovery of basic resources
   b) minimizing the effects of natural hazards
   c) dealing with the human impact on the environment
   d) all of the above

10. This science is divided into two broad divisions—physical and historical.
    a) geology
    b) oceanography
    c) meteorology
    d) astronomy
    e) biology

11. The annual per capita consumption of metallic and nonmetallic mineral resources for the United States is nearly _____ tons.
    a) 2
    b) 6
    c) 8
    d) 11
    e) 15

12. The conditions that surround and influence an organism are referred to as the _____.
    a) resource
    b) social sphere
    c) technosphere
    d) environment

13. Which one of the following is NOT an example of a renewable resource?
    a) water energy
    b) lumber
    c) iron
    d) cotton
    e) chickens

14. In scientific inquiry, when competing hypotheses have been eliminated, a hypothesis may be elevated to the status of a scientific _____.
    a) estimate
    b) theory
    c) idea
    d) understanding
    e) truth

15. The science whose name literally means "study of Earth" is _____.
    a) geology
    b) oceanography
    c) meteorology
    d) astronomy
    e) biology

16. A(n) _____ is a scientific theory that is held with a very high degree of confidence.
    a) explanation
    b) hypothesis
    c) idea
    d) understanding
    e) paradigm

17. Scientific advancements can be made _____.
    a) by applying systematic steps
    b) by accepting purely theoretical ideas that stand up to extensive investigation
    c) from unexpected happenings during an experiment
    d) all of the above

18. Earth is approximately _____ billion years old.
    a) 2.5
    b) 3.0
    c) 4.5
    d) 6.2
    e) 9.8

19. Most natural systems are _____ systems where both energy and matter flow in and out.
    a) closed
    b) inverted
    c) hidden
    d) open
    e) none of the above

20. The _____ subdivides the history of Earth into many different units and provides a meaningful time frame within which the events of the geologic past are arranged.
    a) nebular hypothesis
    b) geologic time scale
    c) light year
    d) hydrologic cycle
    e) none of the above

21. Which one of the following is NOT a layer of Earth's interior as defined by physical properties?
    a) inner core
    b) asthenosphere
    c) troposphere
    d) lithosphere
    e) lower mantle

22. According to the theory of plate tectonics, plate motion is ultimately driven by Earth's _____.
    a) internal heat
    b) rotation
    c) magnetism
    d) ocean currents
    e) none of the above

23. Which one of the following is NOT a feature of ocean basins?
    a) abyssal plain
    b) stable platform
    c) seamount
    d) continental slope
    e) mid-ocean ridge

True/false. For the following true/false questions, if a statement is not completely true, mark it false. For each false statement, change the **italicized** *word to correct the statement.*

1. _____ The life-giving gaseous envelope surrounding Earth is called the *atmosphere.*
2. _____ The *biosphere* is a dynamic mass of water that is continually on the move from the oceans to the atmosphere, precipitating back to the land, and running back to the ocean.
3. _____ Although some *renewable* resources such as aluminum can be used over and over again, others, such as oil, cannot be recycled.
4. _____ The aim of *physical* geology is to understand the origin of Earth and the development of the planet since its formation.
5. _____ The *oceanic* crust is composed of the igneous rock basalt.
6. _____ Science is based on the assumption that the natural world behaves in a(n) *unpredictable* manner.
7. _____ The most obvious difference between the continents and the ocean basins is their relative *levels.*
8. _____ *Divergent* lithospheric plate boundaries are found along oceanic ridges.
9. _____ *Biosphere* refers to everything that surrounds and influences an organism.
10. _____ To determine what is happening in the natural world, scientists collect *facts* through observation and measurement.
11. _____ The most prominent feature on the ocean floor is the *shield.*
12. _____ The light and very thin outer skin of Earth is called the *mantle.*
13. _____ Beneath Earth's lithosphere lies a soft, weak layer known as the *asthenosphere.*
14. _____ By about the year 2015 more than *seven* billion people may inhabit the planet.
15. _____ All life on Earth is included in the *biosphere.*
16. _____ Earth's *inner* core is liquid.
17. _____ A scientific *theory* is a well-tested and widely accepted view that scientists agree best explains certain observable facts.

Word choice. Complete each of the following statements by selecting the most appropriate response.

1. Earth is a [static/dynamic] body with many interacting parts and a long and complex history.
2. The universe began about [14/30] billion years ago.
3. More than 82 percent of Earth's volume is contained in the [mantle/core].
4. The [solar/nebular] hypothesis suggests that the bodies of our solar system evolved from an enormous rotating cloud.
5. In the United States, about [6/20] percent of the world's population uses [10/30] percent of the annual production of mineral and energy resources.
6. The boundary between the continents and the deep-ocean basins lies along the continental [shelf/slope].
7. Earth's crust is [thinnest/thickest] beneath the oceans.
8. The removal of carbon dioxide from the atmosphere has a [positive/negative] impact on global warming.
9. A [hypothesis/paradigm] is held with a very high degree of confidence.

Written questions

1. List the four steps that scientists often use to conduct experiments and gain scientific knowledge.

2. List and briefly describe Earth's four "spheres."

3. Define Earth science. List the four areas commonly included in Earth science.

4. Describe what is meant by the *Earth system*.

5. Distinguish between positive and negative feedback mechanisms.

6. Describe the theory of plate tectonics.

For other interesting and pertinent
information, be sure to visit
the *Earth Science* companion website at

http://www.prenhall.com/tarbuck

MINERALS:

Building Blocks of Rocks

2

Any discussion of Earth materials begins by examining the basic particles of matter, atoms, and how they combine to form minerals. Furthermore, a fundamental knowledge of elements, atoms, compounds, and atomic bonding aids in the understanding of the properties of minerals and how these properties are used for identification. To facilitate their study and understanding, the nearly 4000 minerals have been divided into two groups—silicates and nonsilicates. Each group is characterized by the elements that comprise it and their unique crystalline structure. Silicate minerals, the most abundant of the two mineral groups, commonly combine to form the majority of rocks. Nonsilicate mineral classes, on the other hand, contain members that are prized for their economic value and constitute many mineral resources.

Learning Objectives

After reading, studying, and discussing this chapter, you should be able to:

Explain the difference between a mineral and a rock.

A *mineral* is a naturally occurring inorganic solid that possesses a definite chemical structure, which gives it a unique set of physical properties. Most *rocks* are aggregates composed of two or more minerals.

Describe the basic structure of an atom and explain how atoms combine.

The building blocks of minerals are *elements*. An *atom* is the smallest particle of matter that still retains the characteristics of an element. Each atom has a *nucleus*, which contains *protons* and *neutrons*. Orbiting the nucleus of an atom are *electrons*. The number of protons in an atom's nucleus determines its *atomic number* and the name of the element. Atoms bond together to form a *compound* by either gaining, losing, or sharing electrons with another atom.

Explain isotopes and radioactivity.

Isotopes are variants of the same element but with a different *mass number* (the total number of neutrons plus protons found in an atom's nucleus). Some isotopes are unstable and disintegrate naturally through a process called *radioactive decay*.

Describe the physical properties of minerals and how they can be used for mineral identification.

The properties of minerals include *crystal shape (habit), luster, color, streak, hardness, cleavage, fracture*, and *density* or *specific gravity*. In addition, a number of special physical and chemical properties (*taste, smell, elasticity, feel, magnetism, double refraction*, and *chemical reaction to hydrochloric acid*) are useful in identifying certain minerals. Each mineral has a unique set of properties that can be used for identification.

List the most important elements that compose Earth's continental crust.

The eight most abundant elements found in Earth's continental crust (oxygen, silicon, aluminum, iron, calcium, sodium, potassium, and magnesium) also compose the majority of minerals.

Describe the basic compositions and structures of the silicate minerals.

The most common mineral group is the *silicates.* All silicate minerals have the *silicon-oxygen tetrahedron* as their fundamental building block. In some silicate minerals the tetrahedra are joined in chains; in others, the tetrahedra are arranged into sheets, or three-dimensional networks. Each silicate mineral has a structure and a chemical composition that indicates the conditions under which it was formed.

List the economic use of some nonsilicate minerals.

The *nonsilicate* mineral groups include the *oxides* (e.g., magnetite, mined for iron), *sulfides* (e.g., sphalerite, mined for zinc), *sulfates* (e.g., gypsum, used in plaster and frequently found in sedimentary rocks), *native elements* (e.g., graphite, a dry lubricant), *halides* (e.g., halite, common salt and frequently found in sedimentary rocks), and *carbonates* (e.g., calcite, used in portland cement and a major constituent in two well-known rocks: limestone and marble).

Explain what an ore is and give examples of several ore minerals.

The term *ore* is used to denote useful metallic minerals, like hematite (mined for iron) and galena (mined for lead), that can be mined for a profit, as well as some nonmetallic minerals, such as fluorite and sulfur, that contain useful substances.

Key Terms

atom	habit	periodic table
atomic number	hardness	principal shell
chemical compound	ionic bond	proton
carbonates	ions	quartz
cleavage	isotope	radioactive decay
color	luster	reserve
compound	mass number	rock
covalent bond	mineral	rock-forming minerals
density	mineral resource	silicate
electron	mineralogy	silicon-oxygen tetrahedron
element	Mohs hardness scale	specific gravity
energy levels (shells)	neutron	streak
feldspar	nucleus	tenacity
fracture	ore	valence electrons

Vocabulary Review

Choosing from the list of key terms, furnish the most appropriate response for the following statements.

1. A(n) __________ is an electrically neutral subatomic particle found in the nucleus of an atom.
2. The smallest part of an element that still retains the element's properties is a(n) __________.
3. A(n) __________ is a useful metallic mineral that can be mined at a profit.
4. A naturally occurring, inorganic solid that possesses an orderly crystalline structure and a well-defined chemical composition is a(n) __________.
5. A(n) __________ is an atom that has an electric charge because of a gain or loss of electrons.
6. The tendency of a mineral to break along planes of weak bonding is the property called __________.
7. The subatomic particle that contributes mass and a positive electrical charge to an atom is a(n) __________.
8. A(n) __________ is composed of two or more elements bonded together in definite proportions.

9. Silicon and oxygen combine to form the framework of the most common mineral group, the __________ minerals.
10. An already identified deposit from which minerals can be extracted profitably is called a(n) __________.
11. A(n) __________ is a subatomic particle with a negative electrical charge.
12. A(n) __________ can be defined simply as an aggregate of minerals.
13. Each atom has a central region called the __________.
14. The number of protons in an atom's nucleus determines its __________ and the name of the element.
15. A(n) __________ is a large collection of electrically neutral atoms, all having the same atomic number.
16. The sum of the neutrons and protons in the nucleus is the atom's __________.
17. Electrons are located at a given distance from an atom's nucleus in a region called the __________.
18. When atoms combine by the sharing of electrons, a __________ is produced.
19. A variation of the same element but with a different mass number is called a(n) __________.
20. The term __________ describes a mineral's resistance to breaking or deforming.
21. The process called __________ involves the disintegration of unstable isotopes.
22. The property of a mineral that involves the appearance or quality of light reflected from its surface is termed __________.
23. The fundamental building block of all silicate minerals is the __________.
24. In an __________, one or more valence electrons are transferred from one atom to another.
25. A(n) __________ is a beneficial mineral that can be recovered for use.
26. The property called __________ is the color of a mineral in the powdered form.
27. The resistance of a mineral to abrasion or scratching, called __________, is one of the most useful diagnostic properties.
28. Minerals that do not exhibit cleavage when broken are said to __________.
29. The outermost principal electron shell of an atom contains the __________.
30. Comparing the weight of a mineral to the weight of an equal volume of water determines the mineral's __________.
31. To assign a value to a mineral's resistance to abrasion or scratching, geologists use a standard scale called __________.
32. Due to impurities, a mineral's __________ is often an unreliable diagnostic property.
33. The __________ make up most of Earth's crust.
34. In the __________ elements are organized into rows so that those with similar properties are in the same column.

Comprehensive Review

1. List the four characteristics that any Earth material must exhibit in order to be considered a mineral.

   1) 3)

   2) 4)

2. List the three main subatomic particles found in an atom and describe how they differ from one another. Also indicate where each is located in an atom by placing a corresponding letter on Figure 2.1.

   1) Letter: _____

   2) Letter: _____

   3) Letter: _____

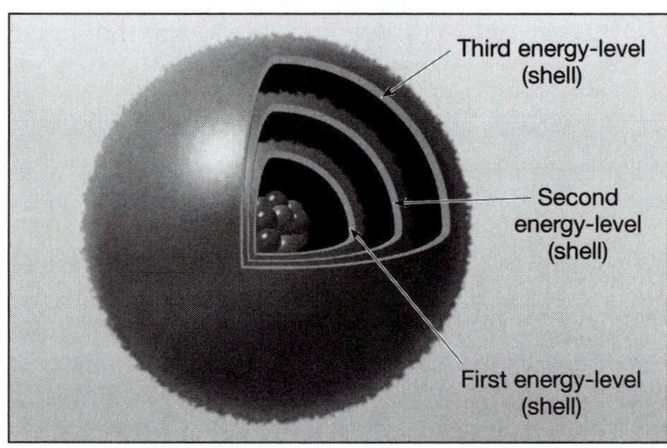

Figure 2.1

3. Briefly explain the difference between a mineral and a rock.

4. If an atom has 6 protons, 6 electrons, and 8 neutrons, what is the atom's

   a) atomic number? _____

   b) mass number? _____

5. If an atom has 17 electrons and its mass number is 35, calculate the following:

   a) Number of protons _____

   b) Atomic number _____

   c) Number of neutrons _____

6. Briefly explain the formation of the chemical bond involving a sodium atom and chlorine atom to produce the compound sodium chloride.

7. Examine the diagram of an atom shown in Figure 2.2A. The atom's nucleus contains 8 protons and 5 neutrons. Is the atom shown in Figure 2.2A an ion? Explain your answer.

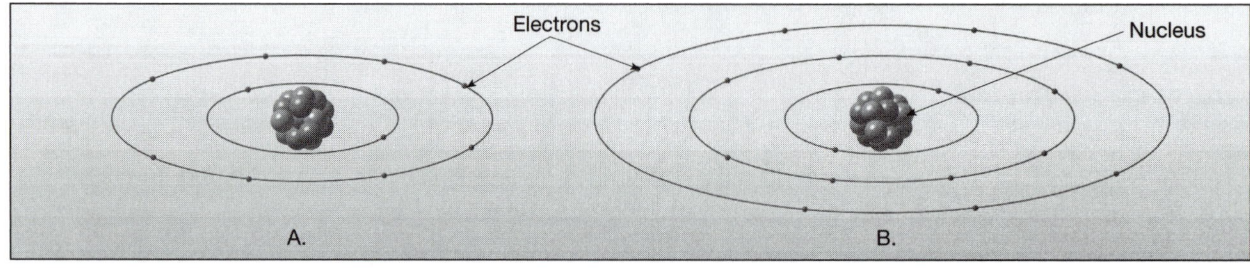

Figure 2.2

a) What is the mass number of the atom shown in Figure 2.2A? ____________

b) What is the atomic number of the atom shown in Figure 2.2A? ____________

8. Examine the diagram of an atom shown in Figure 2.2B. The atom's nucleus contains 17 protons. Is the atom shown in Figure 2.2B an ion? Explain your answer.

9. Define a mineral.

10. In what way does an isotope vary from the common form of the same element?

11. Briefly describe each of the following properties of minerals.

    a) Luster:

    b) Habit:

    c) Streak:

    d) Hardness:

    e) Cleavage:

    f) Fracture:

    g) Specific gravity:

12. The minerals represented by the diagrams in Figure 2.3 exhibit cleavage. Describe the cleavage of each mineral in the space below its diagram.

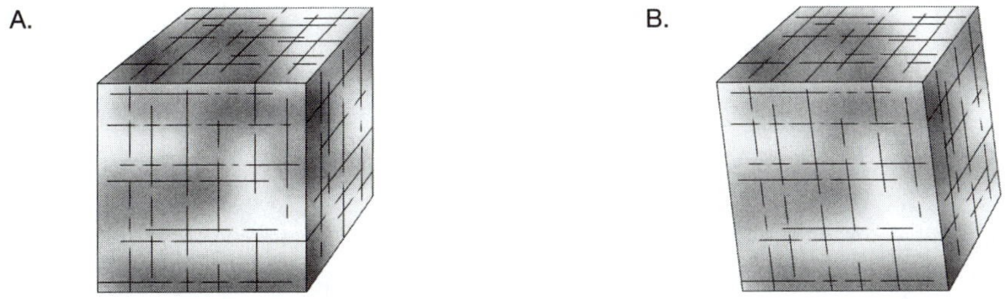

A.

Description:____________________

____

B.

Description:____________________

____

Figure 2.3

13. What are the two most common elements that compose Earth's continental crust? Also list the approximate percentage (by weight) of the continental crust that each element composes, along with its chemical symbol.

14. Figure 2.4 represents the silicon-oxygen tetrahedron. Letter A identifies _____ atoms, while letter B is a _____ atom.

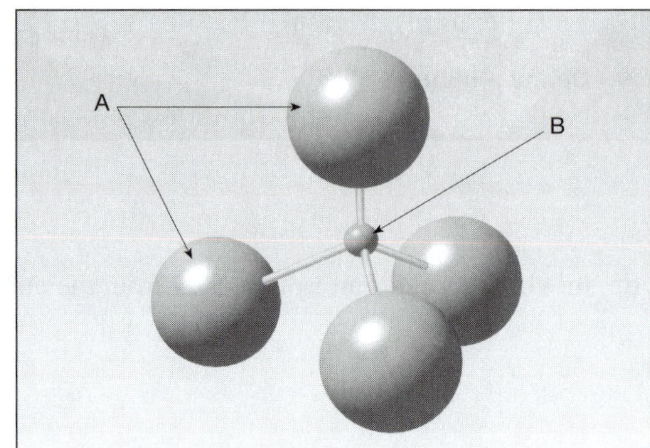

Figure 2.4

15. Name the silicate mineral group(s) that exhibits each of the following silicate structures.

    a) Three-dimensional networks:

    b) Sheets:

    c) Double chains:

    d) Single chains:

    e) Single tetrahedron:

16. Name three common rock-forming silicate minerals and the most common rock-forming nonsilicate mineral.

    a) Three common rock-forming silicate minerals:

    b) Most common rock-forming nonsilicate mineral:

17. Define a rock.

18. How do most silicate minerals form?

19. List a mineral that is an ore of each of the following elements.

    a) Mercury:

    b) Lead:

    c) Zinc:

    d) Iron:

20. Briefly explain what happens to an isotope as it decays.

21. Could a sample of Earth material that contains 10 percent aluminum by weight be profitably mined to extract the aluminum? Explain your answer.

22. Figure 2.5 illustrates two different common silicate mineral structures. Write the name of the silicate mineral group that each structure represents by the corresponding figure.

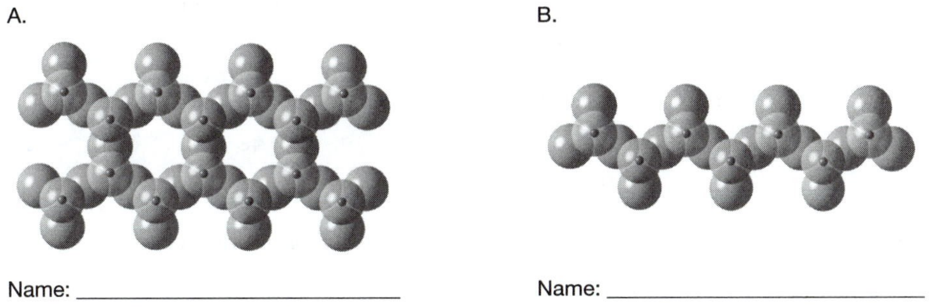

A.

Name: ______________________

B.

Name: ______________________

Figure 2.5

23. What are the primary elements that connect the large silicate structures to one another in the silicate minerals?

Practice Test

Multiple choice. Choose the best answer for the following multiple-choice questions.

1. Which one of the following is NOT a silicate mineral group?

   a) micas
   b) feldspars
   c) carbonates
   d) pyroxenes
   e) quartz

2. The basic building block of the silicate minerals _____.

   a) has the shape of a cube
   b) contains 1 silicon atom and 4 oxygen atoms
   c) always occurs independently
   d) contains 2 iron atoms for each silicon atom
   e) contains 1 oxygen atom and 4 silicon atoms

3. Which one of the following is NOT one of the eight most abundant elements in Earth's continental crust?

   a) oxygen
   b) iron
   c) aluminum
   d) calcium
   e) hydrogen

4. Atoms of the same element possess the same number of _____.

   a) isotopes
   b) ions
   c) protons
   d) neutrons
   e) compounds

5. Atoms containing the same numbers of protons and different numbers of neutrons are _____.

   a) isotopes
   b) ions
   c) protons
   d) neutrons
   e) compounds

6. Which subatomic particle is most involved in chemical bonding?

   a) electrons
   b) ions
   c) protons
   d) neutrons
   e) nucleus

7. The number of _____ in an atom's nucleus determines which element it is.

   a) electrons
   b) ions
   c) protons
   d) neutrons
   e) isotopes

8. Oxygen comprises about _____ percent by weight of Earth's continental crust.

   a) 12
   b) 27
   c) 47
   d) 63
   e) 85

9. The smallest particle of an element that still retains all the element's properties is a(n) _____.

   a) compound
   b) rock
   c) atom
   d) neutron
   e) isotope

10. Atoms that gain or lose electrons become _____.

    a) electrons
    b) ions
    c) protons
    d) neutrons
    e) compounds

11. When two or more elements bond together in definite proportions, they form a(n) _____.

    a) rock
    b) ion
    c) atom
    d) compound
    e) nucleus

12. The sum of the neutrons and protons in an atom's nucleus is the atom's _____.

    a) specific gravity
    b) atomic mass
    c) atomic number
    d) energy-levels
    e) isotope number

13. Radioactive decay occurs when _____.

    a) unstable nuclei disintegrate
    b) ions form
    c) a mineral is cleaved
    d) electrons change energy levels
    e) atoms bond

14. The property that is a measure of the resistance of a mineral to abrasion or scratching is _____.

    a) crystal form
    b) streak
    c) cleavage
    d) fracture
    e) hardness

15. The property that involves the color of a mineral in its powdered form is _____.

    a) crystal form
    b) streak
    c) cleavage
    d) fracture
    e) hardness

16. The tendency of a mineral to break along planes of weak bonding is called _____.

    a) luster
    b) crystal form
    c) cleavage
    d) fracture
    e) bonding

17. Minerals that break into smooth curved surfaces, like those seen in broken glass, have a unique type of fracture, called _____ fracture.

    a) conchoidal
    b) irregular
    c) rounded
    d) vitreous
    e) uneven

18. The basic building block of the silicate minerals is the silicon-oxygen _____.

    a) cube
    b) tetrahedron
    c) octagon
    d) rhombohedron
    e) sphere

19. Which of the following is NOT a primary element that joins silicate structures?

    a) calcium
    b) iron
    c) potassium
    d) magnesium
    e) oxygen

20. Most silicate minerals form from _____.

    a) other minerals
    b) radioactive decay
    c) erosion
    d) molten rock
    e) water

21. Which silicate mineral group has a double-chain silicate structure?

    a) pyroxenes
    b) amphiboles
    c) micas
    d) feldspars
    e) quartz

22. Which major rock-forming mineral is the chief constituent of the rocks limestone and marble?

    a) gypsum
    b) halite
    c) mica
    d) feldspar
    e) calcite

23. The scale used by geologists to measure the hardness of a mineral is called _____ scale.

    a) Mohs
    b) Smiths
    c) Richters
    d) Bowens
    e) Playfairs

24. Which nonsilicate mineral group contains the ores of iron?

a) oxides
b) sulfides
c) sulfates
d) halides
e) carbonates

25. Which one of the following statements concerning minerals is NOT true?

a) naturally occurring
b) solid
c) organic
d) orderly crystalline structure
e) formed of atoms

True/false. For the following true/false questions, if a statement is not completely true, mark it false. For each false statement, change the **italicized** *word to correct the statement.*

1. _____ Ions form when atoms gain or lose *neutrons*.
2. _____ *Protons* have an opposite electrical charge from electrons.
3. _____ Nearly *80* minerals have been named.
4. _____ Only *eight* elements compose the bulk of the rock-forming minerals.
5. _____ The elements silicon and *carbon* comprise nearly three-fourths of Earth's continental crust.
6. _____ The most common mineral group is the *silicate* group.
7. _____ The structure of the silicon-oxygen tetrahedron consists of *four* oxygen atoms surrounding a smaller silicon atom.
8. _____ Each silicate mineral group has a particular silicate *structure*.
9. _____ Silicate minerals tend to cleave *through* the silicon-oxygen structures.
10. _____ Each *silicate* mineral has a structure and chemical composition that indicate the conditions under which it formed.
11. _____ Nonsilicate minerals make up about *one-third* of the continental crust.
12. _____ The mineral calcite is a *sulfide* mineral.
13. _____ *Quartz* is the mineral from which plaster and other similar building materials are composed.
14. _____ *Reserves* are already identified mineral deposits from which minerals can be extracted profitably.
15. _____ *Luster* is the external expression of a mineral's internal orderly arrangement of atoms.
16. _____ Cleavage is defined by the number of directions exhibited and the *angles* at which they meet.
17. _____ Vitreous, pearly, and silky are types of *metallic* lusters.
18. _____ Isotopes of the same element have *different* mass numbers.
19. _____ A(n) *ion* is composed of two or more elements bonded together.
20. _____ The number of *protons* in an atom's nucleus determines its atomic number.
21. _____ Uncombined atoms have the same number of *neutrons* as protons.
22. _____ Individual *electrons* are located at given distances from an atom's nucleus in regions called energy levels.
23. _____ A(n) *compound* is the smallest particle of matter that has all the characteristics of an element.
24. _____ A mineral is a naturally occurring *inorganic* solid that possesses an orderly crystalline structure.
25. _____ Quartz is *softer* than calcite.
26. _____ In an atom, individual electrons move within regions around the nucleus called *principal shells*.

Word choice. Complete each of the following statements by selecting the most appropriate response.

1. The number of protons in an atom's nucleus determines its [atomic number/mass number]; while the [atomic number/mass number] is the sum of the neutrons and protons in the nucleus.
2. The [nucleus/atom] is the smallest particle of matter that has all characteristics of an element.
3. Isotopes of the same element have different [atomic/mass] numbers, the same [atomic/mass] number, and very [similar, different] chemical behavior.
4. Adjectives such as metallic, glassy, and earthy are used to describe the mineral property called [luster/color].
5. Minerals that do not exhibit cleavage when broken are said to [crystallize/fracture].
6. In atoms, a stable configuration occurs when the valence shell contains [6/8] electrons.
7. All silicate minerals contain the elements oxygen and [silicon/aluminum] joined together in the form of a [triangle/tetrahedron].
8. The crystal structure of a mineral has a direct influence on the mineral's [luster/cleavage].
9. The most prominent mineral in the carbonate group of minerals is [calcite/quartz].
10. The mineral name for table salt is [hematite/halite]; and the major constituent of the rock limestone is [calcite/gypsum].
11. An ore of iron is the mineral [pyrite/hematite], while the primary ore of lead is [galena, sphalerite].
12. In an atom, if the electrons are shared, the bond is called a(n) [covalent/ionic] bond; however, if electrons are transferred a(n) [covalent/ionic] bond is formed.

Written questions

1. What are the three main particles of an atom? How do they differ from one another?

2. Suggest a reason as to why some mineral samples may not visibly demonstrate their common habit.

3. Describe the basic building block of all silicate minerals.

4. Define a mineral.

5. Explain the difference between a mineral resource and a mineral reserve.

For other interesting and pertinent
information, be sure to visit
the *Earth Science* companion website at

http://www.prenhall.com/tarbuck

ROCKS:

Materials of the Solid Earth

3

The rock cycle, an integral component of the Earth system, outlines the origins of and processes involved in forming the three rock groups—igneous rock, sedimentary rock, and metamorphic rock. Igneous rocks, formed from magma as it cools and crystallizes within the Earth or on its surface, are classified by their texture and mineral composition. Sedimentary rocks, the most common rocks exposed at Earth's surface, consist of mineral and rock fragments or material once carried by water in solution that has been transformed into rock by compaction or cementation. The single most characteristic feature of these rocks is their layers, called strata, or beds. Metamorphic rocks are produced from preexisting igneous, sedimentary, or other metamorphic rocks by temperature, pressure, and/or chemically active fluids while still essentially solid. A unique arrangement of mineral crystals, called foliation, characterizes many of these rocks. Furthermore, the rocks of the crust host the metallic and nonmetallic mineral resources that are vital to our modern society.

Learning Objectives

After reading, studying, and discussing this chapter, you should be able to:

List the geologic processes involved in the formation of each rock group.

Igneous rock forms from *magma* that cools and solidifies in a process called *crystallization. Sedimentary rock* forms from the *lithification of sediment. Metamorphic rock* forms from rock that has been subjected to great pressure and heat in a process called *metamorphism.*

Briefly describe the crystallization of magma.

The rate of cooling of magma greatly influences the size of mineral crystals in igneous rock—the faster the rate of cooling, the smaller the crystals. The four basic igneous rock textures are (1) *fine-grained*, (2) *coarse-grained*, (3) *porphyritic*, and (4) *glassy.*

List the criteria used to classify igneous rocks.

Igneous rocks are classified by their *texture* and *mineral composition.* Igneous rocks are divided into broad compositional groups based on the percentage of dark and light silicate minerals they contain. *Felsic rocks* (e.g., granite and rhyolite) are composed mostly of the light-colored silicate minerals potassium feldspar and quartz. Rocks of *intermediate* composition (e.g., andesite) contain plagioclase feldspar and amphibole. *Mafic rocks* (e.g., basalt) contain abundant pyroxene, and calcium rich plagioclase feldspar.

Discuss the mineral makeup of igneous rock.

The mineral makeup of an igneous rock is ultimately determined by the chemical composition of the magma from which it crystallized. N. L. Bowen showed that as magma cools, minerals crystallize in an orderly fashion. *Magmatic differentiation* changes the composition of magma and causes more than one rock type to form from a common parent magma.

Explain the difference between detrital and chemical sedimentary rocks.

Detrital sediments are materials that originate and are transported as solid particles derived from weathering. *Chemical sediments* are soluble materials produced largely by chemical weathering that are precipitated by either inorganic or organic processes. *Detrital sedimentary rocks*, which are classified by particle size, contain a variety of mineral and rock fragments, with clay minerals and quartz the chief constituents. *Chemical sedimentary rocks* often contain the products of biological processes such as shells or mineral crystals that form as water evaporates and minerals precipitate. *Lithification* refers to the processes by which sediments are transformed into solid sedimentary rocks.

List the names, textures, and environments of formation for the most common sedimentary rocks.

Common detrital sedimentary rocks include *shale* (the most common sedimentary rock), *sandstone*, and *conglomerate*. The most abundant chemical sedimentary rock is *limestone*, composed chiefly of the mineral *calcite*. *Rock gypsum* and *rock salt* are chemical rocks that form as water evaporates and triggers the deposition of chemical precipitates.

List the common features of sedimentary rocks.

Some of the features of sedimentary rocks that are often used in the interpretation of Earth history and past environments include *strata*, or *beds* (the single most characteristic feature), *fossils, ripple marks*, and *mud cracks*.

Describe the agents of metamorphism.

Two types of metamorphism are 1) *regional metamorphism* and 2) *contact metamorphism*. The agents of metamorphism include *heat, pressure* (*stress*), and *chemically active fluids*. Heat is perhaps the most important because it provides the energy to drive the reactions that result in the *recrystallization* of minerals. Metamorphic processes cause many changes in rocks, including *increased density*, growth of *larger mineral crystals, reorientation of the mineral grains* into a layered or banded appearance known as *foliation*, and the formation of *new minerals*.

List the names and textures of the most common metamorphic rocks.

Some common metamorphic rocks with a *foliated texture* include *slate, schist*, and *gneiss*. Metamorphic rocks with a *nonfoliated texture* include *marble* and *quartzite*.

Discuss metallic mineral resources.

Some of the most important accumulations of *metallic mineral resources* are produced by igneous and metamorphic processes. *Vein deposits* (deposits in fractures or bedding planes) and *disseminated deposits* (deposited distributed throughout the entire rock mass) are produced from *hydrothermal solutions*—hot, metal-rich fluids associated with cooling magma bodies.

Describe nonmetallic mineral resources.

Nonmetallic mineral resources are mined for the nonmetallic elements they contain or for the physical and chemical properties they possess. The two groups of nonmetallic mineral resources are 1) *building materials* (e.g., limestone and gypsum) and 2) *industrial minerals* (e.g., fluorite and corundum).

Key Terms

andesitic (intermediate) composition
basaltic composition
Bowen's reaction series
chemical sedimentary rock
coarse-grained texture
contact (thermal) metamorphism
crystallization
crystal setting
detrital sedimentary rock
disseminated deposit
evaporite deposits
extrusive (volcanic)
felsic
fine-grained texture
fossils
foliated texture
glassy texture
granite composition
hydrothermal solution
igneous rock
intermediate composition
intrusive (plutonic)
lava
lithification
magma
metamorphic rock
metamorphism
nonfoliated
pegmatite
porphyritic texture
regional metamorphism
rock cycle
sediment
sedimentary rock
strata (beds)
texture
thermal metamorphism
ultramafic composition
vein deposit

Vocabulary Review

Choosing from the list of key terms, furnish the most appropriate response for the following statements.

1. The __________ illustrates the origin of the three basic rock types and the role of geologic processes in transforming one rock type into another.
2. A metal-rich accumulation of mineral matter that occurs along a fracture or bedding plane is called a(n) __________.
3. __________ refers to the processes by which sediments are transformed into solid sedimentary rock.
4. Sedimentary rocks, when subjected to great pressures and heat, will turn into the rock type __________, providing they do not melt.
5. __________ is molten material found inside Earth.
6. The process whereby ions begin to arrange themselves into orderly patterns is referred to as __________.
7. Metamorphic rocks that have formed in response to large-scale mountain-building processes are said to have undergone __________.
8. The rock type that forms when sediment is lithified is __________.
9. Magma that reaches Earth's surface is called __________.
10. The term __________ describes the overall appearance of an igneous rock, based on the size and arrangement of its interlocking crystals.
11. The texture of a metamorphic rock that gives it a layered appearance is called __________.
12. Igneous rocks that result when lava solidifies are classified as __________ rocks.
13. Unconsolidated material consisting of particles created by weathering and transported by water, glaciers, wind, or waves is called __________.
14. The rock type that forms from the crystallization of magma is __________.
15. Metamorphic rocks composed of only one mineral that forms equidimensional crystals often have a type of texture called a(n) __________.
16. Igneous rocks that form rapidly at the surface often have a type of texture, called __________, where the individual crystals are too small to be seen with the unaided eye.
17. A(n) __________ is a hot, metal-rich fluid that is associated with a cooling magma body.
18. Detrital sedimentary rocks form from rock fragments; however, a(n) __________ forms when dissolved substances are precipitated back into solids.
19. An igneous rock that has large crystals embedded in a matrix of smaller crystals is said to have a type of texture called a(n) __________.
20. The single most characteristic feature of sedimentary rocks are layers called __________.
21. A(n) __________ is a sedimentary deposit that forms as a shallow arm of the sea evaporates.
22. When rock is in contact with or near a mass of magma, a type of metamorphism, called __________, takes place.
23. When igneous rock cools rapidly and ions do not have time to unite into an orderly crystalline structure, a type of texture called a __________ results.
24. A sedimentary rock that forms from sediments that originate as solid particles from weathered rocks is called a(n) __________.
25. When large masses of magma solidify far below the surface, they form igneous rocks that exhibit a type of texture where the crystals are roughly equal in size and large enough to be identified with the unaided eye, called a(n) __________.

26. A(n) __________ forms when hydrothermal activity distributes minute quantities of ore throughout a large rock mass.
27. An igneous rock composed of unusually large crystals is referred to as a(n) __________.
28. __________ are the traces or remains of prehistoric life.
29. Geologists refer to granitic rocks as being __________, a term derived from feldspar and silica.
30. Igneous rocks such as peridotite that are composed almost entirely of dark silicate minerals have a chemical composition that is referred to as __________.

Comprehensive Review

1. Using Figure 3.1, label the Earth material or chemical/physical process that occurs at each lettered position in the rock cycle.

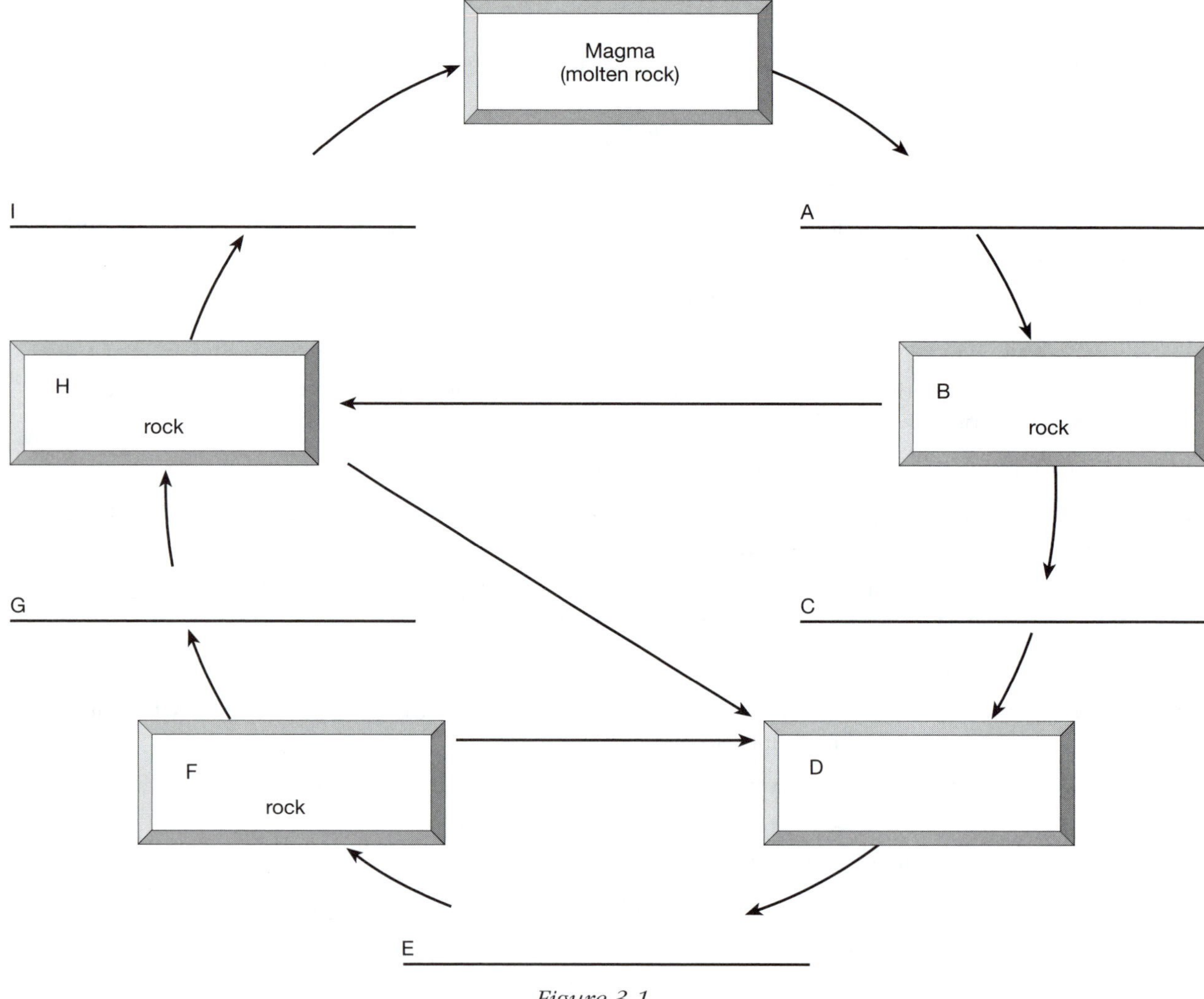

Figure 3.1

2. Why is the rock cycle a good illustration of the Earth system?

3. Referring to the rock cycle, which processes are restricted to Earth's surface?

4. What are the various erosional agents that can pick up, transport, and deposit the products of weathering?

5. Briefly describe the formation of each of the three rock types.

   a) Igneous rocks:

   b) Sedimentary rocks:

   c) Metamorphic rocks:

6. What factor most influences the size of mineral crystals in igneous rocks? How does it influence the size of crystals?

7. What is the difference between magma and lava?

8. List and briefly describe the two criteria used to classify igneous rocks.

   1)

   2)

9. Describe the conditions that will result in the following igneous rock textures. Then, select the igneous rock photograph in Figure 3.2 that illustrates each of the textures.

   a) Fine-grained texture:

      Photograph letter: __________

   b) Glassy texture:

      Photograph letter: __________

   c) Coarse-grained texture:

      Photograph letter: __________

Figure 3.2

d) Porphyritic texture:

Photograph letter: __________

10. What discovery did N.L. Bowen make about the formation of minerals in a cooling magma?

11. According to Bowen's reaction series, which mineral is the first mineral to form from magma? If the reaction series is completed, which mineral will be the last to crystallize?

12. Is it possible for two igneous rocks to have the same mineral constituents but different names? Explain your answer.

13. Using Bowen's reaction series as a guide, list the minerals that will most likely occur in an igneous rock within each of the following rock groups.

    Basaltic group:

    Granitic group:

14. Describe how the solid and liquid components of magma can separate.

15. What are two processes that cause sediment to be lithified into solid sedimentary rock?

16. Describe the two major groups of sedimentary rocks. Give an example of a rock found in each group.

    a) Detrital sedimentary rocks:

        Example:

    b) Chemical sedimentary rocks:

        Example:

17. What criterion is used to subdivide the detrital sedimentary rocks?

18. What does the presence of angular fragments in a detrital sedimentary rock suggest about the transportation of the particles?

19. What basis is used to subdivide the chemical sedimentary rocks?

20. What is the most abundant chemical sedimentary rock? What are the two origins for this rock?

21. Beginning with a swamp environment, list the successive stages in the formation of coal.

22. Selecting from the sedimentary rock photographs in Figure 3.3, answer the following questions.

    a) The rock in Figure 3.3A is made of __________ [chemical, detrital] material.

    b) The rock in Figure 3.3B is __________ [conglomerate, breccia].

    c) The rock in Figure 3.3 __________ [A, B, C] would most likely be associated with a quiet water environment.

A.

B.

C.

Figure 3.3

23. What are fossils? As important tools for interpreting the geologic past, what are some uses of fossils?

24. Briefly describe the two settings in which metamorphism most often occurs.

    a) Regional metamorphism:

    b) Contact metamorphism:

25. What are the three agents of metamorphism?

26. Describe the two textures of metamorphic rocks. Give an example of a rock that exhibits each texture.

    a) Foliated texture:

        Example:

    b) Nonfoliated texture:

        Example:

27. What is the texture of the metamorphic rock in Figure 3.4?

28. How did the gold deposits of the Homestake mine in South Dakota as well as the lead, zinc, and silver ores near Coeur d'Alene originate?

Figure 3.4

Practice Test

Multiple choice. Choose the best answer for the following multiple-choice questions.

1. Which one of the following is NOT an agent of metamorphism?
   a) heat
   b) lithification
   c) chemically active fluids
   d) pressure

2. The great majority of rocks exposed at Earth's surface are of which type?
   a) igneous rocks
   b) sedimentary rocks
   c) metamorphic rocks

3. Molten material found inside Earth is called __________.
   a) lava
   b) rock fluid
   c) plasma
   d) mineraloid
   e) magma

4. The greatest volume of metamorphic rock is produced during __________.
   a) contact metamorphism
   b) lithification
   c) crystallization
   d) regional metamorphism
   e) weathering

5. __________ discovered that as magma cools in the laboratory, certain minerals crystallize first, at very high temperatures.
   a) James Hutton
   b) John Playfair
   c) N.L. Bowen
   d) William Smith
   e) James Smith

6. Which one of the following is NOT a common cement found in sedimentary rocks?
   a) silica
   b) gypsum
   c) calcite
   d) iron oxide

7. The process of igneous rock formation is called __________.
   a) weathering
   b) decomposition
   c) lithification
   d) crystallization
   e) foliation

8. Igneous rocks that contain the last minerals to crystallize from magma and consist mainly of feldspars and quartz are said to have a __________ composition.
   a) granitic
   b) gneissic
   c) basaltic
   d) metamorphic
   e) lithic

9. Fossils found in sedimentary rocks can be used to __________.
   a) interpret past environments
   b) indicate certain periods of time
   c) match rocks of the same age that are found at different places
   d) all of the above

10. When large masses of magma solidify far below the surface, they form igneous rocks that exhibit a(n) __________ texture.
    a) fine-grained
    b) glassy
    c) coarse-grained
    d) porphyritic
    e) fragmental

11. Which one of the following is NOT a detrital sedimentary rock?
    a) shale
    b) limestone
    c) sandstone
    d) conglomerate
    e) siltstone

12. The surface process that slowly disintegrates and decomposes rock is called __________.
    a) erosion
    b) weathering
    c) exfoliation
    d) metamorphism
    e) crystallization

13. Which one of the following is NOT composed of microcrystalline quartz?
    a) travertine
    b) flint
    c) jasper
    d) chert
    e) agate

14. The single most characteristic feature of sedimentary rocks are __________.
    a) crystals
    b) mud cracks
    c) ripple marks
    d) strata

15. Which one of the following is NOT a primary element found in magma?
    a) silicon
    b) aluminum
    c) iron
    d) carbon
    e) oxygen

16. Metamorphic rocks can form from __________.
    a) igneous rocks
    b) sedimentary rocks
    c) other metamorphic rocks
    d) all of the above

17. Vein deposits are usually produced by __________.
    a) cementation and compaction
    b) hydrothermal solutions
    c) foliation
    d) decomposition
    e) weathering

18. Which one of the following metamorphic rocks has a nonfoliated texture?
    a) marble
    b) slate
    c) mica schist
    d) gneiss

19. The term meaning "conversion into rock" is __________.
    a) metamorphism
    b) crystallization
    c) ionization
    d) lithification
    e) weathering

20. Accumulation of silts and clays, and eventually the rocks siltstone and shale, is generally best associated with which one of the following environments?
    a) swiftly flowing river
    b) quiet swamp water
    c) glacier
    d) beach
    e) windy desert

21. Which one of the following is the metamorphic form of coal?
    a) lignite
    b) bituminous
    c) anthracite
    d) peat

22. Most metallic ores are produced by which two processes?
    a) cementation and compaction
    b) igneous and metamorphic
    c) igneous and sedimentary
    d) sedimentary and metamorphic
    e) decomposition and disintegration

23. Why can two igneous rocks have the same minerals but different names?
    a) They may have different colors.
    b) Names of igneous rocks are arbitrary.
    c) The rocks may be of different sizes.
    d) They may have different textures.
    e) They may have been found in different places.

24. Detrital sedimentary rocks are subdivided according to __________.
    a) particle size
    b) color
    c) hardness
    d) the place where they were found
    e) their age

25. Igneous rocks formed from magma that crystallized at depth are called __________ or intrusive rocks.
    a) plutonic
    b) metamorphic
    c) volcanic
    d) porphyritic
    e) foliated

True/false. For the following true/false questions, if a statement is not completely true, mark it false. For each false statement, change the **italicized** *word to correct the statement.*

1. _____ The nonfoliated metamorphic equivalent of limestone is *marble.*
2. _____ The most common extrusive igneous rock is *granite.*
3. _____ Many metallic ore deposits are formed by the precipitation of minerals from *hydrothermal* solutions.
4. _____ The *rock cycle* shows the relations among the three rock types and is essentially an outline of physical geology.
5. _____ About 75 percent of all rock outcrops on the continents are *sedimentary.*
6. _____ *Granite* is a classic example of an igneous rock that exhibits a coarse-grained texture.
7. _____ *Igneous* rocks are the rock type most likely to contain fossils.
8. _____ Igneous rock, when subjected to heat and pressure far below Earth's surface, will change to *sedimentary* rock.
9. _____ Limestone, a common building material, is a *nonmetallic* mineral resource.
10. _____ Because magma's density is *greater* than the surrounding rocks, it works its way to the surface over time spans from thousands to millions of years.
11. _____ Mineral alignment in a metamorphic rock usually gives the rock a *foliated* texture.
12. _____ Most sediment ultimately comes to rest in the *ocean.*
13. _____ The process called *weathering*, whereby magma cools, solidifies, and forms igneous rocks, may take place either beneath Earth's surface or on the surface following a volcanic eruption.
14. _____ Separating *strata* are bedding planes, flat surfaces along which rocks tend to separate or break.
15. _____ *Lithification* literally means to "change form."
16. _____ Molten material found on Earth's surface is called *magma.*
17. _____ Rock salt and rock gypsum form when *evaporation* causes minerals to precipitate from water.
18. _____ The texture of an igneous rock is based on the size and *arrangement* of its interlocking crystals.
19. _____ The two major groups of *metamorphic* rocks are detrital and chemical.
20. _____ The *lithification* of sediment produces sedimentary rock.
21. _____ The agents of *metamorphism* include heat, pressure, and chemically active fluids.
22. _____ According to Bowen's reaction series, quartz is often the *last* mineral to crystallize from a melt.
23. _____ The mineral *sylvite* is the primary ingredient of plaster and wallboard.
24. _____ *Sedimentary* rocks are major sources of energy resources, iron, aluminum, and manganese.
25. _____ Slow cooling of magma results in the formation of *small* mineral crystals.

Word choice. Complete each of the following statements by selecting the most appropriate response.

1. The very first rocks to form on Earth's surface were [igneous/sedimentary/metamorphic].
2. Because a magma body is [less/more] dense than the surrounding rocks, it works its way [toward/away from] the surface over time spans of thousands to millions of years.
3. The molten minerals in magma are primarily members of the [carbonate/silicate/sulfide] group.
4. Detrital sedimentary rocks are primarily classified by particle [size/shape/composition].
5. Coal forms in swamp water that is oxygen-[enriched/deficient]; therefore, complete decay of the plant material is [not/always] possible.
6. During low-grade metamorphism, the sedimentary rock shale becomes the metamorphic rock called [marble/slate].
7. A rock that will effervesce with dilute hydrochloric acid most likely contains the mineral [quartz/calcite/feldspar].
8. The most important agent of metamorphism is [heat/pressure].
9. Sediment derived indirectly through the life processes of water-dwelling organisms is referred to as [detrital/chemical/biochemical] material.
10. Iron oxide, silica, and [calcite/olivine] are the most common cements found in sedimentary rocks.
11. A banded metamorphic rock that contains mostly elongated and granular minerals is called [schist/gneiss].
12. Hydrothermal solutions [can/cannot] migrate great distances through the surrounding rock before they are deposited.

Written questions

1. Referring to the rock cycle, explain why any one rock can be the raw material for another.

2. How is Bowen's reaction series related to the classification of igneous rocks?

3. What are the most common minerals in detrital sedimentary rocks? Why are these minerals so abundant?

4. In what ways do metamorphic rocks differ from the igneous and sedimentary rocks from which they formed?

For other interesting and pertinent information, be sure to visit the *Earth Science* companion website at

http://www.prenhall.com/tarbuck

4

Weathering, Soil, and Mass Wasting

Weathering and mass wasting are the first two steps in the continuous sculpturing of Earth's physical landscape. It is these processes that are responsible for disintegrating and decomposing rock and moving it to lower elevations by gravity where wind, water, and ice can carry the materials away. Chapter four investigates how solid rock deteriorates and why the types and rates of weathering vary from place to place on Earth's surface. Soil, an important product of the weathering process and a vital resource, is also examined. In closing, the factors and mechanisms responsible for moving weathered debris downslope are explored.

Learning Objectives

After reading, studying, and discussing this chapter, you should be able to:

Describe the process of weathering, mass wasting, and erosion.

External processes include (1) *weathering*—the disintegration and decomposition of rock at or near the surface, (2) *mass wasting*—the transfer of rock material downslope under the influence of gravity, and (3) *erosion*—the incorporation and transportation of material by a mobile agent, usually water, wind, or ice. They are called external processes because they occur at or near Earth's surface and are powered by energy from the Sun. By contrast, *internal processes*, such as volcanism and mountain building, derive their energy from Earth's interior.

Explain the difference between mechanical and chemical weathering.

Mechanical weathering is the physical breaking up of rock into smaller pieces. *Chemical weathering* alters a rock's chemistry, changing it into different substances. Rocks can be broken into smaller fragments by *frost wedging, salt crystal growth, unloading*, and *biological activity*. Water is by far the most important agent of chemical weathering. Oxygen in water can *oxidize* some materials, while carbon dioxide (CO_2) dissolved in water forms *carbonic acid*. The chemical weathering of silicate minerals frequently produces (1) soluble products containing sodium, calcium, potassium, and magnesium, (2) insoluble iron oxides, and (3) clay minerals.

List the factors that determine the rate of weathering of a rock.

The rate at which rock weathers depends on such factors as (1) *particle size*—small pieces generally weather faster than large pieces; (2) *mineral makeup*—calcite readily dissolves in mildly acidic solutions, and silicate minerals that form first from magma are least resistant to chemical weathering; and (3) *climatic factors*, particularly temperature and moisture. Frequently, rocks exposed at Earth's surface do not weather at the same rate. This *differential weathering* of rocks is influenced by such factors as mineral makeup and degree of jointing.

Define soil and list the factors that control soil formations.

Soil is a combination of mineral and organic matter, water, and air—that portion of the *regolith* (the layer of rock and mineral fragments produced by weathering) that supports the growth of plants. *Soil texture* refers to the proportions of different particle sizes (clay, silt, and sand) found in soil. The most important *factors that control soil formation are parent material, time, climate, plants, animals, and topography.*

List the typical soil horizons from the surface downward.

Soil-forming process operate from the surface downward and produce zones or layers in the soil that soil scientists call *horizons*. From the surface downward the horizons are designated as O, A, E, B, and C, respectively.

Describe the system of soil classification known as Soil Taxonomy.

In the United States, soils are classified using a system known as the *Soil Taxonomy*. It is based on physical and chemical properties of the soil profile and includes six hierarchical categories. The system is especially useful for agricultural and related land-use purposes.

Discuss soil erosion.

Soil erosion is a natural process; it is part of the constant recycling of Earth materials that we call the rock cycle. *Rates of soil erosion vary* from one place to another and depend on the soil's characteristics as well as such factors as climate, slope, and type of vegetation. Human activities have greatly accelerated the rate of soil erosion in many areas.

Explain how weathering creates mineral deposits.

Weathering creates mineral deposits by consolidating metals into economical concentrations. The process, called *secondary enrichment*, is accomplished by either (1) removing undesirable materials and leaving the desired elements enriched in the upper zones of the soil or (2) removing and carrying the desirable elements to lower soil zones where they are redeposited and become more concentrated. *Bauxite*, the principal ore of aluminum, is one important ore created by secondary enrichment.

Describe the controls of mass wasting.

In the evolution of most landforms, mass wasting is the step that follows weathering. The combined effect of mass wasting and erosion by running water produces stream valleys. *Gravity is the controlling force of mass wasting*. Other factors that influence or trigger downslope movements are saturation of the material with water, oversteepening of slopes beyond the *angle of repose*, removal of anchoring vegetation, and ground vibrations from earthquakes.

List and describe the various types of mass wasting.

The various processes included under the name of mass wasting are classified and described on the basis of (1) the type of material involved (debris, mud, earth, or rock), (2) the kind of motion (fall, slide, or flow), and (3) the rate of the movement (fast, slow). The various kinds of mass wasting include the more rapid forms called *slump, rockslide, debris flow*, and *earthflow*, as well as the slow movements referred to as *creep* and *solifluction*.

Key Terms

angle of repose
chemical weathering
creep
debris flow
differential weathering
earthflow
eluviation
erosion
exfoliation dome
external process
fall
flow
frost wedging
horizon
internal processes
lahar
leaching
mass wasting
mechanical weathering
parent material
permafrost
regolith
rockslide
secondary enrichment
sheeting
slide
slump
soil
soil profile
Soil Taxonomy
soil texture
solifluction
solum
spheroidal weathering
talus slope
weathering

Vocabulary Review

Choosing from the list of key terms, furnish the most appropriate response for the following statements.

1. The process called __________ is the incorporation and transportation of material by a mobile agent, usually water, wind, or ice.
2. The process where water works its way into every crack or void in a rock and, upon freezing, expands and enlarges the opening is called __________.
3. The layer of rock and mineral fragments produced by weathering that nearly everywhere covers Earth's surface is termed __________.
4. __________ refers to the depletion of soluble materials from the upper soil by the downward-percolating water.
5. Stone Mountain, Georgia, is an excellent example of a(n) __________ that has formed as large slabs of rock have separated and spalled off.
6. In the United States, soil scientists have devised a system for classifying soils known as the __________.
7. The downward slipping of mass of rock or unconsolidated material moving as a unit along a curved surface is called __________.
8. The disintegration and decomposition of rock at or near Earth's surface is called __________.
9. Permanently frozen ground, called __________, occurs in Earth's harsh tundra and ice cap climates.
10. __________ occurs when weathering concentrates metals into economically valuable deposits.
11. __________ is that portion of the regolith that supports plant growth.
12. __________ is a type of mass wasting that is common wherever water cannot escape from the saturated surface layer by infiltrating to deeper levels.
13. When a rock undergoes __________ it is broken into smaller and smaller pieces, each retaining the characteristics of the original material.
14. The weathering process called __________ gives the rock a more rounded shape.
15. A(n) __________ occurs when blocks of bedrock break loose and slide down a slope.
16. The complex process that alters the internal structures of minerals by removing and/or adding elements is termed __________.
17. __________ refers to the washing out of fine soil components from the A horizon by downward-percolating water.
18. A(n) __________ is a debris flow that occurs on the slope of a volcano when unstable layers of ash and debris become saturated and flow.
19. The transfer of rock material downslope under the influence of gravity is referred to as __________.
20. __________ relates to the fact that rocks exposed at Earth's surface usually do not weather at the same rate.
21. The relative proportions of clay, silt, and sand in a soil determine the __________.
22. The process called __________ occurs when large masses of igneous rock, particularly granite, begin to break loose like the layers of an onion.
23. A(n) __________ is a layer or zone in a soil profile.
24. __________ is the gradual downhill movement of soil and regolith.
25. The source of the weathered mineral matter from which soils develop is called the __________.
26. A large pile of debris that forms at the base of a slope is called a(n) __________.

27. Together the O, A, E, and B horizons of soil constitute the __________, or "true soil."
28. The stable slope of a material is called its __________.
29. A(n) __________ is a vertical section through a soil showing the succession of soil horizons.

Comprehensive Review

1. List and describe the three general processes that are continually removing materials from higher elevations and transporting them to lower elevations.

   1)

   2)

   3)

2. List and briefly describe the two types of weathering.

   1)

   2)

3. What are three natural processes that break rocks into smaller fragments?

4. Write a brief statement describing the two natural ways that water becomes mildly acidic.

5. What are the three factors that influence the rate of weathering?

   1)

   2)

   3)

6. Referring to the process of weathering, describe what is happening to the rocks in Figure 4.1.

Figure 4.1

7. What are the three products of the weathering of potassium feldspar?

8. What is the difference between regolith and soil?

9. What are the three particle-size classes used for determining soil texture?

10. Referring to Figure 4.2, what are the proportions of clay, silt, and sand of the soil shown at point B?

    Clay: _____ Silt: _____ Sand: _____

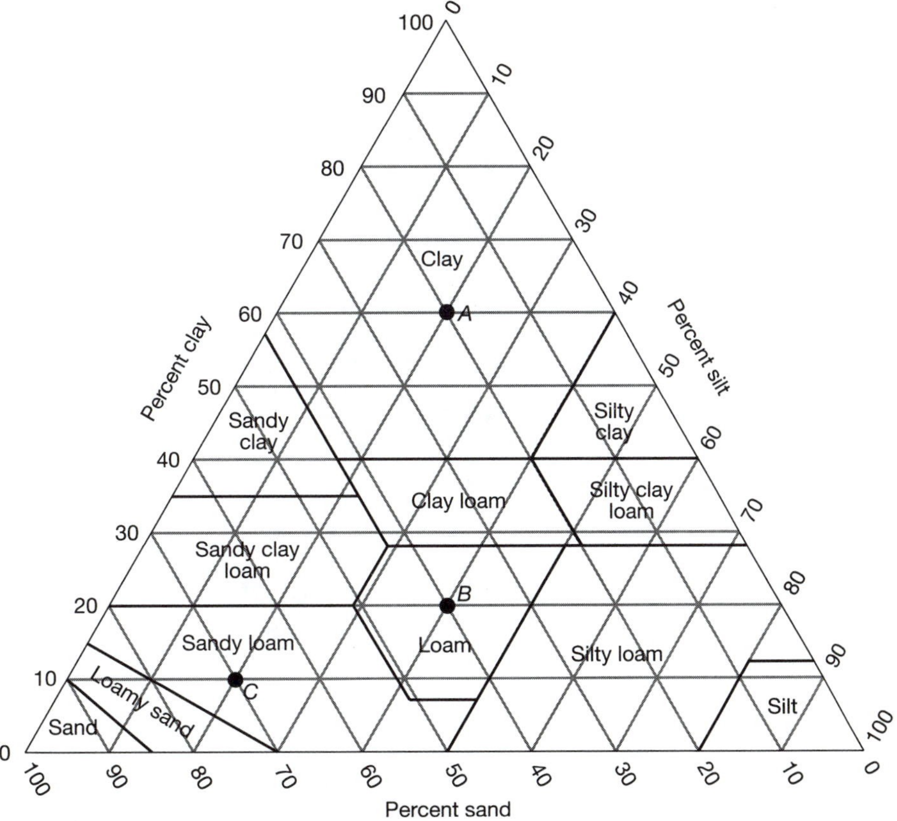

Figure 4.2

11. Referring to Figure 4.2, if a soil has 20 percent clay, 60 percent silt, and 20 percent sand, it is called a(n):

12. What are the four basic soil structures?

13. List the five controls of soil formation.

14. On Figure 4.3, label each of the five soil horizons with the appropriate identifying letter (e.g., O, A, etc.).

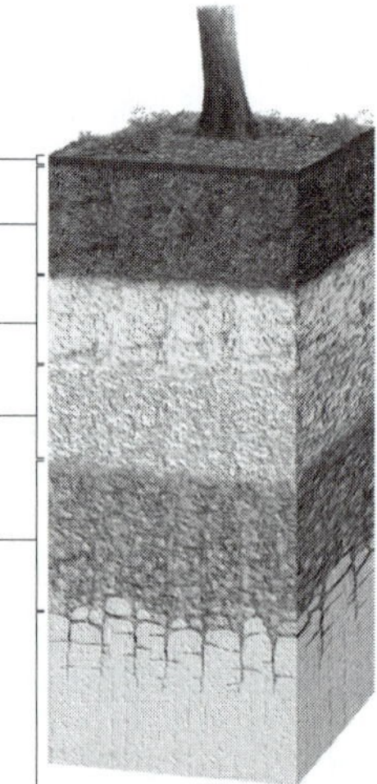

Figure 4.3

15. Use Figure 4.4 to answer the following questions.

    a) At which location would the thickest transported soils be located? _______

    b) Which location would have no soil development? _______

    c) At which location would the thickest residual soil be located? _______

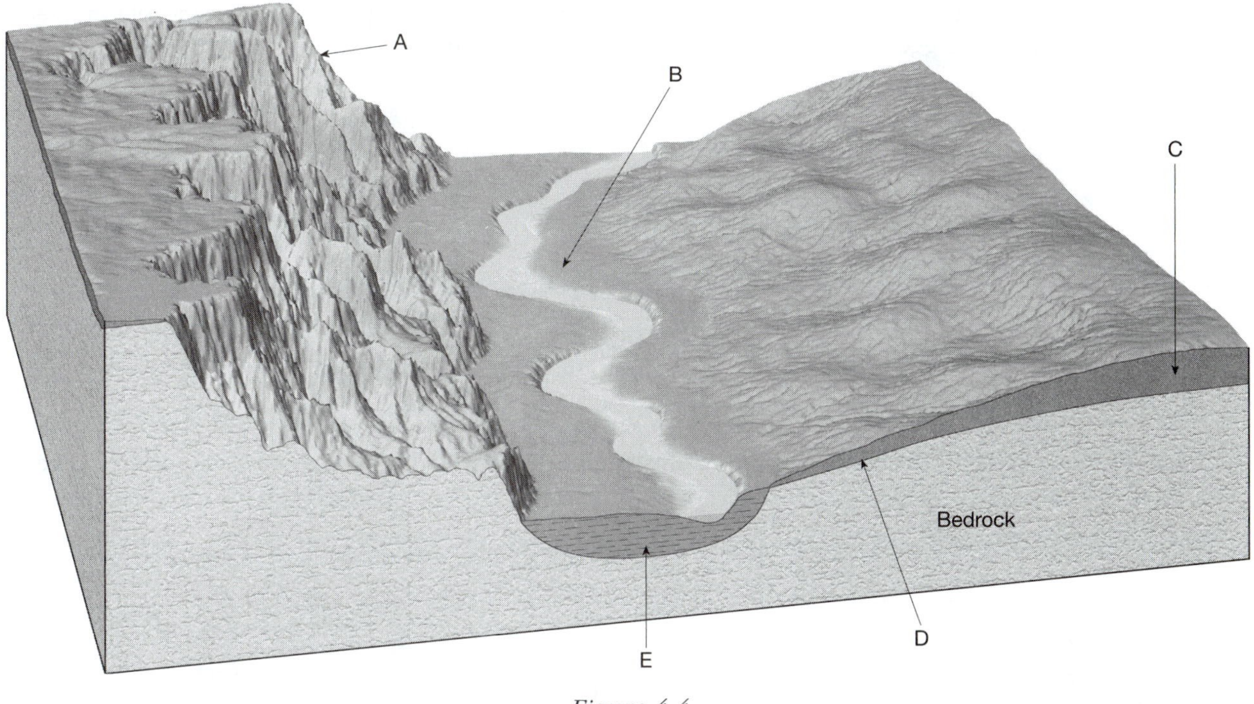

Figure 4.4

16. In addition to prolonged drought, what other factors contributed to the Dust Bowl of the 1930s?

17. Briefly describe how secondary enrichment produces ore deposits.

18. The combined effects of which two processes produce stream valleys?

19. Although gravity is the controlling force of mass wasting, what factors play an important role in overcoming inertia and triggering downslope movements of materials?

20. Identify each of the forms of mass wasting illustrated in Figure 4.5 by writing the name of the process by the diagram.

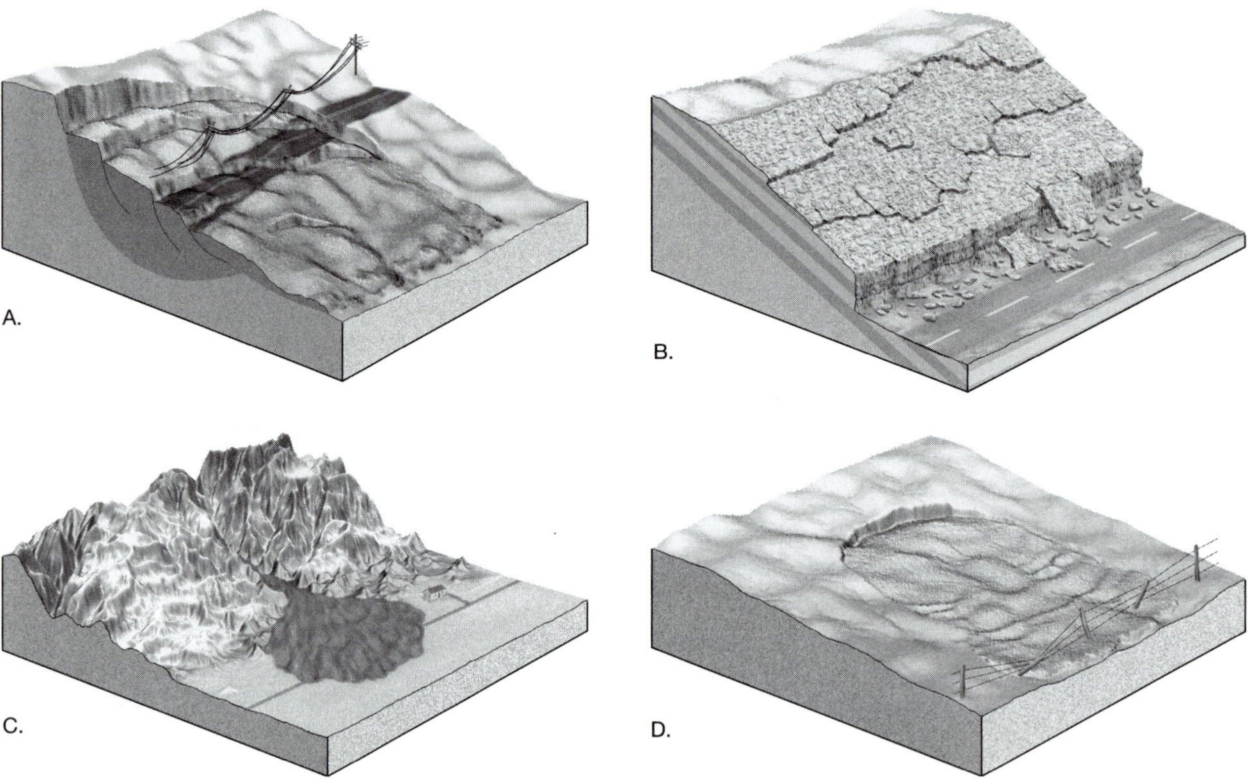

Figure 4.5

Practice Test

Multiple choice. Choose the best answer for the following multiple-choice questions.

1. The most influential control of soil formation is __________.
   a) parent material
   b) climate
   c) topography
   d) time
   e) plants and animals

2. The process responsible for moving material downslope without the aid of running water, wind, or ice is __________.
   a) erosion
   b) mass wasting
   c) weathering
   d) laterization
   e) eluviation

3. When weathering increases the concentration of an economically valuable resource, the process is referred to as __________.
   a) eluviation
   b) solifluction
   c) primary enrichment
   d) deposition
   e) secondary enrichment

4. The disintegration and decomposition of rock at or near Earth's surface is called __________.
   a) erosion
   b) weathering
   c) eluviation
   d) mass wasting
   e) solifluction

5. Compared to the distant past, current rates of soil erosion are __________.
   a) about the same
   b) slower
   c) impossible to determine
   d) faster

6. A soil in which no single particle size predominates over the other two is called __________.
   a) humus
   b) loam
   c) regolith
   d) laterite
   e) caliche

7. Platy, prismatic, blocky, and spheroidal are each a type of __________.
   a) humus
   b) soil structure
   c) mineral composition
   d) weathering
   e) soil texture

8. Chemical weathering will be most rapid in __________ climates.
   a) cold, dry
   b) warm, dry
   c) cold, wet
   d) warm, wet

9. Which soil order is most often found in the tropics and is rich in iron and aluminum oxides?
   a) alfisols
   b) oxisols
   c) andisols
   d) mollisols

10. When a rock is subjected to __________ it breaks into smaller pieces having the same characteristics as the original rock.
    a) mechanical weathering
    b) creep
    c) chemical weathering
    d) oxidation
    e) eluviation

11. Which one of the following sequences is the correct order that water will follow prior to forming a stream channel?
    a) gullies, sheet erosion, rills
    b) sheet erosion, rills, gullies
    c) rills, sheet erosion, gullies
    d) sheet erosion, gullies, rills

12. Clay minerals are produced from the chemical weathering of the __________ minerals.
    a) feldspar
    b) quartz
    c) carbonate
    d) oxide
    e) sulfide

13. The combination of organic matter, rock and mineral fragments, water, and air that covers much of Earth's surface is referred to as __________.
    a) regolith
    b) laterite
    c) soil
    d) caliche
    e) humus

14. Which one of the following minerals is most susceptible to chemical weathering?
    a) pyroxene
    b) biotite
    c) quartz
    d) potassium feldspar

15. Whenever the characteristics and internal structures of weathered minerals have been altered, they have undergone __________.
    a) creep
    b) mechanical weathering
    c) eluviation
    d) chemical weathering
    e) mass wasting

16. Which one of the following is LEAST important in mass wasting?
    a) gravity
    b) slope angle
    c) soil water
    d) slope orientation
    e) vegetation

17. The atmospheric gas that forms a mild acid when dissolved in water is __________.
    a) oxygen
    b) argon
    c) nitrogen
    d) carbon dioxide
    e) silica

18. Soils resting on the bedrock from which they formed are called __________ soils.
    a) transported
    b) static
    c) major
    d) residual
    e) weathered

19. Which of the following is NOT associated with mechanical weathering?
    a) frost wedging
    b) unloading
    c) oxidation
    d) biological activity

20. The layer of weathered rock and mineral fragments that covers much of Earth's surface is called __________.
    a) humus
    b) clay
    c) alluvium
    d) regolith
    e) laterite

21. The most important agent of chemical weathering is __________.
    a) air
    b) water
    c) soil
    d) gravity
    e) erosion

22. When a mass of material moves downslope along a curved surface, the process is called __________.
    a) debris flow
    b) rockslide
    c) creep
    d) slump
    e) solifluction

23. The process of __________ produces a yellow to reddish-brown surface on iron-rich minerals.
    a) lithification
    b) oxidation
    c) acidification
    d) hydrolysis
    e) calcification

24. The process called __________ is responsible for the formation of exfoliation domes.
    a) oxidation
    b) frost wedging
    c) creep
    d) slump
    e) unloading

25. A soil's texture is determined by its __________.
    a) mineral composition
    b) type of humus
    c) particle sizes
    d) water content
    e) age

26. In a well-developed soil profile the __________ horizon is the topmost layer.
    a) A
    b) B
    c) C
    d) O
    e) E

27. The accumulation of rock fragments found at the base of many cliffs is referred to as the __________.
    a) talus slope
    b) lahar
    c) transported residue
    d) earthflow
    e) base horizon

28. Which is the least effective form of mechanical weathering?
    a) thermal expansion
    b) unloading
    c) organic activity
    d) frost wedging

29. The removal of soluble materials from soil by percolating water is called __________.
    a) leaching
    b) erosion
    c) laterization
    d) eluviation
    e) degradation

30. The most widespread form of mass wasting is __________.
    a) rockslide
    b) creep
    c) earthflow
    d) mudflow
    e) slump

True/false. For the following true/false questions, if a statement is not completely true, mark it false. For each false statement, change the **italicized** *word to correct the statement.*

1. _____ Soils formed from bedrock are called *residual* soils.
2. _____ The B horizon of a soil is also called the *topsoil*.
3. _____ Removing vegetation will *increase* the rate of soil erosion.
4. _____ In Bowen's reaction series, the minerals that form *last* are the most susceptible to chemical weathering.
5. _____ The free descent of individual rock pieces of any size is called *fall*.
6. _____ A vertical section through all the layers of a soil is called the soil *horizon*.
7. _____ *Lahars* are debris flows on the sides of volcanoes.
8. _____ About 50 percent of the total volume of a good-quality soil consists of pore spaces.
9. _____ Eluviation and leaching are most active in the *E horizon*.
10. _____ In general, the greater the slope, the *thicker* the soil.
11. _____ The soil order that is most common in dry places is *aridosols*.
12. _____ *Climate* is the most influential control of soil formation.
13. _____ The soil layer composed of partially altered parent material is called the *C* horizon.
14. _____ A soil layer (or zone) is referred to as a(n) *strata*.
15. _____ Tiny, threadlike water channels are called *rills*.
16. _____ The downward movement of material as a viscous fluid is termed *slide*.
17. _____ The B horizon of a soil is also called the zone of *accumulation*.
18. _____ The greatest amount of soil erosion is caused by *wind*.
19. _____ The depletion of soluble materials from the upper soil is termed *leaching*.

Word choice. Complete each of the following statements by selecting the most appropriate response.

1. Quartz is [more/less] resistant to chemical weathering than feldspar.
2. Soils lacking clearly defined horizons are called [mature/hardpan/immature] soils.
3. Atmospheric nitrogen is fixed (altered) into soil nitrogen by [earthworms/microorganisms].
4. Frost wedging is most effective in the [tropical/middle/polar] latitudes.
5. The O and A horizons of a soil make up what is commonly called the [topsoil/subsoil].
6. Excessive clay in soil will result in [good/poor] drainage.
7. The primary ore of aluminum is [pyrite/bauxite].
8. When minerals containing iron are chemically weathered, the iron is usually [dissolved/oxidized/reduced].
9. Increasing the amount of water in the pore spaces of soil and regolith will usually [increase/decrease] the likelihood of downslope movement.
10. A compact and impermeable layer of accumulated clay in soil is called [caliche/hardpan].
11. When carbon dioxide (CO_2) dissolves in water, a [weak/strong] [acid/base] is produced.
12. Mass wasting will most likely occur when materials are at an angle that is [greater/less] than their angle of repose.
13. The same parent material [will/will not] always produce the same type of soil.
14. The poor soil quality of ultisols is due to [evaporation/leaching].
15. Most talus slopes form as the result of [falls/flows/slides].
16. Water [expands/contracts] when it freezes.
17. During slump, movement of the material is along [flat/curved] surfaces.
18. Sheeting is believed to occur due to the [reduction/increase] in pressure when overlying rock is [removed/deposited].
19. The most common form of mass wasting is [slump/creep].

Written questions

1. Briefly describe the two types of weathering and how they are related.

2. What is soil? How does a soil profile develop?

3. What is the controlling force of mass wasting? What other factors are important?

For other interesting and pertinent
information, be sure to visit
the *Earth Science* companion website at

http://www.prenhall.com/tarbuck

Running Water and Groundwater

5

Water is constantly on the move from the ocean to the land and back again in an endless cycle. Chapter five investigates that part of the hydrologic cycle that returns water back to the sea, rushing over the surface as running water or slowly below the ground as groundwater. When viewed as integral parts of the Earth system, running water is responsible for shaping most of Earth's surface landforms and groundwater represents the largest reservoir of freshwater that is readily available to humans. The chapter examines the factors that influence the movement of water in a channel, as well as the work of streams and the erosional and depositional features associated with them and their valleys. Also explored are the importance, distribution, storage, and movement of groundwater. The chapter closes with a discussion of a few of the environmental problems associated with groundwater and a look at the geologic work of groundwater.

Learning Objectives

After reading, studying, and discussing this chapter, you should be able to:

Describe the movement of water through the hydrologic cycle.

The *hydrologic cycle* describes the continuous interchange of water among the oceans, atmosphere, and continents. Powered by energy from the Sun, it is a global system in which the atmosphere provides the link between the oceans and continents. The processes involved in the water cycle include *precipitation, evaporation, infiltration* (the movement of water into rocks or soil through cracks and pore spaces), *runoff* (water that flows over the land rather than infiltrating into the ground), and *transpiration* (the release of water vapor to the atmosphere by plants). *Running water is the single most important agent sculpturing Earth's land surface.*

Name the land area that contributes water to a stream.

The land area that contributes water to a stream is its *drainage basin*. Drainage basins are separated by imaginary lines called *divides*.

List the three main parts of a river system.

River systems consist of three main parts: the zones of erosion, transportation, and deposition.

List the factors that determine a stream's velocity and gradient.

The factors that determine a stream's *velocity* are *gradient* (slope of the stream channel), *shape, size* and *roughness* of the channel, and the stream's *discharge* (amount of water passing a given point per unit of time, frequently measured in cubic feet per second). Most often, the gradient and roughness of a stream decrease downstream, while width, depth, discharge, and velocity increase.

Describe how streams transport their load.

Streams transport their load of sediment in solution (*dissolved load*), in suspension (*suspended load*), and along the bottom of the channel (*bed load*). Much of the dissolved load is contributed by groundwater. Most streams carry the greatest part of their load in suspension. The bed load moves only intermittently and is usually the smallest portion of a stream's load.

Discuss the factors that influence a stream's ability to transport solid particles.

A stream's ability to transport solid particles is described using two criteria: *capacity* (the maximum load of solid particles a stream can carry) and *competence* (the maximum particle size a stream can transport). Competence increases as the square of stream velocity, so if velocity doubles, water's force increases fourfold.

Describe stream deposition and list several features of stream deposition.

Streams deposit sediment when velocity slows and competence is reduced. This results in *sorting*, the process by which like-sized particles are deposited together. Stream deposits are called *alluvium* and may occur as channel deposits called *bars*, as floodplain deposits, which include *natural levees*, and as *deltas* or *alluvial fans* at the mouths of streams.

List and describe the basic types of stream channels.

Stream channels are of two basic types: *bedrock channels* and *alluvial channels*. Bedrock channels are most common in headwaters regions where gradients are steep. Rapids and waterfalls are common features. Two types of alluvial channels are *meandering channels* and *braided channels*.

List and describe the two general types of base level.

The two general types of *base level* (the lowest point to which a stream may erode its channel) are (1) *ultimate base level* and (2) *temporary*, or *local, base level*. Any change in base level will cause a stream to adjust and establish a new balance. Lowering base level will cause a stream to downcut, while raising base level results in deposition of material in the channel.

Discuss the origin of floodplains and list some common features associated with them.

When a stream has cut its channel closer to base level, its energy is directed from side to side, and erosion produces a flat valley floor, or *floodplain*. Streams that flow upon floodplains often move in sweeping bends called *meanders*. Widespread meandering may result in shorter channel segments, called *cutoffs*, and/or abandoned bends, called *oxbow lakes*.

Describe some common flood-control measures.

Floods are triggered by heavy rains and/or snowmelt. Sometimes human interference can worsen or even cause floods. Flood-control measures include the building of *artificial levees* and dams, as well as *channelization*, which could involve creating *artificial cutoffs*. Many scientists and engineers advocate a nonstructural approach to flood control that involves more appropriate land use.

List the common drainage patterns produced by streams.

Common *drainage patterns* produced by streams include (1) *dendritic*, (2) *radial*, (3) *rectangular*, and (4) *trellis*.

Describe the significance of groundwater as a resource.

As a resource, *groundwater* represents the largest reservoir of freshwater that is readily available to humans. Geologically, the dissolving action of groundwater produces *caves* and *sinkholes*. Groundwater is also an equalizer of streamflow.

Discuss the occurence of groundwater.

Groundwater is water that occupies the pore spaces in sediment and rock in a zone beneath the surface called the *zone of saturation*. The upper limit of this zone is the *water table*. The *unsaturated zone* is above the water table, where the soil, sediment, and rock are not saturated.

List the factors that determine the quantity and movement of groundwater.

The quantity of water that can be stored depends on the *porosity* (the volume of open spaces) of the material. The *permeability* (the ability to transmit a fluid through interconnected pore spaces) of a material is a very important factor controlling the movement of groundwater.

Distinguish between aquitards and aquifers.

Materials with very small pore spaces (such as clay) hinder or prevent groundwater movement and are called *aquitards*. *Aquifers* consist of materials with larger pore spaces (such as sand) that are permeable and transmit groundwater freely.

Describe springs, wells, and artesian wells.

Springs occur whenever the water table intersects the land surface and a natural flow of groundwater results. *Wells*, openings bored into the zone of saturation, withdraw groundwater and create roughly conical depressions in the water table known as *cones of depression*. *Artesian wells* occur when water rises above the level at which it was initially encountered.

Describe hot springs and geysers.

When groundwater circulates at great depths, it becomes heated. If it rises, the water may emerge as a *hot spring*. *Geysers* occur when groundwater is heated in underground chambers, expands, and some water quickly changes to steam, causing the geyser to erupt. The source of heat for most hot springs and geysers is hot igneous rock.

List some of the current environmental problems associated with groundwater.

Some of the current environmental problems involving groundwater include (1) *overuse* by intense irrigation, (2) *land subsidence* caused by groundwater withdrawal, and (3) *contamination* by pollutants.

Discuss the formation of caverns and karst topography.

Most *caverns* form in limestone at or below the water table when acidic groundwater dissolves rock along lines of weakness, such as joints and bedding planes. *Karst topography* exhibits an irregular terrain punctuated with many depressions, called *sinkholes*.

Key Terms

alluvial fan	delta	incised meander	sinkhole (sink)
alluvium	dendritic pattern	infiltration	sorting
aquifer	discharge	infiltration capacity	spring
aquitard	dissolved load	karst topography	stalactite
artesian well	distributary	laminar flow	stalagmite
backswamp	divide	meander	stream valley
bar	drainage basin	natural levee	suspended load
base level	drawdown	oxbow lake	transpiration
bed load	evapotranspiration	permeability	trellis pattern
braided stream	flood	point bar	turbulent flow
capacity	floodplain	porosity	unsaturated zone
cavern	geyser	radial pattern	water table
competence	gradient	rectangular pattern	well
cone of depression	groundwater	runoff	yazoo tributary
cut bank	hot spring	saltation	zone of saturation
cutoff	hydrologic cycle	settling velocity	

Vocabulary Review

Choosing from the list of key terms, furnish the most appropriate response for the following statements.

1. __________ is the movement of surface water into rocks or soil through cracks or pore spaces.
2. The land area that contributes water to a stream is referred to as the stream's __________.
3. The zone beneath Earth's surface where all open spaces in sediment and rock are completely filled with water is the __________.

4. The __________ of a material is the percentage of the total volume that consists of pore spaces.
5. __________ is the slope of a stream channel expressed as the vertical drop of a stream over a specified distance.
6. The general term for any well-sorted stream-deposited sediment is __________.
7. The __________ of a stream is the maximum load it can carry.
8. Water beneath Earth's surface in the zone of saturation is known as __________.
9. A tributary that flows parallel to the main stream because a natural levee is present is called a __________.
10. The release of water vapor to the atmosphere by plants is referred to as __________.
11. A "looplike," sweeping bend in a stream is called a(n) __________.
12. The __________ is the upper limit of the zone of saturation.
13. The volume of water flowing past a certain point per unit of time is called a stream's __________.
14. The term __________ is applied to any well in which groundwater rises above the level where it was initially encountered.
15. The elevated landform that parallels some streams, called a(n) __________, acts to confine the stream's waters, except during flood stage.
16. The __________ of a material indicates its ability to transmit water through interconnected pore spaces.
17. The unending circulation of Earth's water supply from the oceans, to the atmosphere, to the land, and back to the oceans is called the __________.
18. A(n) __________ is the flat, low-lying portion of a stream valley that is subject to inundation during flooding.
19. The __________ of a stream measures the maximum size of particles it is capable of transporting.
20. The drainage basin of one stream is separated from the drainage basin of another by an imaginary line called a(n) __________.
21. An impermeable layer that hinders or prevents the movement of groundwater is termed a(n) __________.
22. The combined amount of water that is evaporated from the land and transpired by plants is called __________.
23. __________ is the lowest point to which a stream can erode its channel.
24. Whenever the water table intersects the ground surface, a natural flow of groundwater, called a __________, results.
25. A rock strata or sediment that transmits groundwater freely is called a(n) __________.
26. __________ occurs when water particles move in roughly straight-line paths that parallel the stream channel.
27. The fine sediment transported within the body of flowing water is called a stream's __________.
28. Landscapes that to a large extent have been shaped by the dissolving power of groundwater are said to exhibit __________.
29. A __________ has channels with an interwoven appearance.
30. Water that flows over the land rather than infiltrating into the ground is referred to as __________.

Comprehensive Review

1. The combined effects of which processes produce stream valleys?

2. Figure 5.1 illustrates the hydrologic cycle. Select the letter in the figure that represents each of the following processes.

    a) Runoff: _____

    b) Evaporation: _____

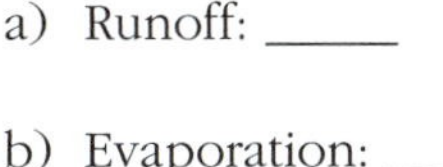

    c) Precipitation: _____

    d) Infiltration: _____

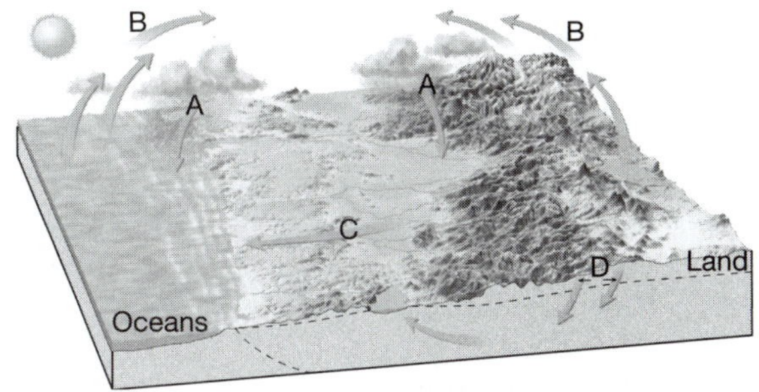

Figure 5.1

3. Briefly define the following measurements that are used to assess streamflow.

    a) Velocity:

    b) Gradient:

    c) Discharge:

4. What is the average gradient of a stream whose headwaters are 10 kilometers from the sea, at an altitude of 40 meters?

5. By circling the correct response, indicate how each of the following measurements changes from a stream's head to its mouth.

    a) Gradient [increases, decreases]

    b) Velocity [increases, decreases]

    c) Channel roughness [increases, decreases]

    d) Discharge [increases, decreases]

    e) Width and depth [increase, decrease]

6. Describe the two general types of base level.

    a) Ultimate base level:

    b) Temporary, or local, base level:

7. List and describe the three ways streams transport their load of sediment.

    1)

    2)

    3)

8. Why does the greatest erosion and transportation of sediment by a stream occur during a flood?

9. Figure 5.2 illustrates a wide stream valley. Select the letter in the figure that identifies each of the following features.

    a) Yazoo tributary: _____

    b) Backswamp: _____

    c) Natural levees: _____

    d) Oxbow lake: _____

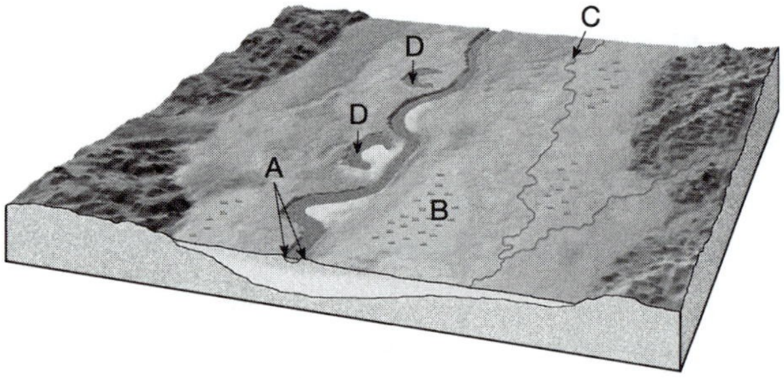

Figure 5.2

10. Briefly describe the appearance of the following types of valleys. Also list the major features you would expect to find associated with each type.

    1) Narrow, V-shaped, valley:

        Features:

    2) Wide valley:

        Features:

11. Describe the type of drainage pattern that would develop on an isolated volcanic cone or domal uplift.

12. What are some strategies that have been devised to eliminate or lessen the effects of floods?

13. List some key words or features that describe or are associated with alluvial channels.

14. Using Figure 5.3, select the letter that illustrates each of the following.

    a) Zone of saturation: _____

    b) Aquitard: _____

    c) Spring: _____

    d) Water table: _____

    e) Unsaturated zone: _____

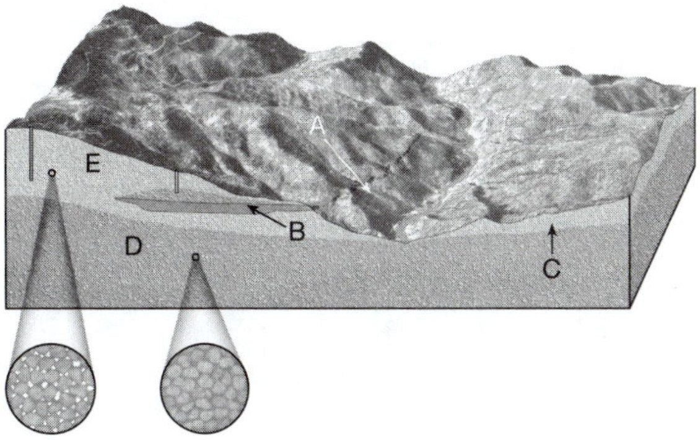

Figure 5.3

15. In general, how is the shape of the water table related to the shape of the surface?

16. What is the source of heat for most hot springs and geysers?

17. Using the terms *porosity* and *permeability*, describe an aquifer.

18. List two environmental problems that are associated with groundwater.

    1)

    2)

19. What are two ways that sinkholes form?

    1)

    2)

20. Identify the type of drainage pattern illustrated in each of the two diagrams in Figure 5.4 by writing the correct response by the diagram. Describe how each pattern forms.

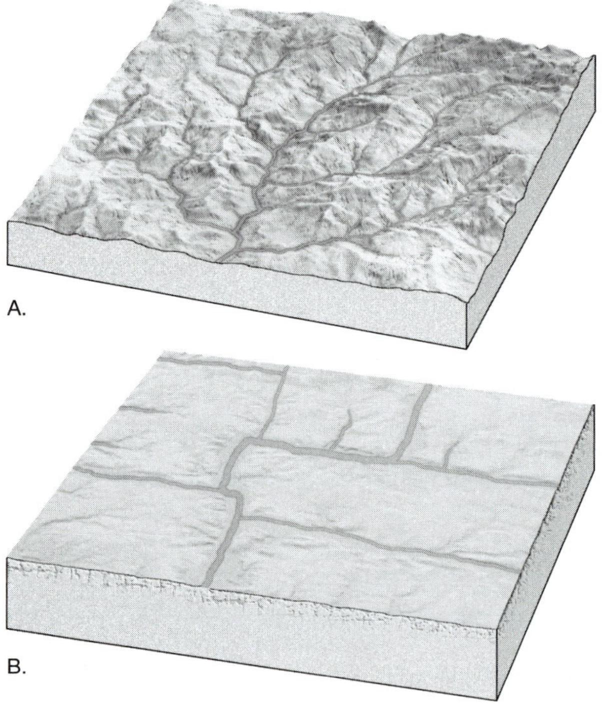

Figure 5.4

Practice Test

Multiple choice. Choose the best answer for the following multiple-choice questions.

1. The largest reservoir of freshwater that is available to humans is __________.
   a) lakes and reservoirs
   b) glaciers
   c) groundwater
   d) river water
   e) atmospheric water vapor

2. The single most important agent sculpturing Earth's land surface is __________.
   a) ice
   b) wind
   c) running water
   d) waves

3. Lowering a stream's base level will cause the stream to __________.
   a) deposit
   b) meander
   c) change course
   d) downcut
   e) stop flowing

4. Along a stream meander, the maximum velocity of water occurs __________.
   a) near the inner bank of the meander
   b) in the center
   c) near the outer bank of the meander
   d) near the stream's bed

5. A typical profile of a stream exhibits a constantly __________ gradient from the head to the mouth.
   a) increasing
   b) decreasing
   c) uniform
   d) arbitrary

6. The flat portion of a valley floor adjacent to a stream channel is called a __________.
   a) meander
   b) floodplain
   c) yazoo
   d) divide
   e) tributary

7. Which type of drainage pattern will most likely develop where the underlying material is relatively uniform?
   a) dendritic
   b) radial
   c) rectangular
   d) trellis

8. Ice sheets and glaciers represent approximately __________ percent of the total volume of freshwater.
   a) 15
   b) 35
   c) 55
   d) 85

9. The function of artificial levees built along a river is to control __________.
   a) erosion
   b) flooding
   c) gradient
   d) meandering
   e) oxbow lakes

10. Plants release water into the atmosphere through a process called __________.
    a) evaporation
    b) infiltration
    c) erosion
    d) transpiration
    e) precipitation

11. Occasionally, deposition causes the main channel of a stream to divide into several smaller channels called __________.
    a) deltas
    b) meanders
    c) oxbows
    d) yazoos
    e) distributaries

12. The average annual precipitation worldwide must equal the quantity of water __________.
    a) evaporated
    b) transpired
    c) that becomes runoff
    d) infiltrated
    e) locked in glaciers

13. During periods when rain does not fall, rivers are sustained by __________.
    a) transpirated water
    b) groundwater
    c) atmospheric moisture
    d) runoff

14. Which one of the following would least likely be found in a wide valley?
    a) oxbow lake
    b) floodplain
    c) rapids
    d) cutoff
    e) meanders

15. Along straight stretches, the highest velocities of water are near the __________ of the channel, just below the surface.
    a) banks
    b) center

16. The combined effects of weathering, __________, and running water produce stream valleys.
    a) weathering
    b) tectonics
    c) mass wasting
    d) discharge
    e) capacity

17. Which one of the following is a measure of a material's ability to transmit water through interconnected pore spaces?
    a) competence
    b) gradient
    c) capacity
    d) porosity
    e) permeability

18. The capacity of a stream is directly related to its __________.
    a) velocity
    b) competence
    c) gradient
    d) meandering
    e) discharge

19. Drawdown of a well creates a roughly conical lowering of the water table called the cone of __________.
    a) saturation
    b) lowering
    c) permeability
    d) depression
    e) aeration

20. In a typical stream, where gradient is steep, discharge is __________.
    a) large
    b) small
    c) variable
    d) impossible to calculate
    e) reversed

21. Most natural water contains the weak acid, __________ acid, which makes it possible for groundwater to dissolve limestone and form caverns.
    a) acetic
    b) hydrochloric
    c) carbonic
    d) sulfuric
    e) nitric

22. The area above the water table where soil is not saturated is called the __________ zone.
    a) unsaturated
    b) soil moisture
    c) saturation
    d) porosity
    e) aquifer

23. Most streams carry the largest part of their load __________.
    a) as bed load
    b) in suspension
    c) in solution
    d) near their head

24. The great majority of hot springs and geysers in the United States are found in the __________.
    a) North
    b) South
    c) East
    d) West
    e) Midwest

25. Whenever the water table intersects the ground surface, a natural flow of water, called a __________, results.
    a) spring
    b) stalagmite
    c) aquitard
    d) sinkhole
    e) geyser

True/false. For the following true/false questions, if a statement is not completely true, mark it false. For each false statement, change the **italicized** *word to correct the statement.*

1. _____ Most of a stream's dissolved load is brought to it by *groundwater.*
2. _____ *Ultimate* base levels include lakes and resistant layers of rock.
3. _____ The unending circulation of Earth's water supply is called the *hydrologic* cycle.
4. _____ The reduced current velocity at the inside of a meander results in the deposition of coarse sediment, especially sand, called a *point bar.*
5. _____ Landscapes shaped by the dissolving power of groundwater are often said to exhibit *karst* topography.
6. _____ A flat valley floor, or floodplain, often occurs in *narrow* valleys.
7. _____ Since the water cycle is balanced, the average annual *precipitation* worldwide must equal the quantity of water evaporated.
8. _____ The gradient of a stream *increases* downstream.
9. _____ The drainage pattern that develops when the bedrock is crisscrossed by many right-angle joints and/or faults is the *radial* pattern.
10. _____ *Discharge* is one factor that determines the velocity of a stream.
11. _____ Much of the water that flows in rivers is not direct runoff from rain or snowmelt but originates as *groundwater.*
12. _____ *Permeable* layers, such as clay, that hinder or prevent the movement of water beneath the surface are termed aquitards.
13. _____ The term *artesian* is applied to any situation in which groundwater rises above the level where it was initially encountered in a well.
14. _____ A channel's shape, size, and *roughness* affect the amount of friction with the water in the channel.
15. _____ When rivers become rejuvenated, meanders often stay in the same place but become *deeper.*
16. _____ Most streams carry the largest part of their load in *solution.*
17. _____ The *competence* of a stream measures the maximum size of particles it is capable of transporting.
18. _____ A stream's velocity *increases* downstream.
19. _____ *Lowering* base level will cause a stream to gain energy and downcut.
20. _____ Marshes called *backswamps* form because water on a floodplain cannot flow up the levee and into the river.
21. _____ The formation of caverns takes place in the zone of *saturation.*
22. _____ *Groundwater* represents the largest reservoir of freshwater that is readily available to humans.
23. _____ Some of the water that infiltrates the ground surface is absorbed by plants, which then release it into the atmosphere through a process called *infiltration.*

24. _____ Sorting of sediment occurs because each particle *size* has a critical settling velocity.
25. _____ Near the *mouth* of a river where discharge is great, gradient is small.
26. _____ The zone in the ground where all of the open spaces in sediment and rock are completely filled with water is called the zone of *saturation*.
27. _____ *Geysers* are intermittent hot springs that eject water with great force at various intervals.
28. _____ Two types of stream channels are bedrock and *alluvial* channels.
29. _____ Land *subsidence* is an environmental problem that can be caused by groundwater withdrawal.
30. _____ *Ice* is the single most important agent sculpturing Earth's land surface.

Word choice. Complete each of the following statements by selecting the most appropriate response.

1. The velocity of a stream in a straight channel is greatest at the [center/sides] of the channel.
2. A region with alternating bands of resistant and less resistant rock exposed at the surface will probably develop a [dendritic/radial/trellis] drainage pattern.
3. On a delta, the main channel of a river often divides into smaller channels called [tributaries/distributaries/rills].
4. The upper limit of the zone of saturation is called the [water table/aquifer].
5. A stream's gradient usually [increases/decreases/remains unchanged] downstream.
6. A permeable rock layer that transmits water freely is referred to as an [aquifer/aquitard].
7. Urbanization [decreases/increases] infiltration and [decreases/increases] runoff.
8. Karst topography is most likely to develop in regions with [basalt/limestone/sandstone] bedrock with a [dry/wet] climate.
9. Infiltration will be greatest on [steep/gentle] slopes following a [heavy/light] rain.
10. The amount of water passing a given point per unit of time is a stream's [velocity/discharge/competence].
11. Building a dam will [raise/lower] the base level of a stream upstream from the dam.
12. The velocity of a stream is greatest at the [inside/outside] of a meander.
13. The acid that does most of the work of groundwater is [nitric/carbonic/sulfuric] acid.
14. A stream's discharge [increases/decreases] downstream.
15. The amount of groundwater a material can hold is controlled by the material's [permeability/porosity].

Written questions

1. Briefly describe the movement of water through the hydrologic cycle.

2. Discuss the role of base level in determining the work of streams.

3. What is the difference between the porosity of a material and its permeability?

4. What are the conditions required for the development of karst topography?

For other interesting and pertinent
information, be sure to visit
the *Earth Science* companion website at

http://www.prenhall.com/tarbuck

6

Glaciers, Deserts, and Wind

Glaciers and wind, like running water and groundwater in the previous chapter, are significant erosional and depositional processes responsible for creating a wide variety of landforms. They are important parts of the Earth system and play integral roles in the rock cycle by transporting and depositing the products of weathering. Today glaciers cover approximately 10 percent of Earth's land surface. However, many regions of the north and central United States bear the markings of much more extensive ice sheets that once covered vast areas with ice several thousand meters thick.

Chapter six begins with an examination of how glaciers form and move. Also investigated are the erosional and depositional features produced by ice. The second part of the chapter is devoted to dry lands and the geologic work of wind. Because desert and near-desert conditions prevail over about 30 percent of Earth's land surface, the nature of such landscapes is indeed worth exploring.

Learning Objectives

After reading, studying, and discussing this chapter, you should be able to:

Describe the types and locations of glaciers.

A *glacier* is a thick mass of ice originating on the land from the compaction and recrystallization of snow, and shows evidence of past or present flow. Today, *valley* or *alpine glaciers* are found in mountain areas, where they usually follow valleys that were originally occupied by streams. *Ice sheets* exist on a much larger scale, covering most of Greenland and Antarctica.

Discuss glacial movement.

Near the surface of a glacier, in the *zone of fracture*, ice is brittle. However, below about 50 meters, pressure is great, causing ice to *flow* like a *plastic material*. A second important mechanism of glacial movement consists of the whole ice mass *slipping* along the ground.

Describe the budget of a glacier.

Glaciers form in areas where more snow falls in winter than melts in summer. Snow accumulation and ice formation occur in the *zone of accumulation*. Beyond this area is the *zone of wastage*, where there is a net loss to the glacier. The glacial budget is the balance, or lack of balance, between accumulation at the upper end of the glacier and loss at the lower end.

Describe how glaciers erode and list several features of glacial erosion.

Glaciers erode land by *plucking* (lifting pieces of bedrock out of place) and *abrasion* (grinding and scraping of a rock surface). Erosional features produced by valley glaciers include *glacial troughs, hanging valleys, cirques, arêtes, horns, and fiords*.

List the types of glacial drift.

Any sediment of glacial origin is called *drift*. The two distinct types of glacial drift are (1) *till*, which is unsorted sediment deposited directly by the ice; and (2) *stratified drift*, which is relatively well-sorted sediment laid down by glacial meltwater.

Discuss the most widespread features created by glacial deposition.

The most widespread features created by glacial deposition are layers or ridges of till, called *moraines*. Associated with valley glaciers are *lateral moraines*, formed along the sides of the valley, and *medial moraines*, formed between two valley glaciers that have joined. *End moraines*, which mark the former position of the front of a glacier, and *ground moraine*, an undulating layer of till deposited as the ice front retreats, are common to both valley glaciers and ice sheets.

List several effects of Ice Age glaciers.

Perhaps the most convincing evidence for the occurrence of several glacial advances during the *Ice Age* is the widespread existence of *multiple layers of drift* and an uninterrupted record of climate cycles preserved in *seafloor sediments*. In addition to massive erosional and depositional work, other effects of Ice Age glaciers included the *migration* of organisms, *changes in stream courses, adjustment of the crust* by rebounding after the removal of the immense load of ice, and *climate changes* caused by the existence of the glaciers themselves. In the sea, the most far-reaching effect of the Ice Age was the *worldwide change in sea level* that accompanied each advance and retreat of the ice sheets.

List the theories for the causes of glacial ages.

Any theory that attempts to explain the causes of glacial ages must answer the two basic questions: (1) What causes the onset of glacial conditions? and (2) What caused the alternating glacial and interglacial stages that have been documented for the Pleistocene epoch? Two of the many hypotheses for the cause of glacial ages involve (1) plate tectonics and (2) variations in Earth's orbit. Other factors that are related to climate change during glacial ages include changes in atmospheric composition, variations in the amount of sunlight reflected by Earth's surface, and changes in ocean circulation.

Discuss the roles of wind and water in arid climates.

Practically all desert streams are dry most of the time and are said to be *ephemeral*. Nevertheless, *running water is responsible for most of the erosional work in a desert*. Although wind erosion is more significant in dry areas than elsewhere, the main role of wind in a desert is in the transportation and deposition of sediment.

Describe the geologic evolution of the Basin and Range region.

Many of the landscapes of the Basin and Range region of the western and southwestern United States are the result of streams eroding uplifted mountain blocks and depositing the sediment in interior basins. *Alluvial fans, playas*, and *playa lakes* are features often associated with these landscapes.

Discuss the mechanisms of wind erosion.

For wind erosion to be effective, dryness and scant vegetation are essential. *Deflation*, the lifting and removal of loose material, often produces shallow depressions called *blowouts*.

Describe the features produced by wind erosion.

Desert pavement is a thin layer of coarse pebbles and cobbles that covers some desert surfaces. Once established, it protects the surface from further deflation. Depending on circumstances, it may develop as a result of deflation or deposition of fine particles.

Abrasion, the sandblasting effect of wind, is often given too much credit for producing desert features. However, abrasion does cut and polish rock near the surface.

List the types of wind deposits and describe the features of wind deposition.

Wind deposits are of two distinct types: (1) extensive *blankets of silt*, called *loess*, that is carried by wind in *suspension*; and (2) *mounds and ridges of sand*, called *dunes*, which are formed from sediment that is carried as part of the wind's *bed load*. The *types of sand dunes* include (1) *barchan dunes*, solitary dunes shaped like crescents with their tips pointing downwind; (2) *transverse dunes*, which form a series of long ridges orientated at right angles to the prevailing wind; (3) *longitudinal dunes*, long ridges that are more or less parallel to the prevailing wind; (4) *parabolic dunes*, similar in shape to barchans except that their tips point into the wind; and (5) *star dunes*, isolated hills of sand that exhibit a complex form.

Key Terms

ablation
abrasion
alluvial fan
alpine glacier
arête
barchan dune
barchanoid dune
blowout
cirque
crevasse
cross beds
deflation
desert pavement
drumlin
end moraine
ephemeral stream
esker
fiord
glacial drift
glacial erratic
glacial striations
glacial trough
glacier
ground moraine
hanging valley
horn
ice cap
ice sheet
ice shelf
interior drainage
kame
kettle
lateral moraine
loess
longitudinal dune
medial moraine
outwash plain
parabolic dune
piedmont glacier
playa lake
Pleistocene epoch
plucking
pluvial lake
rock flour
slip face
star dune
stratified drift
till
transverse dune
valley glacier
valley train
zone of accumulation
zone of wastage

Vocabulary Review

Choosing from the list of key terms, furnish the most appropriate response for the following statements.

1. The largest type of glacier is the __________, which often covers a large portion of a continent.
2. A(n) __________ is a type of stream that carries water only in response to a specific episode of rainfall.
3. The all-embracing term for sediments of glacial origin, no matter how, where, or in what form they were deposited is __________.
4. A(n) __________ is a streamlined asymmetrical hill composed of till.
5. Most of the major glacial episodes occurred during a division of the geologic time scale known as the __________.
6. Embedded material in a glacier may gouge long scratches and grooves called __________ in the bedrock as the ice passes over.
7. Arid regions typically lack permanent streams and are often characterized by a type of drainage called __________.
8. A(n) __________ is a type of moraine that forms when two valley glaciers coalesce to form a single ice stream.
9. The area where there is net loss to a glacier due to melting is known as the __________.
10. A(n) __________ is a U-shaped valley produced by the erosion of a valley glacier.
11. A(n) __________ is a hollowed-out, bowl-shaped depression at the head of a valley produced by glacial erosion.
12. A sand dune shaped like a crescent with its tips pointing downwind is called a(n) __________.
13. The part of a glacier where snow accumulates and ice forms is called the __________.
14. As a stream emerges from a canyon and quickly loses velocity, it often deposits a cone of debris known as a(n) __________.
15. A long ridge of sand that forms more or less parallel to the prevailing wind direction is known as a(n) __________.
16. Deposits of windblown silt are called __________.

17. Materials deposited directly by glacial ice are known as __________.
18. A broad ramplike surface of stratified drift that is built adjacent to the downstream edge of an ice sheet is termed a(n) __________.
19. Deflation by wind often forms a shallow depression called a(n) __________.
20. The term applied to the sloping layers of sand that compose a dune is __________.
21. A(n) __________ is a thick mass of ice that forms over land from the accumulation, compaction, and re-crystallization of snow.
22. A glacier that occupies the broad lowland at the base of a steep mountain is called a(n) __________.
23. The budget of a glacier depends on the balance between accumulation and __________.
24. A large mass of floating ice that extends seaward from the coast, but remains attached to the land is called a(n) __________.

Comprehensive Review

1. List and describe the two major types of glaciers.

    1)

    2)

2. Why are the uppermost 50 meters of a glacier appropriately referred to as the zone of fracture?

3. Briefly describe the following two ways that glaciers erode land.

    a) Plucking:

    b) Abrasion:

4. The area represented in Figure 6.1 was subjected to alpine glaciation. Select the appropriate letter in the figure that identifies each of the following features.

    a) Cirque: _____

    b) Glacial trough: _____

    c) Hanging valley: _____

    d) Horn: _____

    e) Arête: _____

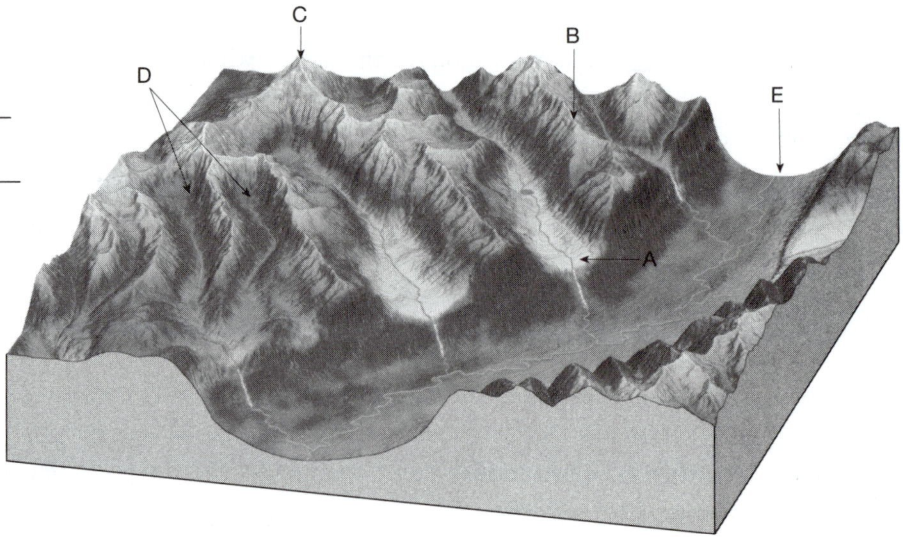

Figure 6.1

5. Distinguish between the two terms *till* and *stratified drift*.

6. What are two hypotheses relating to the possible causes of glacial periods?

7. In Figure 6.2, diagram __________ [A, B] illustrates the early stage in the evolution of the desert landscape. Select the appropriate letter in the figure that identifies each of the following features.

   a) Playa lake: __________

   b) Alluvial fan: __________

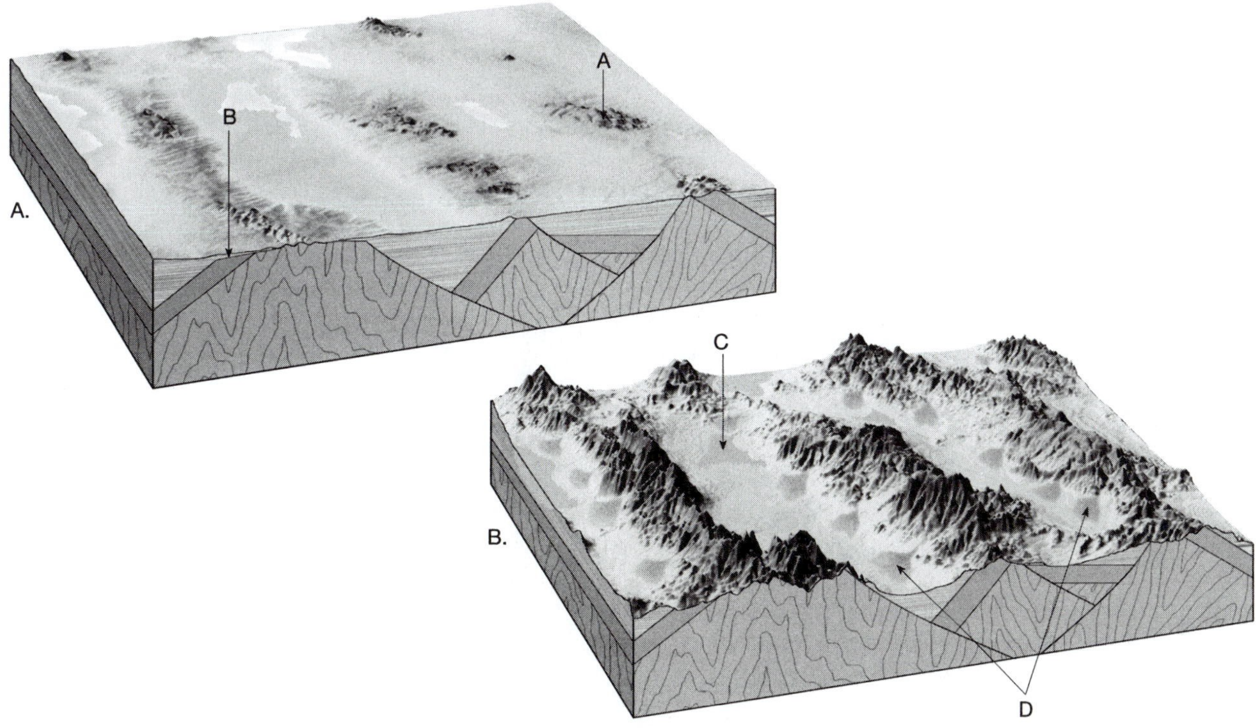

Figure 6.2

8. Describe each of the following types of moraines.

   a) End moraine:

   b) Lateral moraine:

   c) Ground moraine:

9. The area represented in Figure 6.3 is being subjected to glaciation by an ice sheet. Select the appropriate letter in the figure that identifies each of the following features.

   a) Drumlin: _____ c) Esker: _____

   b) Outwash plain: _____ d) End moraine: _____

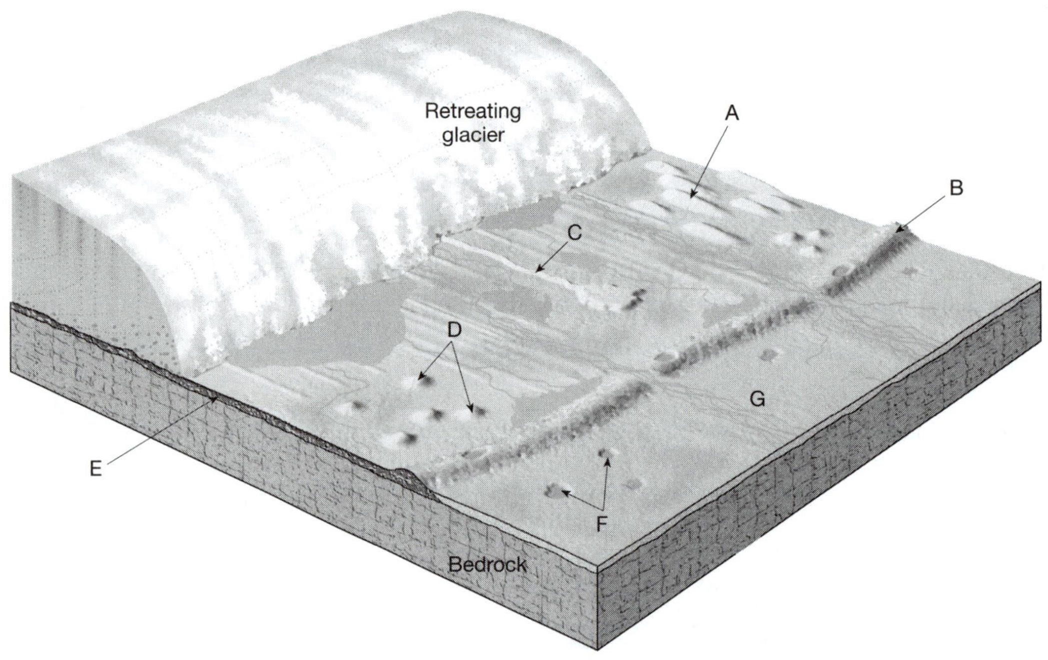

Figure 6.3

10. Briefly describe the process responsible for the migration of sand dunes.

11. What were some indirect effects of Ice Age glaciers?

12. What features would you look for in a mountainous region to determine if the area had been glaciated by alpine glaciers?

13. What evidence suggests that there were several glacial advances during the Ice Age?

14. In what two ways does the transport of sediment by wind differ from that by running water?

15. Referring to Figure 6.4, write the name of the type of dune illustrated by each diagram.

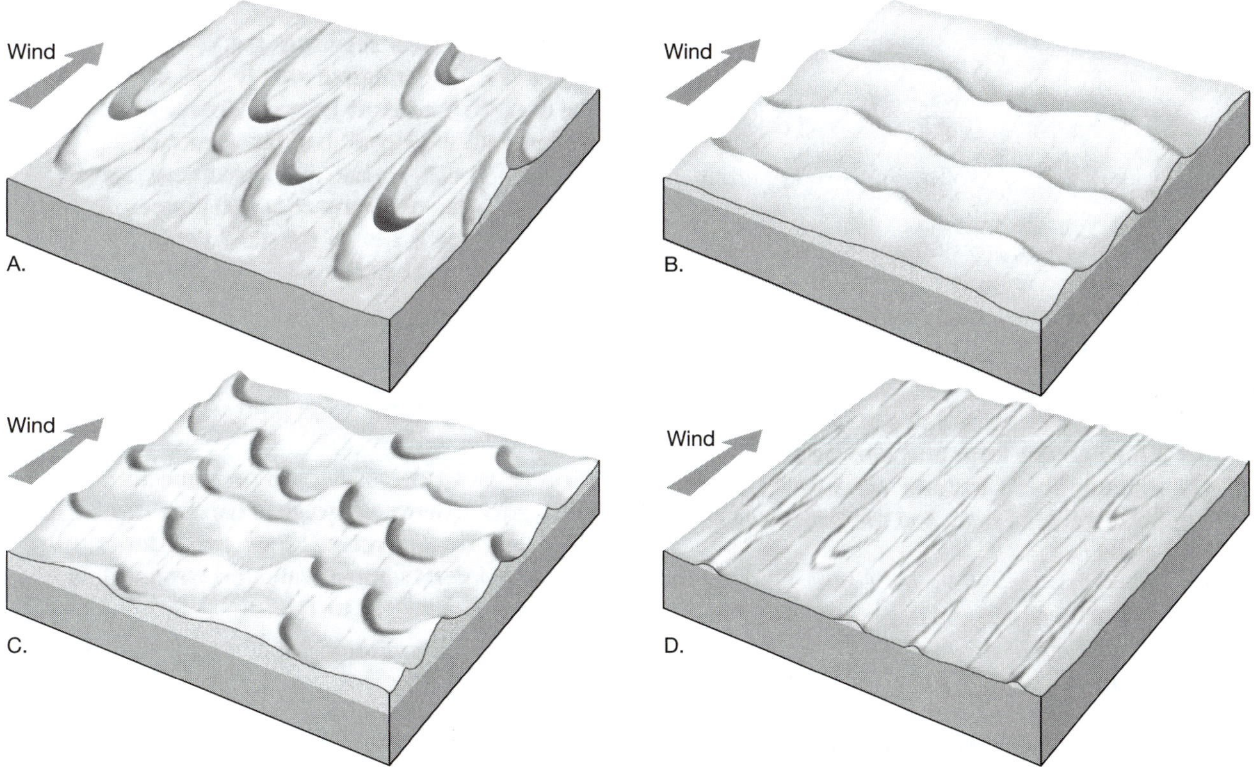

Figure 6.4

Practice Test

Multiple choice. Choose the best answer for the following multiple-choice questions.

1. A broad, ramplike surface of stratified drift built adjacent to the downstream edge of most end moraines is called a(n) __________.
   a) moraine
   b) valley train
   c) outwash plain
   d) fiord
   e) kettle

2. Icebergs are produced when large pieces of ice break off from the front of a glacier during a process termed __________.
   a) ablation
   b) deflation
   c) calving
   d) abrasion
   e) plucking

3. Which type of dune will form at a right angle to the prevailing wind when there is abundant sand, little or no vegetation, and a constant wind direction?
   a) barchan dune
   b) transverse dune
   c) longitudinal dune
   d) parabolic dune
   e) star dune

4. Most of the major glacial episodes during the Ice Age occurred during a division of the geologic time scale called the __________ epoch.
   a) Pliocene
   b) Eocene
   c) Paleocene
   d) Pleistocene
   e) Miocene

5. The two major ways that glaciers erode land are abrasion and __________.
   a) plucking
   b) slipping
   c) deflation
   d) gouging
   e) scouring

6. Which one of the following is NOT an effect that Pleistocene glaciers had upon the landscape?
   a) mass extinctions
   b) crustal depression and rebounding
   c) animal and plant migration
   d) adjustments of stream courses
   e) worldwide change in sea level

7. A thick ice mass that forms over the land from the accumulation, compaction, and recrystallization of snow is a __________.
   a) wadi
   b) fiord
   c) drumlin
   d) glacier
   e) loess

8. The __________ drainage of arid regions is characterized by intermittent streams that do not flow out of the desert to the ocean.
   a) radial
   b) interior
   c) exterior
   d) excellent
   e) well-developed

9. The all-embracing term for sediments of glacial origin is __________.
   a) outwash
   b) glacial drift
   c) loess
   d) silt
   e) erratic

10. The end moraine that marks the farthest advance of a glacier is called the __________ moraine.
    a) lateral
    b) terminal
    c) medial
    d) ground
    e) recessional

11. Cracks that form in the zone of fracture of a glacier are called __________.
    a) fractures
    b) crevasses
    c) faults
    d) gouges
    e) kettles

12. Desert streams that carry water only in response to specific episodes of rainfall are said to be __________.
    a) episodic
    b) flowing
    c) occasional
    d) ephemeral
    e) youthful

13. Sinuous ridges composed of sand and gravel deposited by streams flowing in tunnels beneath glacial ice are __________.
    a) eskers
    b) cirques
    c) drumlins
    d) horns
    e) moraines

14. Materials that have been deposited directly by a glacier are called __________.
    a) outwash
    b) sediment
    c) till
    d) loess
    e) stratified drift

15. Valleys that are left standing high above the main trough of a receding valley glacier are termed __________ valleys.
    a) ephemeral
    b) wadis
    c) glacial
    d) fiord
    e) hanging

16. The thickest and most extensive loess deposits occur in western and northern __________.
    a) Canada
    b) China
    c) Australia
    d) Libya
    e) Egypt

17. Glacial troughs that have become deep, steep-sided inlets of the sea are called __________.
    a) fiords
    b) hanging valleys
    c) cirques
    d) washes
    e) kames

18. Moraines that form when two valley glaciers coalesce to form a single ice stream are termed __________ moraines.
    a) lateral
    b) medial
    c) terminal
    d) ground
    e) recessional

19. Boulders found in glacial till or lying free on the surface are called glacial __________.
    a) remnants
    b) striations
    c) drumlins
    d) moraines
    e) erratics

20. Which one of the following is NOT a feature associated with valley glaciers?
    a) arête
    b) glacial trough
    c) horn
    d) arroyo
    e) cirque

21. Dry, flat lake beds located in the center of basins in arid areas are called __________.
    a) playas
    b) arroyos
    c) kames
    d) deltas
    e) alluvial fans

22. The most noticeable result of deflation in some places are shallow depressions called __________.
    a) kettles
    b) blowouts
    c) dunes
    d) drumlins
    e) sinkholes

23. Evidence indicates that in addition to the Pleistocene epoch, there were at least __________ earlier periods of glacial activity.
    a) two
    b) three
    c) four
    d) five
    e) six

24. The leeward slope of a dune is called the __________ face.
    a) cross
    b) slope
    c) wind
    d) slip
    e) gradual

25. Which one of the following states is included in the Basin and Range region?
    a) Nevada
    b) Missouri
    c) Texas
    d) New York
    e) Florida

True/false. For the following true/false questions, if a statement is not completely true, mark it false. For each false statement, change the **italicized** *word to correct the statement.*

1. _____ During the Ice Age, ice sheets and alpine glaciers were far *more* extensive than they are today.
2. _____ Snow accumulation and glacial ice formation occur in an area known as the zone of *wastage*.
3. _____ Much of the weathered debris in deserts is the result of *mechanical* weathering processes.
4. _____ *Medial* moraines form along the sides of valley glaciers.
5. _____ The two major types of glaciers are *valley* glaciers and ice sheets.
6. _____ Under pressure equivalent to more than the weight of *10* meters of ice, ice will behave as a plastic and flow.
7. _____ The Ice Age began between two and three *million* years ago.
8. _____ The position of the front of a glacier depends on the balance between *accumulation* and wastage.
9. _____ The combined areas of present-day continental ice sheets represents almost *30* percent of Earth's land area.
10. _____ Dry regions of the world encompass nearly *60* percent of Earth's land surface.
11. _____ The rates that glaciers advance vary considerably from one glacier to another but can be as great as several *kilometers* per day.
12. _____ Desert floods arrive suddenly and subside *quickly*.
13. _____ Glaciated valleys typically exhibit a *V-shaped* appearance.
14. _____ Streamlined asymmetrical hills composed of till occurring in clusters are called *eskers*.
15. _____ The movement of glacial ice is generally referred to as *flow*.
16. _____ *Glaciers* are capable of carrying huge blocks of material that no other erosional agent could budge.
17. _____ In *humid* regions, moisture binds particles together and vegetation anchors the soil so that wind erosion is negligible.
18. _____ The Matterhorn in the Swiss Alps is the most famous example of a *cirque*.
19. _____ Sediments laid down by glacial meltwater are called stratified *drift*.
20. _____ *Moraines* are bowl-shaped erosional features found at the heads of glacial valleys.

21. _____ In the United States, *loess* deposits are an indirect product of glaciation.
22. _____ Continued sand accumulation, coupled with periodic slides down the *slip* face, causes the slow migration of a sand dune in the direction of air movement.
23. _____ The tips of *parabolic* dunes point into the wind.

Word choice. Complete each of the following statements by selecting the most appropriate response.

1. Moraines are composed of [stratified drift/till/loess].
2. The pulverized rock produced by glacial abrasion is called rock [till/loess/flour].
3. Most of the weathered debris in deserts is the result of [mechanical/chemical] weathering.
4. Most deserts have [exterior/interior] drainage.
5. [Kettles/Fiords] are deep, steep-sided ocean inlets formed by glacial erosion.
6. An individual boulder, different from the local bedrock, that was deposited by a glacier is called a glacial [erratic/kame].
7. The breaking off of large pieces of ice at the front of a glacier is a process called [calving/crevassing].
8. Stratified drift is [sorted/unsorted] sediment.
9. Desert streams often have [many/few] tributaries.
10. Erosion by ice sheets tends to [accentuate/subdue] the topography of a region.
11. Desert pavement is created by [abrasion/deflation].
12. Stream erosion in most desert regions [is/is not] highly influenced by sea level.
13. Loess deposits [have/lack] visible layers.

Written questions

1. List and briefly describe at least four glacial depositional features.

2. List and briefly describe at least three erosional features you might expect to find in an area where valley glaciers exist, or have recently existed.

3. List three indirect effects of Ice Age glaciers.

4. Describe how sand dunes migrate.

For other interesting and pertinent information, be sure to visit the *Earth Science* companion website at

http://www.prenhall.com/tarbuck

7

Plate Tectonics: A Scientific Theory Unfolds

One of the most significant scientific revelations of the past century was the discovery of the fact that continents gradually migrate across the globe. Where landmasses split apart, new ocean basins are created between the diverging plates. Meanwhile, older portions of the seafloor are carried back into Earth's mantle in regions where trenches occur in the deep ocean floor. Because of these movements, blocks of continental crust eventually collide and form Earth's great mountain ranges. Although the vast majority of the scientific community rejected the original hypothesis, called continental drift, proposed in the early 1900s, beginning in the mid-1900s new evidence led to a far more encompassing and acceptable explanation, a theory known as plate tectonics.

Chapter seven briefly traces the development of the concept of continental drift, the reasons why the hypothesis was rejected, and the evidence that led to the acceptance of the theory of plate tectonics. Also investigated are the driving forces for plate motions and the future implications of plate tectonics.

Learning Objectives

After reading, studying, and discussing this chapter, you should be able to:

List the evidence that was used to support the continental drift hypothesis.

In the early 1900s *Alfred Wegener* set forth his *continental drift* hypothesis. One of its major tenets was that a supercontinent called *Pangaea* began breaking apart into smaller continents about 200 million years ago. The smaller continental fragments then "drifted" to their present positions. To support the claim that the now-separate continents were once joined, Wegener and others used the *fit of South America and Africa, ancient climatic similarities, fossil evidence,* and *rock structures.*

Describe the main objection to the continental drift hypothesis.

One of the main objections to the continental drift hypothesis was its inability to provide an acceptable mechanism for the movement of continents.

Explain the differences between the continental drift hypothesis and the theory of plate tectonics.

The theory of *plate tectonics,* a far more encompassing theory than continental drift, holds that Earth's rigid outer shell, called the *lithosphere,* consists of numerous segments called *plates* that are in motion relative to one another. Most of Earth's *seismic activity, volcanism,* and *mountain building* occur along the dynamic margins of these plates.

Describe the major departure of the plate tectonics theory from continental drift.

A major departure of the plate tectonics theory from the continental drift hypothesis is that large plates contain both continental and ocean crust and the entire plate moves. By contrast, in continental drift, Wegener proposed that the sturdier continents "drifted" by breaking through the oceanic crust, much like ice breakers cut through ice.

Describe divergent plate boundaries.

Divergent plate boundaries occur where plates move apart, resulting in upwelling of material from the mantle to create new seafloor. Most divergent boundaries occur along the axis of the oceanic ridge system and are associated with seafloor spreading, which occurs at rates of 2 to 15 centimeters (1 and 6 inches) per year. New divergent boundaries may form within a continent (for example, the East African Rift Valleys), where they may fragment a landmass and develop a new ocean basin.

Describe convergent plate boundaries.

Convergent plate boundaries occur where plates move together, resulting in the subduction of oceanic lithosphere into the mantle along a deep oceanic trench. Convergence between an oceanic and continental block results in subduction of the oceanic slab and the formation of a *continental volcanic arc* such as the Andes of South America. Oceanic-oceanic convergence results in an arc-shaped chain of volcanic islands called a *volcanic island arc*. When two plates carrying continental crust converge, both plates are too buoyant to be subducted. The result is a "collision" resulting in the formation of a mountain belt such as the Himalayas.

Describe transform fault boundaries.

Transform fault boundaries occur where plates grind past each other without the production or destruction of lithosphere. Most transform faults join two segments of a mid-oceanic ridge. Others connect spreading centers to subduction zones and thus facilitate the transport of oceanic crust created at a ridge crest to its site of destruction, at a deep-ocean trench. Still others, like the San Andreas Fault, cut through continental crust.

List and describe the evidence used to support the plate tectonics theory.

The theory of plate tectonics is supported by (1) *paleomagnetism,* the direction and intensity of Earth's magnetism in the geologic past, (2) the ages of *sediments* from the floors of the deep-ocean basins, and (3) the existence of island groups that formed over *hot spots* and provide a frame of reference for tracing the direction of plate motion.

Discuss the models that have been proposed to explain the driving mechanism for plate motion.

Mechanisms that contribute to this convective flow are slab pull, ridge push, and mantle plumes. *Slab pull* occurs where cold, dense oceanic lithosphere is subducted and pulls the trailing lithosphere along. *Ridge push* results when gravity sets the elevated slabs astride ocean ridges in motion. Hot, buoyant *mantle plumes* are considered the upward flowing arms on mantle convection. One model suggests that mantle convection occurs in two layers separated at a depth of 660 kilometers (410 miles). Another model proposes whole-mantle convection that stirs the entire 2900-kilometer-thick (1800-miles thick) rocky mantle. Yet another model suggests that the bottom third of the mantle gradually bulges upward in some areas and sinks in others without appreciable mixing.

Key Terms

asthenosphere
continental drift
continental volcanic arc
convergent plate boundary
Curie point
deep-ocean trench
divergent plate boundary
fossil magnetism
fracture zone
hot spot
island arc
lithosphere
magnetic reversal
mantle plume
normal polarity
oceanic ridge system
paleomagnetism
Pangaea
partial melting
plate
plate tectonics
reverse polarity
ridge-push
rift (rift valley)
seafloor spreading
slab-pull
slab suction
subduction zone
transform boundary
volcanic island arc

Vocabulary Review

Choosing from the list of key terms, furnish the most appropriate response for the following statements.

1. In his 1915 book, *The Origin of Continents and Oceans,* Wegener set forth his radical hypothesis of __________.
2. Each of the rigid slabs of Earth's lithosphere is referred to as a(n) __________.
3. __________ is the mechanism responsible for producing the new seafloor between two diverging plates.
4. The type of plate boundary where plates move together, causing one of the slabs to be consumed into the mantle as it descends beneath an overriding plate, is referred to as a(n) __________.
5. Remnant magnetism in rock bodies is referred to as __________.
6. The theory of __________ holds that Earth's rigid outer shell consists of seven large and numerous smaller rigid segments.
7. When rocks exhibit the same magnetism as the present magnetic field, they are said to possess __________.
8. The type of plate boundary where plates move apart, resulting in upwelling of material from the mantle to create new seafloor, is referred to as a(n) __________.
9. A rising plume of mantle material often causes a(n) __________, such as the one responsible for the intraplate volcanism that produced the Hawaiian Islands.
10. Wegener hypothesized that about 200 million years ago a supercontinent that he called __________ began breaking into smaller continents, which then drifted to their present positions.
11. The region where an oceanic plate descends into the asthenosphere because of convergence is called a(n) __________.
12. Material displaced downward along spreading centers on continents often creates a downfaulted valley called a(n) __________.
13. A(n) __________ occurs where plates grind past each other without creating or destroying lithosphere.
14. When heated above a temperature known as the __________, magnetic minerals lose their magnetism.
15. Rocks that exhibit magnetism that is opposite of the present magnetic field are said to possess __________.
16. A chain of small volcanic islands that forms when two oceanic slabs converge, one descending beneath the other, is called a(n) __________.
17. As an oceanic plate slides beneath an overriding plate, a deep, linear feature, called a(n) __________, often forms on the ocean floor adjacent to the zone of subduction.
18. The __________ is an elevated portion of the seafloor that forms along well-developed divergent plate boundaries.

Comprehensive Review

1. List four lines of evidence that were used to support the continental drift hypothesis.

   1)

   2)

   3)

   4)

2. What was one of the main objections to Wegener's continental drift hypothesis?

3. Using Figure 7.1, select the letter of the diagram that portrays each of the following types of plate boundaries.

   a) Transform boundary: ________

   b) Divergent boundary: ________

   c) Convergent boundary: ________

4. Briefly explain the theory of plate tectonics.

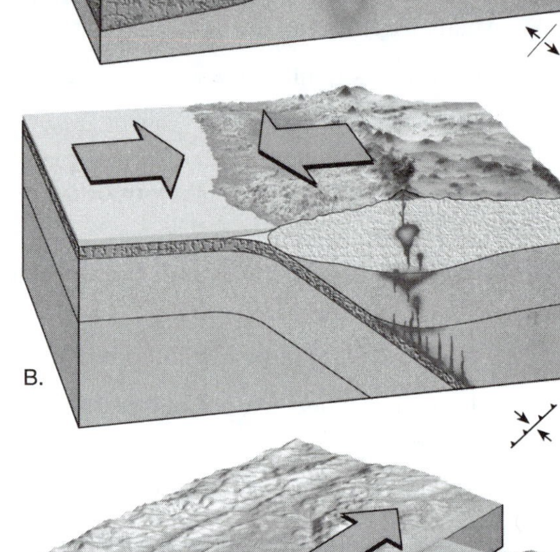

Figure 7.1

5. What are the three types of convergent plate boundaries?

   1)

   2)

   3)

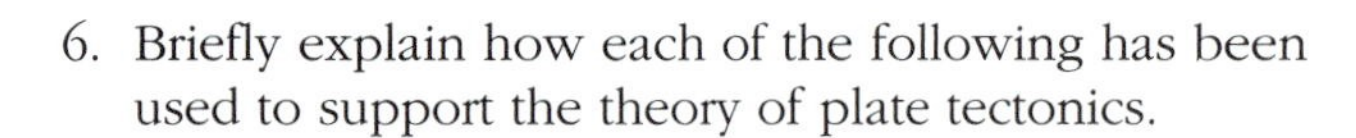

6. Briefly explain how each of the following has been used to support the theory of plate tectonics.

   a) Paleomagnetism:

   b) Ocean drilling:

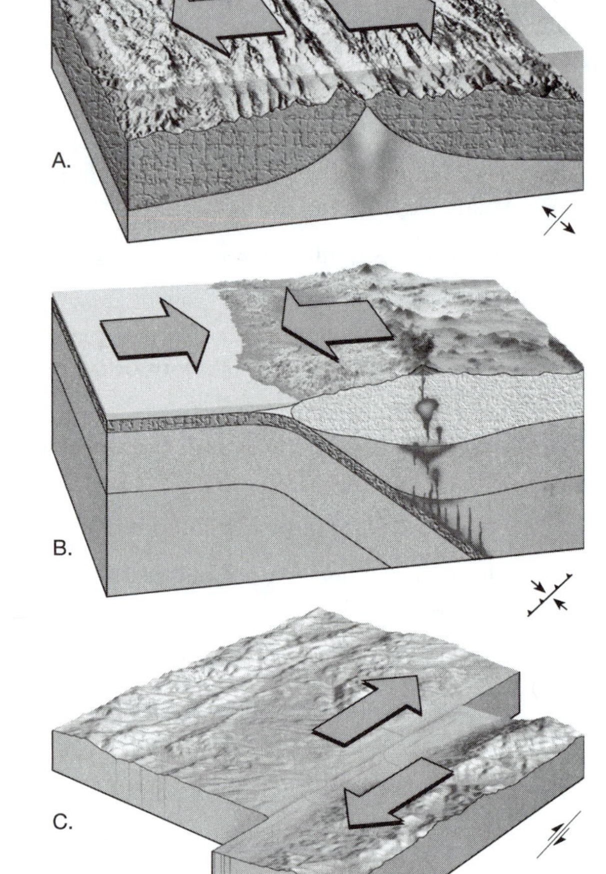

7. Using Figure 7.2, select the appropriate letter that identifies each of the following features.

   a) Subducting oceanic lithosphere: _____

   b) Trench: _____

   c) Continental volcanic arc: _____

   d) Continental lithosphere: _____

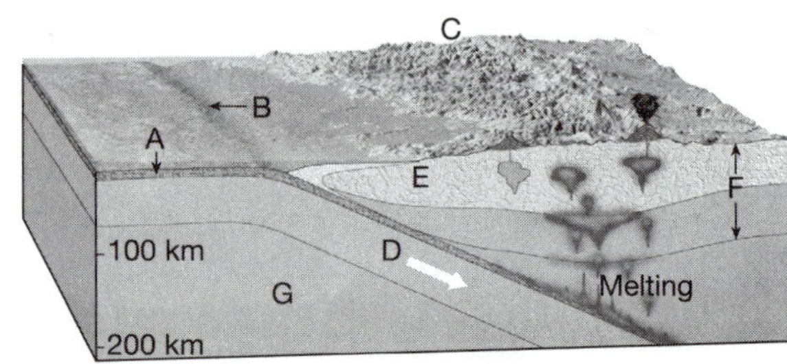

Figure 7.2

8. Briefly describe the relationship between the Himalaya mountains and plate tectonics.

9. List and briefly explain the hypotheses that have been proposed for the driving mechanism of plate motion.

10. Figure 7.3 illustrates the ancient supercontinent of Pangaea as it is thought to have existed about 300 million years ago. Using the figure, select the appropriate letter that identifies each of the following current-day landmasses.

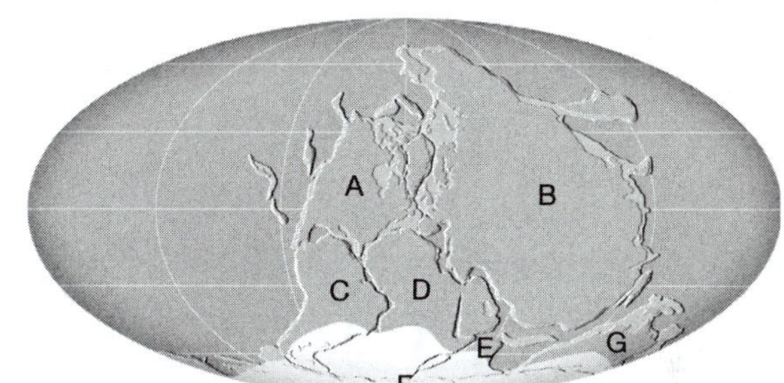

Figure 7.3

    a) Antarctica: __________

    b) Eurasia: __________

    c) South America: __________

    d) India: __________

    e) North America: __________

    f) Africa: __________

    g) Australia: __________

11. How are satellites being used to directly test the theory of plate tectonics?

Practice Test

Multiple choice. Choose the best answer for the following multiple-choice questions.

1. The general term that refers to the deformation of Earth's crust and results in the formation of structural features such as mountains is _____.
   a) erosion
   b) tectonics
   c) mass wasting
   d) volcanism
   e) subduction

2. During the last 4 million years, Earth's magnetic field has reversed _____.
   a) twice
   b) several times
   c) hundreds of times
   d) thousands of times

3. Most of Earth's seismic activity, volcanism, and mountain building occur along _____.
   a) lines of magnetism
   b) plate boundaries
   c) parallels of latitude
   d) random trends
   e) hot spots

4. The apparent movement of Earth's magnetic poles through time is referred to as _____.
   a) magnetic drift
   b) perturbation
   c) subduction
   d) polar spreading
   e) polar wandering

5. Alfred Wegener is best known for his hypothesis of _____.
   a) continental drift
   b) natural selection
   c) atoll formation
   d) subduction
   e) seafloor spreading

6. Complex mountain systems such as the Himalayas are the result of _____.
   a) oceanic-oceanic convergence
   b) hot spots
   c) oceanic-continental convergence
   d) continental-continental convergence
   e) continental volcanic arcs

7. The type of plate boundary where plates move together, causing one of the slabs of lithosphere to be consumed into the mantle as it descends beneath an overriding plate, is called a _____ boundary.
   a) divergent
   b) transitional
   c) transform
   d) gradational
   e) convergent

8. The spreading rate for the North Atlantic Ridge is about _____ centimeters per year.
   a) 2
   b) 10
   c) 15
   d) 20
   e) 25

9. The name given by Wegener to the supercontinent he believed existed prior to the current continents was _____.
   a) Pangaea
   b) Atlantis
   c) Euroamerica
   d) Pantheon
   e) Panamerica

10. Magnetic minerals lose their magnetism when heated above a certain temperature, called the _____.
    a) magnetopoint
    b) Curie point
    c) melting point
    d) break point
    e) Bechtol point

11. The best approximation of the true outer boundary of the continents is the seaward edge of the _____.
    a) deep-ocean trench
    b) abyssal plain
    c) mid-ocean ridge
    d) present-day shorelines
    e) continental shelf

12. The Aleutian, Mariana, and Tonga islands are examples of _____.
    a) volcanic island arcs
    b) abyssal plains
    c) transform boundaries
    d) hot spots
    e) mid-ocean ridges

13. Beneath Earth's lithosphere, the hotter, weaker zone known as the _____ allows for motion of Earth's rigid outer shell.
    a) crust
    b) asthenosphere
    c) outer core
    d) Moho
    e) oceanic crust

14. Which one of the following mountain systems is NOT the result of continental collisions?
    a) Alps
    b) Urals
    c) Andes
    d) Himalayas
    e) Appalachians

15. Which one of the following was NOT used in support of Wegener's continental drift hypothesis?
    a) fossil evidence
    b) paleomagnetism
    c) fit of South America and Africa
    d) ancient climates
    e) rock structures

16. The Red Sea is believed to be the site of a recently formed _____.
    a) divergent boundary
    b) ocean trench
    c) convergent boundary
    d) hot spot
    e) gradational boundary

17. The type of plate boundary where plates move apart, resulting in upwelling of material from the mantle to create new seafloor, is a _____ boundary.
    a) divergent
    b) transitional
    c) transform
    d) gradational
    e) convergent

18. One of the main objections to Wegener's hypothesis was his inability to provide an acceptable _____ for continental drift.
    a) time
    b) rate
    c) mechanism
    d) direction
    e) place

19. The spreading rate for the Mid-Atlantic Ridge is _____ than the rate for most of the East Pacific Rise.
    a) greater
    b) less

20. Earth's rigid outer shell is called the _____.
    a) asthenosphere
    b) mantle
    c) outer core
    d) continental mass
    e) lithosphere

21. Most deep-focus earthquakes occur in association with _____.
    a) hot spots
    b) ocean trenches
    c) oceanic ridge systems
    d) abyssal plains
    e) transform boundaries

22. The theory of plate tectonics holds that Earth's rigid outer shell consists of _____ major plates.
    a) seven
    b) ten
    c) fifteen
    d) twenty
    e) twenty-five

23. The age of the deepest sediment in an ocean basin _____ with increasing distance from the oceanic ridge.
    a) increases
    b) decreases
    c) remains the same
    d) varies

24. During oceanic-continental convergence, as the oceanic plate slides beneath the overriding plate, a _____ is often produced adjacent to the zone of subduction.
    a) deep-ocean terrace
    b) deep-ocean ridge
    c) transform fault
    d) deep-ocean trench
    e) divergent boundary

25. The chain of volcanic structures extending from the Hawaiian Islands to Midway Island and then continuing northward toward the Aleutian trench has formed over a _____ as the Pacific plate moved.
    a) subduction zone
    b) hot spot
    c) volcanic island arc
    d) convergent boundary
    e) divergent boundary

26. Which one of the following is NOT a hypothesis that has been proposed for the mechanism of plate motion?
    a) slab pull hypothesis
    b) ridge push hypothesis
    c) slab suction hypothesis
    d) mantle density hypothesis

True/false. For the following true/false questions, if a statement is not completely true, mark it false. For each false statement, change the **italicized** *word to correct the statement.*

1. _____ The *lithosphere* consists of both crustal rocks and a portion of the upper mantle.
2. _____ *Transform* faults connect convergent and divergent plate boundaries in various combinations.
3. _____ The rate of plate movement is measured in *kilometers* per year.
4. _____ The Red Sea is believed to be the site of a recently formed *convergent* plate boundary.
5. _____ In North America, the *Cascade Range* and Sierra Nevada system are continental volcanic arcs associated with the subduction of oceanic lithosphere.
6. _____ Older portions of the seafloor are carried into Earth's *core* in regions where trenches occur in the deep-ocean floor.
7. _____ The island of Hawaii is *older* than Midway Island.
8. _____ The age of the deepest ocean sediments *increases* with increasing distance from a ridge.
9. _____ *Transform* faults are roughly parallel to the direction of plate movement.
10. _____ To explain continental drift, Wegener proposed that the *continents* broke through the oceanic crust, much like ice breakers cut through ice.
11. _____ There is a close association between deep-focus earthquakes and ocean *ridges*.
12. _____ Beneath Earth's lithosphere is the hotter and weaker zone known as the *asthenosphere*.
13. _____ Seafloor spreading is the mechanism that has produced the floor of the *Atlantic* Ocean during the past 165 million years.
14. _____ The unequal distribution of heat inside Earth generates some type of thermal convection in the *crust* that ultimately drives plate motion.
15. _____ According to Wegener, the supercontinent *Pangaea* began breaking apart about 200 million years ago.
16. _____ When rocks exhibit the same magnetism as the present magnetic field, they are said to possess *reverse* polarity.
17. _____ Deep-ocean trenches are located adjacent to *subduction* zones.
18. _____ The *oldest* oceanic crust is located at the oceanic ridge crests.
19. _____ The largest single rigid slab of Earth's outer shell is the *Pacific* plate.
20. _____ The oldest sediments found in the ocean basins are *less* than 180 million years old.

21. _____ Along a *transform* plate boundary, plates grind past each other without creating or destroying lithosphere.
22. _____ The Aleutian, Mariana, and Tonga islands are volcanic island arcs associated with *oceanic-oceanic* plate convergence.
23. _____ Lithospheric plates are *thickest* in the ocean basins.
24. _____ The idea that Earth's *magnetic* poles had migrated through time is known as polar wandering.
25. _____ Using the dates of the most recent *magnetic* reversals, the rate at which spreading occurs at the various ridges can be determined.

Word choice. Complete each of the following statements by selecting the most appropriate response.

1. Tectonic plates are large segments of Earth's [lithosphere/asthenosphere].
2. The best way to determine the true shape of a continent is to trace the outer boundary of its [continental shelf/shoreline/mountain ranges].
3. At convergent plate boundaries, oceanic lithosphere is being [created/consumed].
4. Most large tectonic plates containing continental crust [also/do not] contain oceanic crust.
5. Most divergent plate boundaries are associated with [continental/oceanic] ridges.
6. Tectonic plates are [flexible/rigid] slabs of Earth materials.
7. The supercontinent Pangaea began breaking apart about [200/500] million years ago.
8. The coal fields of North America contain fossil evidence that these regions were once located in [tropical/polar] climates.
9. The region where one plate descends into the asthenosphere below another plate is called a [rifting/subduction] zone.
10. Most plates have [only/more than] one type of plate boundary.
11. At divergent plate boundaries, lithosphere is being [created/destroyed].
12. Wegener proposed that the portion of the supercontinent that is now South Africa was once centered over the [equator/South Pole].
13. The primary driving force for plate movement comes from the unequal distribution of [heat/gravity] within Earth.
14. Continental crust is [thicker/thinner] than oceanic crust.
15. As oceanic crust moves away from a spreading center, it becomes [warmer/cooler] and [less/more] dense.
16. Most deep-focus earthquakes are associated with [hot spots/subduction zones].
17. The average positions of the magnetic poles correspond closely to the [equator/geographic poles].
18. The study of paleomagnetism has provided evidence about the [rate/depth] of seafloor spreading.
19. Hot spots are believed to be created by [converging plates/mantle plumes].
20. The age of the deepest ocean sediments [increases/decreases] with increasing distance from an oceanic ridge crest.
21. Once a rock forms, changing its position [can/will not] change the magnetic alignment of its minerals.
22. Transform faults are roughly [perpendicular/parallel] to the direction of plate movement.
23. Earthquakes associated with divergent and transform boundaries have [deep/shallow] foci.
24. When spreading centers develop within a continent, valleys called [trenches/rifts] form.
25. The magma produced in a subduction zone often produces [oceanic ridges/volcanic island arcs].

26. A [faster/slower] spreading rate exists for the Pacific as compared to the Atlantic.
27. The increasing density of a slab of oceanic crust as it cools and moves away from a spreading center explains the [ridge push/slab pull] model for the driving force of plate tectonics.

Written questions

1. What relation exists between the ages of the Hawaiian Islands, hot spots, and plate tectonics?

2. List and briefly describe the three major types of plate boundaries.

3. List the lines of evidence used to support the theory of plate tectonics.

For other interesting and pertinent
information, be sure to visit
the *Earth Science* companion website at

http://www.prenhall.com/tarbuck

8

Earthquakes and Earth's Interior

Each year over 30,000 earthquakes strong enough to be felt occur worldwide. Fortunately, most are minor and do little damage. On the other hand, large earthquakes rank near the top of all destructive forces on Earth. Ongoing research helps reduce the devastation caused by major earthquakes and their associated perils and moves scientists closer to reliable earthquake prediction. Furthermore, using their knowledge of the causes, methods of propagation, and worldwide distribution of earthquakes, seismologists also utilize seismic vibrations to provide us with a more detailed understanding of Earth's interior structure and the processes that operate deep within the planet.

Learning Objectives

After reading, studying, and discussing this chapter, you should be able to:

Describe the cause of earthquakes.

Earthquakes are vibrations of Earth produced by the rapid release of energy from rocks that rupture because they have been subjected to stresses beyond their limit. This energy, which takes the form of waves, radiates in all directions from the earthquake's source, called the *focus*. The movements that produce most earthquakes occur along large fractures, called *faults*, that are associated with plate boundaries.

List the types of seismic waves and describe their propagation.

Two main groups of *seismic waves* are generated during an earthquake: (1) *surface waves*, which travel along the outer layer of Earth; and (2) *body waves*, which travel through Earth's interior. Body waves are further divided into *primary*, or *P, waves*, which push (compress) and pull (dilate) rocks in the direction the wave is traveling, and *secondary*, or *S, waves*, which "shake" the particles in rock at right angles to their direction of travel. P waves can travel through solids, liquids, and gases. Fluids (gases and liquids) will not transmit S waves. In any solid material, P waves travel about 1.7 times faster than S waves.

Describe how an earthquake epicenter is located.

The location on Earth's surface directly above the focus of an earthquake is the *epicenter*. An epicenter is determined using the difference in velocities of P and S waves.

Describe the worldwide distribution of earthquake epicenters.

There is a close correlation between earthquake epicenters and plate boundaries. The principal earthquake epicenter zones are along the outer margin of the Pacific Ocean, known as the *circum-Pacific belt*, and through the world's oceans along the *oceanic ridge system*.

Explain how the magnitude of an earthquake is determined.

Seismologists use two fundamentally different measures to describe the size of an earthquake—intensity and magnitude. *Intensity* is a measure of the degree of ground shaking at a given locale based on the amount of damage. The *Modified Mercalli Intensity Scale* uses damage to buildings in California to estimate the intensity of ground shaking for a local earthquake. *Magnitude* is calculated from seismic records and estimates the amount of energy released at the source of an earthquake. Using the *Richter scale*, the magnitude of an earthquake is estimated by measuring the *amplitude* (maximum displacement) of the largest seismic wave recorded. A logarithmic scale is used to express magnitude, in which a tenfold increase in ground shaking corresponds to an increase of 1 on the magnitude scale. *Moment magnitude* is currently used to estimate the size of moderate and large earthquakes. It is calculated using the average displacement of the fault, the area of the fault surface, and the sheer strength of the faulted rock.

Discuss the destruction that often accompanies an earthquake.

The most obvious factors that determine the amount of destruction accompanying an earthquake are the *magnitude* of the earthquake and the *proximity* of the quake to a populated area. *Structural damage* attributable to earthquake vibrations depends on several factors, including (1) *intensity*, (2) *duration* of the vibrations, (3) *nature of the material* upon which the structure rests, and (4) the *design* of the structure. Secondary effects of earthquakes include *tsunamis, landslides, ground subsidence*, and *fire*.

Discuss the status of earthquake prediction.

Substantial research to predict earthquakes is under way in Japan, the United States, China, and Russia—countries where earthquake risk is high. No consistent method of short-range prediction has yet been devised. Long-range forecasts are based on the premise that earthquakes are repetitive or cyclical. Seismologists study the history of earthquakes for patterns so their occurrences might be predicted.

List the major zones of Earth's interior.

Earth's internal structure is divided into layers based on differences in chemical composition and on the basis of changes in physical properties. Compositionally, Earth is divided into a thin outer *crust,* a solid, rocky *mantle,* and a dense *core.* Based on physical properties, the layers of Earth are (1) the *lithosphere*—the cool, rigid, outermost layer that averages about 100 kilometers thick, (2) the *asthenosphere,* a relatively weak layer located in the mantle beneath the lithosphere, (3) the more rigid *lower mantle* where rocks are very hot and capable of very gradual flow, (4) the liquid *outer core*, where Earth's magnetic field is generated, and (5) the solid *inner core.*

Describe the composition of Earth's interior.

The *continental crust* has an average composition of a *granitic* rock called *granodiorite*, whereas the *oceanic crust* has a *basaltic* composition. Rocks similar to *peridotite* make up the *mantle.* The *core* is made up mainly of *iron and nickel.* An iron core explains the high density of Earth's interior as well as Earth's magnetic field.

Key Terms

aftershock	foreshock	outer core
asthenosphere	inner core	primary (P) wave
body wave	intensity	Richter scale
crust	liquefaction	secondary (S) wave
earthquake	lithosphere	seismic sea wave (tsunami)
elastic rebound	magnitude	seismogram
epicenter	mantle	seismograph
fault	Modified Mercalli Intensity scale	seismology
fault creep	Mohorovicic discontinuity (Moho)	shadow zone
focus	moment magnitude	surface wave

Vocabulary Review

Choosing from the list of key terms, furnish the most appropriate response for the following statements.

1. The vibration of Earth produced by a rapid release of energy is called a(n) __________.
2. A large ocean wave that often results from vertical displacement of the ocean floor during an earthquake is called a(n) __________.

3. A small earthquake, called a(n) __________, often precedes a major earthquake by days or, in some cases, by as much as several years.
4. The source of an earthquake is called the __________.
5. The "springing back" of rock after it has been deformed, much like a stretched rubber band does when released, is termed __________.
6. The type of earthquake wave that travels along the outer layer of Earth is the __________.
7. Earth's __________ is a layer of the interior that exhibits the characteristics of a mobile liquid composed of iron and nickel.
8. The __________ of an earthquake is the location on the surface directly above the focus.
9. The zone between about 100 kilometers and 350 kilometers within Earth that consists of hot, weak rock that is easily deformed is called the __________.
10. A fracture in rock along which displacement has occurred is termed a(n) __________.
11. Earth's very thin outer layer is called the __________.
12. The type of earthquake wave that "shakes" the particles at right angles to the direction it is traveling is the __________.
13. A recording, or trace, of an earthquake is called a(n) __________.
14. The boundary that separates the crust from the mantle is known as the __________.
15. Using the Richter scale, the __________ of an earthquake is determined by measuring the amplitude of the largest wave recorded on the seismogram.
16. Earth's __________ is a solid metallic sphere about 1216 kilometers in radius.
17. The type of earthquake wave that pushes and pulls rock in the direction it is traveling is the __________.
18. The cool, rigid layer of Earth, which includes the entire crust as well as the uppermost mantle, is the __________.
19. The __________ determines the magnitude of an earthquake by measuring the amplitude of the largest wave recorded on a seismogram.
20. The __________ is a belt from about 105 degrees to 140 degrees distance from an earthquake epicenter in which direct P waves are absent because of refraction by Earth's core.
21. The rocky layer of Earth located beneath the crust and having a thickness of about 2900 kilometers is the __________.
22. An instrument that records earthquake waves is called a(n) __________.
23. __________ is currently used to estimate the size of moderate and large earthquakes.

Comprehensive Review

1. List and describe the three basic types of earthquake waves.

   1)

   2)

   3)

2. Using Figure 8.1, select the letter of the diagram that illustrates the characteristic motion of the following types of earthquake waves.

    a) P wave: __________

    b) S wave: __________

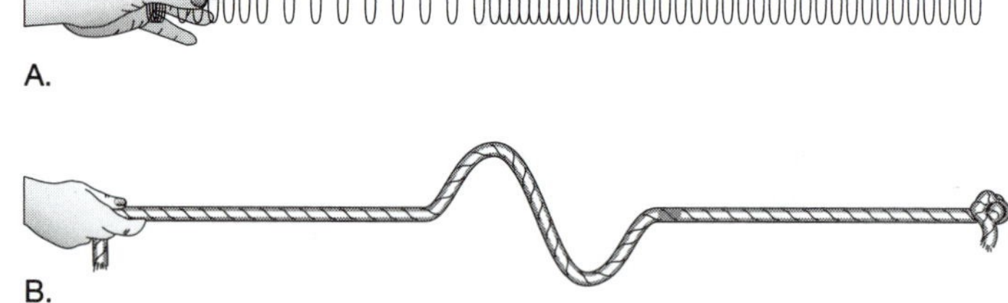

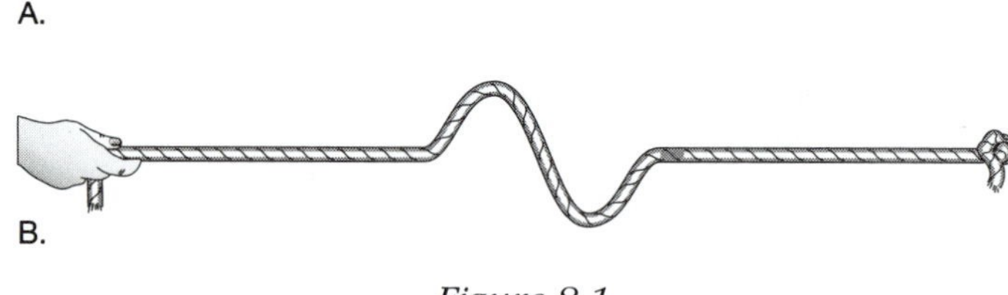

Figure 8.1

3. Figure 8.2 is a typical recording of an earthquake, called a seismogram. Using Figure 8.2, select the letter on the figure that identifies the recording of each of the following types of earthquake waves.

    a) Surface waves: __________

    b) S waves: __________

    c) P waves: __________

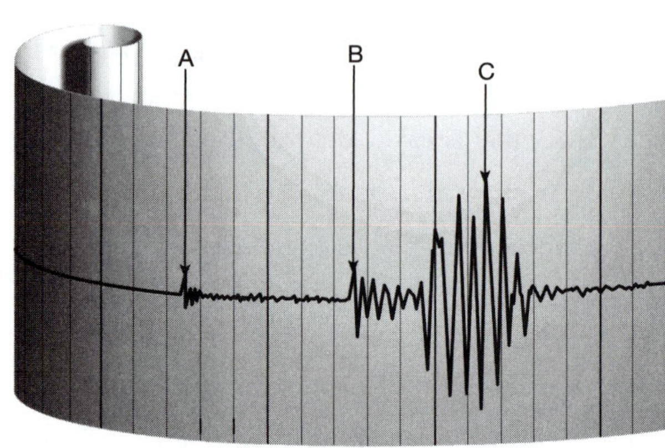

Figure 8.2

4. Describe the differences in velocity and mode of travel between P waves and S waves.

5. Briefly describe how the epicenter of an earthquake is located.

6. What are two factors that determine the degree of destruction that accompanies an earthquake?

    1)

    2)

7. List two major, continuous earthquake belts. With what general Earth feature are most earthquake epicenters closely correlated?

8. Using Figure 8.3, select the letter that identifies each of the following parts of Earth's interior.

   a) Mantle: __________

   b) Inner core: __________

   c) Continental crust: __________

   d) Oceanic crust: __________

   e) Asthenosphere: __________

   f) Outer core: __________

   g) Lithosphere: __________

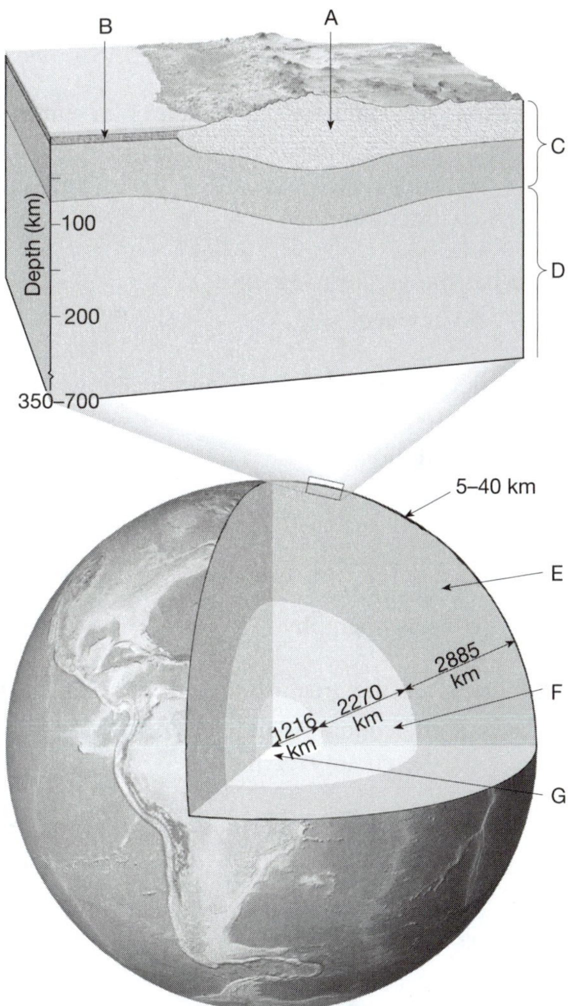

Figure 8.3

9. Briefly describe the composition of the following zones of Earth's interior.

   a) Mantle:

   b) Outer core:

Practice Test

Multiple choice. Choose the best answer for the following multiple-choice questions.

1. It is estimated that over __________ earthquakes that are strong enough to be felt occur worldwide annually.
   a) 500
   b) 1000
   c) 10,000
   d) 20,000
   e) 30,000

2. The location on the surface directly above the earthquake focus is called the __________.
   a) ephemeral
   b) epicycle
   c) epicenter
   d) epinode
   e) epitaph

3. The cool, rigid layer of Earth that includes the entire crust as well as the uppermost mantle is called the __________.
   a) asthenosphere
   b) lower crust
   c) oceanic crust
   d) Moho
   e) lithosphere

4. Underground storage tanks rising to the surface as the result of an earthquake is evidence of __________.
   a) tsunamis
   b) liquefaction
   c) subsidence
   d) fracturing
   e) fault creep

5. Which earthquake body wave has the greatest velocity?
   a) P wave
   b) S wave

6. The belt from about 105 to 140 degrees away from an earthquake where no P waves are recorded is known as the __________.
   a) shadow zone
   b) absent zone
   c) Moho zone
   d) reflective zone
   e) low-velocity zone

7. The study of earthquakes is called __________.
   a) seismogram
   b) seismicity
   c) seismology
   d) seismography
   e) seismogony

8. Which one of the following regions has the greatest amount of seismic activity?
   a) central Europe
   b) southern Russia
   c) the central Atlantic basin
   d) the circum-Pacific belt
   e) the eastern United States

9. The difference in __________ of P and S waves provides a method for determining the epicenter of an earthquake.
   a) magnitudes
   b) velocities
   c) sizes
   d) modes of travel
   e) foci

10. The source of an earthquake is called the __________.
    a) fulcrum
    b) ephemeral
    c) epicenter
    d) foreshock
    e) focus

11. Long-range earthquake forecasts are based on the premise that earthquakes are __________.
    a) random
    b) destructive
    c) fully understood
    d) always occurring
    e) repetitive

12. The epicenter of an earthquake is located using the distances from a minimum of __________ seismic stations.

    a) three
    b) four
    c) five
    d) six
    e) seven

13. Which of the earthquake body waves cannot be transmitted through fluids?

    a) P waves
    b) S waves

14. Dense rocks like __________ are thought to make up the mantle and provide the lava for oceanic eruptions.

    a) limestone
    b) granite
    c) peridotite
    d) sandstone
    e) rhyolite

15. In areas where unconsolidated materials are saturated with water, earthquakes can turn stable soil into a fluid during a phenomenon called __________.

    a) libation
    b) lithification
    c) leaching
    d) liquefaction
    e) localization

16. The adjustments of materials that follow a major earthquake often generate smaller earthquakes called __________.

    a) tremors
    b) aftershocks
    c) foreshocks
    d) surface waves
    e) body waves

17. An earthquake with a magnitude of 6.5 releases __________ times more energy than one with a magnitude of 5.5.

    a) 10
    b) 20
    c) 30
    d) 40
    e) 50

18. Earthquake epicenters are most closely correlated with __________.

    a) continental interiors
    b) plate boundaries
    c) population centers
    d) continental shelves
    e) high latitudes

19. The amount of damage caused by an earthquake at a specific location is used to determine the intensity of an earthquake in the __________.

    a) Modified Mercalli scale
    b) Richter scale
    c) Mohs scale
    d) seismic scale
    e) Gutenberg scale

20. The greatest concentration of metals occurs in Earth's __________.

    a) crust
    b) asthenosphere
    c) mantle
    d) core
    e) lithosphere

True/false. For the following true/false questions, if a statement is not completely true, mark it false. For each false statement, change the **italicized** *word to correct the statement.*

1. _____ Earthquake waves that travel through Earth's interior are called *body* waves.
2. _____ The adjustments that follow a major earthquake often generate smaller earthquakes called *foreshocks*.
3. _____ Earthquakes with a Richter magnitude less than *eight* are usually not felt by humans.
4. _____ Most of our knowledge of Earth's interior comes from the study of *earthquakes*.
5. _____ Vibrations known as earthquakes occur as rock slips and *elastically* returns to its original shape.

6. _____ Earthquake *body* waves are divided into two types called primary (P) waves and secondary (S) waves.
7. _____ Most tsumanis result from *horizontal* displacement of the ocean floor during an earthquake.
8. _____ The study of earthquakes is called *seismography.*
9. _____ No reliable method of *short-range* earthquake prediction has yet been devised.
10. _____ Fluids (gases and liquids) *cannot* transmit P waves.
11. _____ P waves arrive at a recording station *after* S waves.
12. _____ The boundary that separates the crust from the underlying mantle is known as the *shadow* discontinuity.
13. _____ Most earthquakes occur along faults associated with *plate* boundaries.
14. _____ The lithosphere is situated *below* the asthenosphere.
15. _____ Earth's *inner* core is a solid metallic sphere.
16. _____ The *epicenter* of an earthquake is the location on the surface directly above the focus.
17. _____ To locate an epicenter, the distance from *three* or more different seismic stations must be known.
18. _____ The mantle is *solid* because both P and S waves travel through it.
19. _____ The farther an earthquake recording station is from an earthquake, the *greater* the difference in arrival times of the P and S waves.
20. _____ Earthquakes in the central and eastern United States occur *more* frequently than along plate-boundary areas.
21. _____ The continental crust is mostly made of *granitic* rocks.
22. _____ A refined Richter scale is used to describe earthquake *magnitude.*

Word choice. Complete each of the following statements by selecting the most appropriate response.

1. [S/P] waves cannot pass through Earth's outer core.
2. The location on Earth's surface above the source of an earthquake is called the [epicenter/focus].
3. Seismic waves that travel along Earth's outer layer are called [body/surface] waves.
4. The energy of an earthquake is released during a process called elastic [deformation/rebound].
5. The boundary between the crust and the mantle is called the [Moho/shadow zone].
6. Most of our knowledge about Earth's interior comes from the study of [rocks/seismic waves].
7. Large fractures in Earth's crust along which movement occurs are referred to as [epicenters/faults].
8. A scale used to measure earthquake magnitude is the [Mercalli/Richter] scale.
9. The [lithosphere/asthenosphere] consists of partially melted rock.
10. The instruments used to record earthquakes are called [seismographs/seismograms], and the records they produce are called [seismographs/seismograms].
11. The greatest damage done by a tsunami is [at sea/on land].
12. In general, the most earthquake-resistant structures are [rigid/flexible].
13. [Inertia/Gravity] is the tendency of a stationary object to hold still, or a moving object to stay in motion.
14. Earth's [lithosphere/asthenosphere] is rigid.
15. A seismograph records the relative motion between a [stationary/moving] weight and a base that vibrates.

Written questions

1. Describe an earthquake and the circumstances that cause it to occur.

2. What are the major differences between P and S earthquake waves?

3. Describe the composition (mineral/rock makeup) of Earth's crust, mantle, and core.

For other interesting and pertinent
information, be sure to visit
the *Earth Science* companion website at

http://www.prenhall.com/tarbuck

9

Volcanoes and Other Igneous Activity

At first glance the significance of igneous activity may not be readily obvious. However, because of the nature of volcanic eruptions, volcanoes provide the only window for direct observation of the materials and processes that occur many kilometers inside our planet. Furthermore, gases emitted by volcanoes very early in Earth's history contributed to the evolution of the primitive atmosphere and oceans. Although these facts alone are reasons enough for igneous activity to warrant our attention, other questions also come to mind and merit consideration. Why do some volcanoes erupt violently while others are quiescent? What mechanism is responsible for the countless number of volcanoes located on the ocean floor? And can volcanoes change Earth's climate? Chapter nine explores these and other concerns as the nature of volcanic eruptions, the materials produced during eruptions, and the various types of surface landforms and intrusive features produced by igneous activity are investigated.

Learning Objectives

After reading, studying, and discussing this chapter, you should be able to:

List the factors that determine the violence of volcanic eruptions.

The primary factors that determine the nature of volcanic eruptions include the magma's *temperature,* its *composition,* and the *amount of dissolved gases* it contains. As lava cools, it begins to congeal, and as *viscosity* increases, its mobility decreases. *The viscosity of magma is directly related to its silica content. Rhyolitic* lava, with its high silica content, is very viscous and forms short, thick flows. *Basaltic* lava, with a lower silica content, is more fluid and may travel a long distance before congealing. Dissolved gases provide the force that propels molten rock from the vent of a volcano.

List the materials that are extruded from volcanoes.

The materials associated with a volcanic eruption include *lava flows (pahoehoe* and *aa* flows for basaltic lavas), *gases* (primarily in the form of *water vapor*), and *pyroclastic material* (pulverized rock and lava fragments blown from the volcano's vent, which include *ash, pumice, lapilli, cinders, blocks,* and *bombs*).

Describe the major features produced by volcanic activity.

Successive eruptions of lava from a central vent result in a mountainous accumulation of material known as a *volcano.* Located at the summit of many volcanoes is a steep-walled depression called a *crater. Shield cones* are broad, slightly domed volcanoes built primarily of fluid, basaltic lava. *Cinder cones* have very steep slopes composed of pyroclastic material. *Composite cones,* or stratovolcanoes, are large, nearly symmetrical structures built of interbedded lavas and pyroclastic deposits. Composite cones produce some of the most violent volcanic activity. Often associated with a violent eruption is a *nuée ardente,* a fiery cloud of hot gases infused with incandescent ash that races down steep volcanic slopes. Large composite cones may also generate a type of mudflow known as a *lahar.*

Describe the major features typically found in volcanic regions.

Most volcanoes are fed by *conduits* or *pipes.* As erosion progresses, the rock occupying the pipe is often more resistant and may remain standing above the surrounding terrain as a *volcanic neck.* The summits of some volcanoes have large, nearly circular depressions called *calderas* that result from collapse. Calderas also form our shield volcanoes by subterranean drainage from a central magma chamber, and the largest calderas form by the discharge of collossal volumes of silica-rich pumice along ring fractures. Although volcanic eruptions from a central vent are the most familiar, by far the largest amounts of volcanic material are extruded from cracks in the crust called *fissures*. The term *flood basalts* describes the fluid, water-like, basaltic lava flows that cover an extensive region in the northwestern United States known as the Columbia Plateau. When silica-rich magma is extruded, *pyroclastic flows* consisting largely of ash and pumice fragments usually result.

Discuss the classification of igneous intrusive bodies.

Igneous intrusive bodies are classified according to their *shape* and by their *orientation with respect to the host rock,* generally sedimentary rock. The two general shapes are *tabular* (sheetlike) and *massive*. Intrusive igneous bodies that cut across existing sedimentary beds are said to be *discordant,* whereas those that form parallel to existing sedimentary beds are *concordant*.

List and describe the major intrusive igneous features.

Dikes are tabular, discordant igneous bodies produced when magma is injected into fractures that cut across rock layers. Tabular, concordant bodies, called *sills,* form when magma is injected along the bedding surfaces of sedimentary rocks. In many respects sills closely resemble buried lava flows. *Laccoliths* are similar to sills but form from less-fluid magma that collects as a lens-shaped mass that arches the overlying strata upward. *Batholiths,* the largest intrusive igneous bodies with surface exposures of more than 100 square kilometers (40 square miles), frequently compose the cores of mountains.

Discuss the origin of magma.

Magma originates from essentially solid rock of the crust and mantle. In addition to a rock's composition, its temperature, depth (confining pressure), and water content determine whether it exists as a solid or liquid. Thus, magma can be generated by *raising a rock's temperature,* as occurs when a hot mantle plume "ponds" beneath crustal rocks. A *decrease in pressure* can cause *decompression melting*. Further, the *introduction of volatiles* (water) can lower a rock's melting point sufficiently to generate magma. Because melting is generally not complete, a process called *partial melting* produces a melt made of the lowest-melting-temperature minerals, which are higher in silica than the original rock. Thus, magmas generated by partial melting are nearer to the granitic (felsic) end of the compositional spectrum than are the rocks from which they formed.

Describe the relation between igneous activity and plate tectonics.

Most active volcanoes are associated with plate boundaries. Active areas of volcanism are found along mid-ocean ridges where seafloor spreading is occurring (*divergent plate boundaries*), in the vicinity of ocean trenches where one plate is being subducted beneath another (*convergent plate boundaries*), and in the interiors of plates themselves (*intraplate volcanism*). Rising plumes of hot mantle rock are the source of most intraplate volcanism.

Key Terms

aa flow
batholith
caldera
cinder cone
columnar joints
composite cone
conduit
continental volcanic arc
crater
decompression melting
dike
fissure
fissure eruption
flood basalt
fumarole
geothermal gradient
hot spot
intraplate volcanism
island arc
laccolith
lahar
lava tube
mantle plume
nuée ardente
pahoehoe flow
parasitic cone
partial melting
pipe
pluton
pyroclastic flow
pyroclastic materials
scoria cone
shield volcano
sill
stratovolcanoes
vent
viscosity
volatiles
volcanic island arc
volcanic neck
volcano

Vocabulary Review

Choosing from the list of key terms, furnish the most appropriate response for the following statements.

1. Particles of pulverized rock, lava, and glass fragments blown from the vent of a volcano are referred to as __________.
2. A(n) __________ is an unusually large volcanic summit depression that exceeds 1 kilometer in diameter.
3. A lava flow that has a surface of rough, jagged blocks is called a(n) __________.
4. Ship Rock, New Mexico, a feature produced by erosion, is an example of a(n) __________.
5. A(n) __________ is a sheetlike body that is produced when magma is injected into a fracture that cuts across rock layers.
6. The crater of a volcano is connected to a magma chamber via a pipelike conduit called a(n) __________.
7. A(n) __________ is a rather small volcano with steep slopes built from ejected lava fragments.
8. By far the largest intrusive igneous body is a(n) __________.
9. A volcano that takes the shape of a broad, slightly domed structure is called a(n) __________.
10. A flow of fluid basaltic lava that issues from cracks or fissures and commonly covers an extensive area to a thickness of hundreds of meters is called a(n) __________.
11. A lava flow with a smooth-to-ropy appearance that is produced from fluid basaltic lava is called a(n) __________.
12. A hot plume, which may extend to Earth's core-mantle boundary, produces a(n) __________, a volcanic region a few hundred kilometers across.
13. A(n) __________ is a large, nearly symmetrical volcano built of interbedded strata of lavas and pyroclastic deposits.
14. A(n) __________ is a mountainous accumulation of material formed by successive eruptions from a central vent.
15. A(n) __________ is an igneous intrusive feature that forms from a lens-shaped mass of magma that arches the overlying strata upward.
16. An eruption in which volcanic material is extruded from a long, narrow crack in the crust is known as a(n) __________.
17. A(n) __________ is a tabular pluton formed when magma is injected along sedimentary bedding surfaces.
18. Hot gases infused with incandescent ash ejected from a volcano produce a fiery cloud called a(n) __________, which flows down steep slopes at high speed.
19. Most igneous rocks melt over a temperature range of a few hundred degrees, a process known as __________, which produces most, if not all, magma.
20. A material's __________ is a measure of its resistance to flow.
21. A type of mudflow associated with violent eruptions is known as a(n) __________.
22. The change in temperature with depth inside Earth is referred to as the __________.
23. __________ are structures that result from the emplacement of igneous materials at depth.
24. A cave-like void or tunnel that once served as a conduit carrying lava from a volcanic vent to the flow's leading edge is called a(n) __________.

Comprehensive Review

1. List three factors that determine whether a volcano extrudes magma violently or "gently."

   1)

   2)

   3)

2. Using Figure 9.1, select the letter of the diagram that illustrates each of the following types of volcanoes. Name a volcano that is an example of each type. Also describe the eruptive pattern most commonly associated with each type.

   a) Shield volcano: ________ (Example:)

   b) Cinder cone: ________ (Example:)

   c) Composite cone (stratovolcano): ________ (Example:)

   Eruptive pattern:

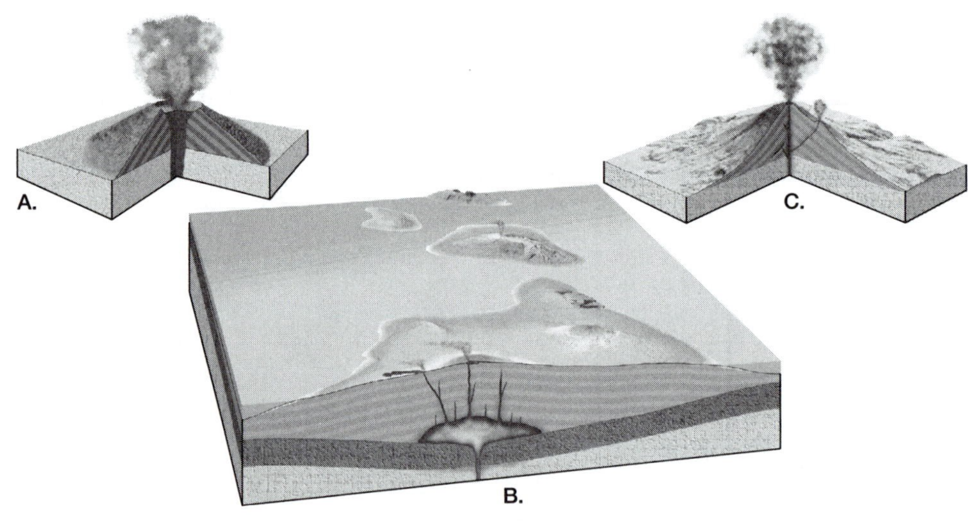

Figure 9.1

3. What are the major gases released during a volcanic eruption?

4. Describe the meaning of the following terms as they are related to igneous plutons.

   a) Discordant:

   b) Concordant:

5. What factors affect the viscosity of magma?

6. Using Figure 9.2, select the letter that illustrates each of the following igneous intrusive features.

   a) Sill: _________

   b) Batholith: _________

   c) Laccolith: _________

   d) Dike: __________

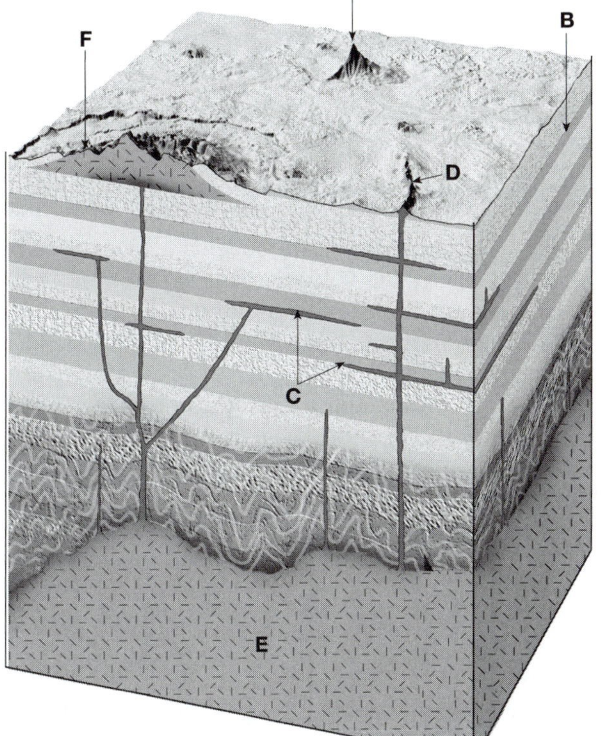

Figure 9.2

7. Briefly comment on the role of each of the following in the origin of magma.

   a) Temperature:

   b) Pressure and/or volatiles:

   c) Partial melting:

8. Compare basaltic magma to rhyolitic/felsic magma in terms of

   a) Silica content:

   b) Viscosity:

   c) Melting temperature:

9. List three major zones of volcanic activity and relate each to global tectonics.

   1)

   2)

   3)

Practice Test

Multiple choice. Choose the best answer for the following multiple-choice questions.

1. Underground igneous rock bodies are called __________.
   a) aquifers
   b) plutons
   c) playas
   d) pluvials
   e) placers

2. The greatest volume of volcanic material is produced by __________.
   a) cinder cones
   b) fissure eruptions
   c) laccoliths
   d) shield cones
   e) explosive eruptions

3. Highly viscous magmas tend to impede the upward migration of expanding gases, which often results in __________ eruptions.
   a) relatively quiet
   b) explosive

4. The most violent type of volcanic activity is associated with __________.
   a) cinder cones
   b) sills
   c) composite cones
   d) intermediate cones
   e) shield cones

5. Which one of the following is NOT a factor that determines the violence of a volcanic eruption?
   a) temperature of the magma
   b) size of the volcanic cone
   c) the magma's composition
   d) amount of dissolved gases in the magma

6. The most abundant gas produced during Hawaiian eruptions is __________.
   a) nitrogen
   b) carbon dioxide
   c) oxygen
   d) chlorine
   e) water vapor

7. The force that extrudes magma from a volcanic vent is provided by __________.
   a) dissolved gases
   b) gravity
   c) the magma's heat
   d) the volcano's slope
   e) discordant plutons

8. The area of igneous activity commonly called the Ring of Fire surrounds the __________.
   a) Indian Ocean
   b) Coral Sea
   c) Atlantic Ocean
   d) Sea of Japan
   e) Pacific Ocean

9. A magma's viscosity is directly related to its __________.
   a) depth
   b) age
   c) volcanic cone
   d) silica content
   e) color

10. Pulverized rock, lava, and glass fragments produced from the vent of a volcano are known as _____.
    a) nuée ardentes
    b) sills
    c) pyroclastic materials
    d) craters
    e) pahoehoes

11. Which type of volcano consists of interbedded strata of lavas and pyroclastic material?
    a) cinder cone
    b) composite cone
    c) intermediate cone
    d) shield cone
    e) pyro-cone

12. Fluid basaltic lavas of the Hawaiian type commonly form __________.
    a) aa flows
    b) pahoehoe flows
    c) pyroclastic flows
    d) nuée ardentes
    e) lapilli

13. Basaltic lava tends to be __________ fluid than rhyolitic lava.
    a) more
    b) less

14. When silica-rich magma is extruded, ash and pumice fragments may be propelled from the vent at high speeds and produce __________.
    a) flood basalts
    b) pahoehoe flows
    c) batholiths
    d) a shield volcano
    e) pyroclastic flows

15. Unusually large volcanic summit depressions that exceed 1 kilometer in diameter are known as __________.
    a) calderas
    b) vents
    c) craters
    d) sills
    e) laccoliths

16. Large particles of hardened lava ejected from a volcano are termed __________.
    a) bombs
    b) lapilli
    c) welded tuff
    d) cinders
    e) blocks

17. Eruptions of fluid basaltic lavas, such as those that occur in Hawaii, tend to be __________.
    a) relatively quiet
    b) unpredictable
    c) extremely violent
    d) explosive

18. Hot gases infused with incandescent ash ejected from a volcano often produce a fiery cloud called a __________.
    a) nuée ardente
    b) laccolith
    c) lapilli
    d) pyroclastic
    e) volcanic bomb

19. The type of volcano produced almost entirely of pyroclastic material is the __________.
    a) shield volcano
    b) cinder cone
    c) composite cone
    d) pyro-cone
    e) batholith

20. A __________ is a tabular, concordant pluton.
    a) dike
    b) laccolith
    c) stock
    d) sill
    e) batholith

21. Which type of volcanoes are generally small and occur in groups?
    a) composite cones
    b) laccoliths
    c) shield cones
    d) cinder cones
    e) intermediate cones

22. In a near-surface environment, silica-rich rocks of rhyolitic (felsic) composition melt at a __________ temperature than basaltic rocks.
    a) higher
    b) lower

23. The largest intrusive igneous bodies are __________.
    a) dikes
    b) stocks
    c) batholiths
    d) laccoliths
    e) sills

24. Intraplate volcanism may be associated with the formation of __________ over rising plumes of hot mantle material.
    a) hot spots
    b) dikes
    c) subduction zones
    d) ocean ridges
    e) volcanic necks

25. In general, an increase in the confining pressure __________ a rock's melting temperature.
    a) increases
    b) decreases
    c) stabilizes

True/false. For the following true/false questions, if a statement is not completely true, mark it false. For each false statement, change the **italicized** *word to correct the statement.*

1. _____ Most of Earth's active volcanoes are near *divergent* plate margins.
2. _____ The more viscous a material, the *greater* its resistance to flow.
3. _____ Plutons similar to but smaller than batholiths are termed *stocks.*
4. _____ Water vapor is the *most* abundant gas in magma.
5. _____ The smallest volcanoes are *composite* cones.
6. _____ The greatest volume of volcanic rock is produced along the oceanic *ridge* system.
7. _____ Magmas that produce basaltic rocks contain *more* silica than those that form granitic rocks.
8. _____ Reducing confining pressure *lowers* a rock's melting temperature.
9. _____ Located at the summit of most volcanoes is a steep-walled depression called a *sill.*
10. _____ An important consequence of partial melting is the production of a melt with a *higher* silica content than the parent rock.
11. _____ *Dikes* are tabular, discordant plutons.
12. _____ With increasing depth in Earth's interior, there is a gradual *decrease* in temperature.
13. _____ A magma's viscosity is directly related to its *iron* content.
14. _____ Most of the volcanoes of the Cascade Range in the northwestern United States are *shield* cones.
15. _____ When *subduction* volcanism occurs in the ocean, a chain of volcanoes called a volcanic island arc is produced.
16. _____ The large expanse of granitic rock exposed in the interior of North America is called the *Canadian* Shield.
17. _____ The viscosity of magma, plus the quantity of dissolved *gases* and the ease with which they can escape, determines the nature of volcanic eruptions.
18. _____ One of the best known *batholiths* in North America is along the Hudson River near New York.
19. _____ Secondary volcanic vents that emit only gases are called *fumaroles.*
20. _____ The Columbia River Plateau in the northwestern United States formed from very fluid *basaltic* lava that erupted from numerous fissures.

Word choice. Complete each of the following statements by selecting the most appropriate response.

1. [Increasing/Decreasing] the temperature of magma will increase its viscosity.
2. Most calderas form when the [side/summit] of a volcano collapses into a partially emptied [magma chamber/vent].
3. Most of the pungent odor associated with a volcano is from [carbon/nitrogen/sulfur] compounds.
4. The most explosive volcanoes are produced by [high/low] viscosity magma containing a [large/small] quantity of dissolved gases.
5. Partial melting of a rock usually results in a magma with a [higher/lower] silica content than the parent rock.
6. The greatest volume of volcanic material is produced along the [ocean trenches/oceanic ridge system].
7. The magma of spreading-center volcanism is produced primarily by the partial melting of [continental crust/upper mantle rock/oceanic crust] and is mostly of [basaltic/granitic] composition.

8. A volcano's affect on climate, if any, would be caused by ash and [lava/gases] being ejected into the [upper/lower] atmosphere.
9. In subduction-zone volcanism, rocks melt due to a(n) [increase/decrease] in [temperature/pressure].
10. [Increasing/Decreasing] the silica content of magma will increase its viscosity.
11. The smallest volcanic cones are typically [cinder/composite/shield] cones.
12. At divergent plate boundaries, rocks melt due to a(n) [increase/decrease] in [temperature/pressure].
13. The most fluid lavas have a [basaltic/rhyolitic] composition.
14. The most explosive volcanism is associated with [convergent plate/divergent plate] volcanism due to the relatively high [crystal/basalt/water] content of the associated magma.
15. Rocks can melt without changing temperature when their confining pressure is [increased/reduced].

Written questions

1. What is the difference between magma and lava?

2. List the three main types of volcanoes, and describe the appearance of each.

3. List and describe three igneous intrusive features.

For other interesting and pertinent information, be sure to visit the *Earth Science* companion website at

http://www.prenhall.com/tarbuck

10

Mountain Building

Geologists believe that at some time all continental regions were mountainous masses and have concluded that the continents grow by the addition of new mountains to their flanks. Consequently, as earth scientists unravel the process of mountain formation, they also gain an understanding of the evolution of Earth's continental crust. However, if continents do indeed expand by adding material to their borders, how do we explain the existence of mountains such as the Urals that are located in the interior of a landmass? To answer this and related questions, the chapter begins by examining rock deformation and the structures that result from the process. Also extensively investigated is the sequence of events responsible for mountain building during the recent geologic past along convergent plate boundaries and other locations around the world.

Learning Objectives

After reading, studying, and discussing this chapter, you should be able to:

Discuss rock deformation and list the factors that influence the strength of a rock.

Deformation refers to changes in the shape and/or volume of a rock body. Rocks first respond by deforming *elastically* and will return to their original shape when the stress is removed. Once their elastic limit (strength) is surpassed, rocks either deform by ductile flow or they fracture. *Ductile deformation* is a solid state flow that results in a change in size and shape of rocks without fracturing. Ductile deformation occurs in a high-temperature/high-pressure environment. In a near-surface environment, when stress is applied rapidly, most rocks deform by *brittle failure.*

Describe the major types of folds and how they form.

One of the most basic geologic structures associated with rock deformation is *folds* (flat-lying sedimentary and volcanic rocks bent into a series of wavelike undulations). The two most common types of folds are *anticlines,* formed by the upfolding, or arching, of rock layers, and *synclines,* which are downfolds. Most folds are the result of horizontal *compressional stresses. Domes* (upwarped structures) and *basins* (downwarped structures) are circular or somewhat elongated folds formed by vertical displacements of strata.

Describe the major types of faults and how they form.

Faults are fractures in the crust along which appreciable displacement has occurred. Faults in which the movement is primarily vertical are called *dip-slip faults.* Dip-slip faults include both *normal* and *reverse faults.* Low-angle reverse faults are called *thrust faults.* Normal faults indicate *tensional stresses* that pull the crust apart. Along spreading centers, divergence can cause a central block called a *graben,* bounded by normal faults, to drop as the plates separate. Reverse and thrust faulting indicate that *compressional forces* are at work. Large thrust faults are found along subduction zones and other convergent boundaries where plates are colliding. *Strike-slip faults* exhibit mainly horizontal displacement parallel to the fault surface. Large strike-slip faults, called *transform faults,* accommodate displacement between plate boundaries. Most transform faults cut the oceanic lithosphere and link spreading centers. The San Andreas Fault cuts the continental lithosphere and accommodates the northward displacement of southwestern California.

Discuss the occurrence of joints in rock.

Joints are fractures along which no appreciable displacement has occurred. Joints generally occur in groups with roughly parallel orientations and are the result of brittle failure of rock units located in the outermost crust.

Give the name for the processes that collectively produce a mountain system.

The name for the processes that collectively produce a mountain system is *orogenesis*. Most mountains consist of roughly parallel ridges of folded and faulted sedimentary and volcanic rocks, portions of which have been strongly metamorphosed and intruded by younger igneous bodies.

Describe the formation of an Andean-type mountain belt.

Subduction of oceanic lithosphere under a continental block gives rise to an *Andean-type plate margin* that is characterized by a continental volcanic arc and associated igneous plutons. In addition, sediment derived from the land, as well as material scraped from the subducting plate, becomes plastered against the landward side of the trench, forming an *accretionary wedge*. An excellent example of an inactive Andean-type mountain belt is found in the western United States and includes the Sierra Nevada and the Coast Range in California.

Discuss continental collisions and the development of compressional mountains.

Continued subduction of oceanic lithosphere beneath an Andean-type continental margin will eventually close an ocean basin. The result will be a *continental collision* and the development of compressional mountains that are characterized by shortened and thickened crust as exhibited by the Himalayas. The development of a major mountain belt is often complex involving two or more distinct episodes of mountain building. A common feature of compressional mountains are *fold-and-thrust belts*. Continental collisions have generated many mountain belts, including the Alps, Urals, and Appalachians.

Describe the formation of the mountain belts of Alaska and British Columbia.

Mountain belts can develop as a result of the collision and merger of an island arc, oceanic plateau, or some other small crustal fragment to a continental block. Many of the mountain belts of the North American Cordillera, principally those in Alaska and British Columbia, were generated in this manner.

Explain the formation of fault-block mountains.

Although most mountains form along convergent plate boundaries, other tectonic processes, such as continental rifting, can produce uplift and the formation of topographic mountains. The mountains that form in these settings, termed *fault-block mountains,* are bounded by high-angle normal faults that gradually flatten with depth. The Basin and Range Province in the western United States consists of hundreds of faulted blocks that give rise to nearly parallel mountain ranges that stand above sediment-laden basins.

Describe the concept of isostasy.

Earth's less dense crust floats on top of the denser and deformable rocks of the mantle, much like wooden blocks floating in water. The concept of a floating crust in gravitational balance is called *isostasy*. Most mountainous topography is located where the crust has been shortened and thickened. Therefore, mountains have deep crustal roots that isostatically support them. As erosion lowers the peaks, *isostatic adjustment* gradually raises the mountains in response. The processes of uplifting and erosion will continue until the mountain block reaches "normal" crustal thickness. Gravity also causes elevated mountainous structures to collapse under their own "weight."

Key Terms

accretionary wedge	fault scarp	normal fault
active continental margin	fault-block mountains	orogenesis
anticline	fold	passive continental margin
basin	graben	reverse fault
brittle failure (deformation)	gravitational collapse	strike-slip fault
deformation	horst	syncline
dip-slip fault	isostasy	terrane
dome	isostatic adjustment	thrust fault
ductile deformation	joint	transform fault
fault	monocline	

Vocabulary Review

Choosing from the list of key terms, furnish the most appropriate response for the following statements.

1. A fault in which the dominant displacement is along the trend of the fault is called a(n) __________.
2. A(n) __________ is an elongated, uplifted block of crust bounded by faults.
3. __________ is the concept that Earth's crust is "floating" in gravitational balance upon the material of the mantle.
4. A wavelike layer of rock that was originally horizontal and subsequently deformed is a(n) __________.
5. A(n) __________ is a circular or somewhat elongated downfolded structure.
6. __________ is the name for the processes that collectively produce a mountain system.
7. A fracture in rock along which displacement has occurred is termed a(n) __________.
8. A reverse fault having a very low angle is also called a(n) __________.
9. A(n) __________ is a valley formed by the downward displacement of a fault-bounded block.
10. The East Coast of North America is a good example of a(n) __________.
11. A(n) __________ is the type of fold most commonly formed by the upfolding, or arching, of rock layers.
12. A downfold, or trough, of rock layers forms a type of fold called a(n) __________.
13. Faults in which the dominant displacement is horizontal and parallel to the trend are called __________.
14. ___________ are large, steplike folds in otherwise horizontal sedimentary strata.
15. A fault where the primary movement is vertical is called a(n) __________.
16. ___________ is a general term that refers to all changes in the original shape and/or size of a rock body.
17. A dip-slip fault is classified as a(n) __________ when the hanging wall moves down relative to the footwall.
18. When weight is removed from the crust and uplifting occurs, the response is called a(n) ___________.
19. A(n) __________ is a fracture in rock along which no appreciable displacement has occurred.
20. An accumulation of sedimentary and metamorphic rocks, including occasional scraps of ocean crust, that forms in association with a subduction zone is called a(n) __________.

21. A crustal block whose geologic history is distinct from the histories of adjoining crustal blocks is termed a(n) __________.
22. A circular or somewhat elongated upfolded structure is called a(n) __________.
23. A(n) __________ is the type of fault that is produced when the hanging wall moves upward relative to the footwall.

Comprehensive Review

1. What evidence supports the fact that sedimentary rocks found at high elevations in mountains were once below sea level?

2. Why is the crust beneath the oceans at a lower elevation than the continental crust?

3. Describe the concept of isostasy.

4. Write a brief paragraph describing elastic deformation of rock.

5. Briefly describe each of the following types of folds and select the letter of the diagram in Figure 10.1 that illustrates it.

   a) Anticline: __________

   b) Syncline: __________

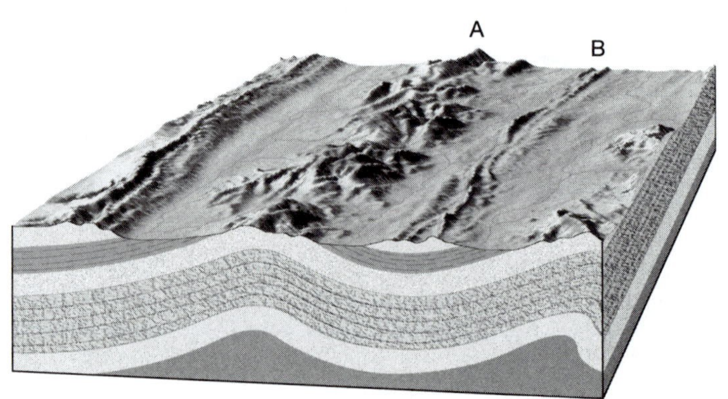

Figure 10.1

6. Write the name of the geologic structure illustrated by each diagram in Figure 10.2 below the appropriate illustration.

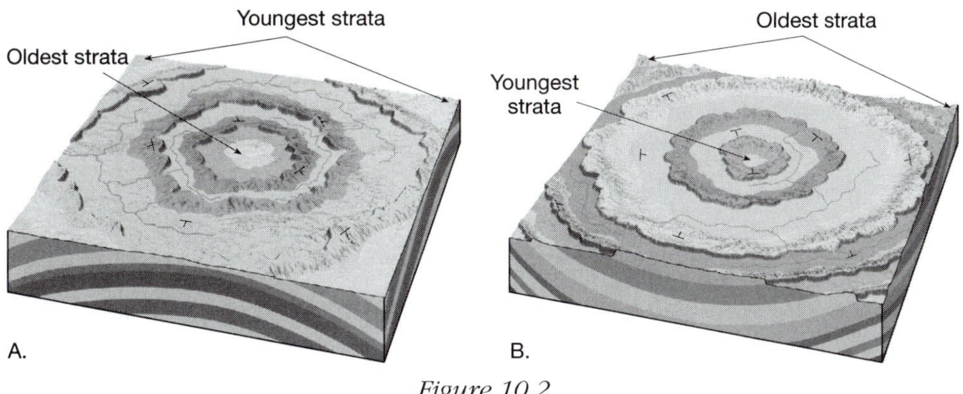

Figure 10.2

7. Briefly describe each of the following types of faults and select the letter of the diagram in Figure 10.3 that illustrates it.

   a) Reverse fault: __________

   b) Strike-slip fault: __________

   c) Normal fault: __________

   d) Thrust fault: __________

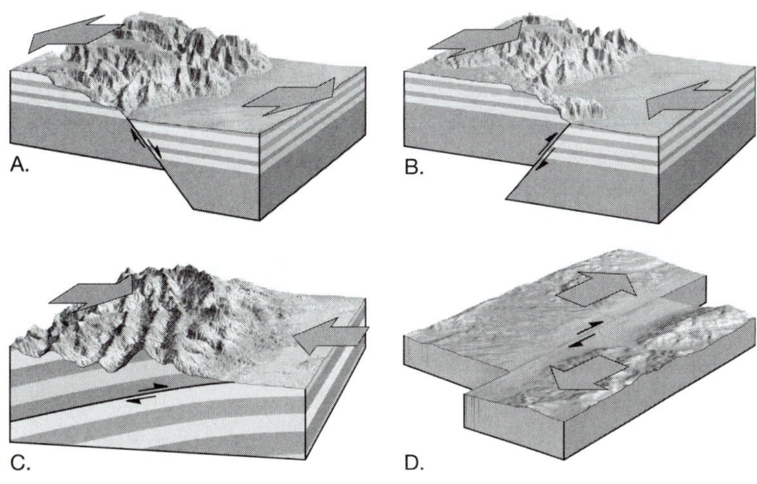

Figure 10.3

8. Using Figure 10.4, select the letter that illustrates each of the following.

   a) Graben: __________

   b) Horst: __________

9. Which type of fault are those illustrated in Figure 10.4?

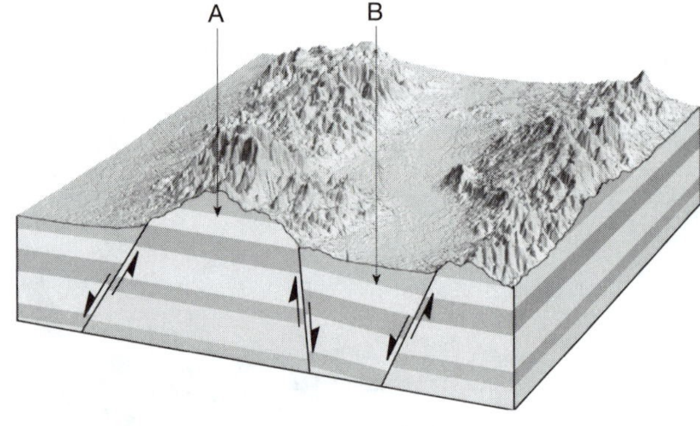

Figure 10.4

10. What are the two types of faults that are most often produced by compressional forces?

11. Briefly explain the events that produce mountains at convergent boundaries where

    a) Oceanic and continental crusts converge:

    b) Continental crusts converge:

12. Figure 10.5 illustrates mountain building along an Andean-type subduction zone. Select the appropriate letter in the figure that identifies each of the following features.

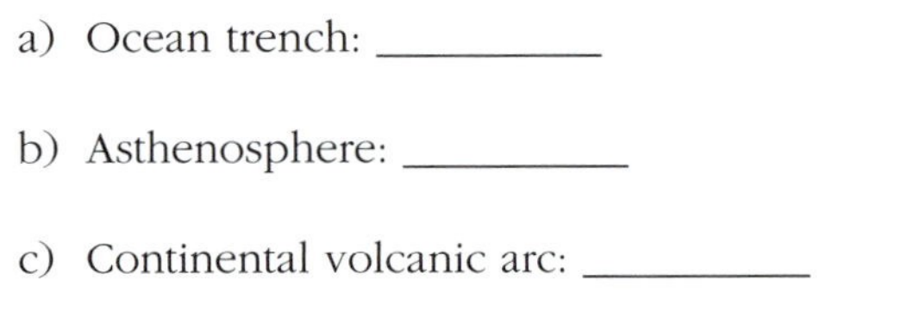

    a) Ocean trench: __________

    b) Asthenosphere: __________

    c) Continental volcanic arc: __________

    d) Accretionary wedge: __________

    e) Subducting oceanic lithosphere: __________

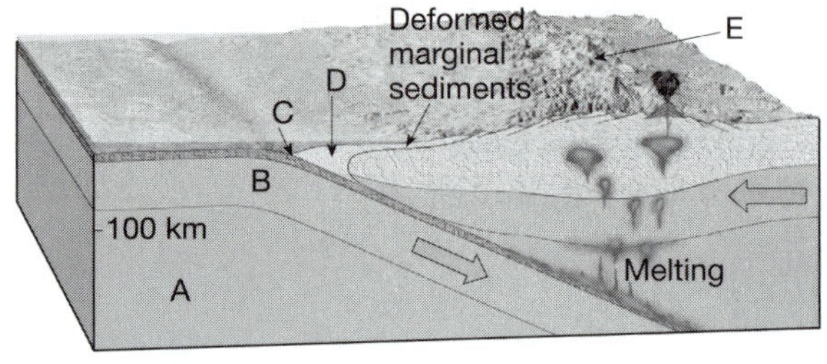

Figure 10.5

13. What are two mountain systems that have formed as the result of continental collision?

14. Why is the San Andreas Fault more appropriately referred to as the San Andreas fault system?

Practice Test

Multiple choice. Choose the best answer for the following multiple-choice questions.

1. Which one of the following is NOT a form of rock deformation?
 a) elastic deformation
 b) ductile deformation
 c) brittle failure
 d) erosion

2. The thickest part of the crust occurs in __________.
 a) the asthenosphere
 b) the ocean basin
 c) young mountain ranges
 d) eastern Canada
 e) old eroded mountains

3. Which one of the following is NOT a factor that influences the strength of a rock?
 a) time
 b) rock type
 c) temperature
 d) confining pressure
 e) age

4. The two most common types of folds are anticlines and __________.
 a) domes
 b) synclines
 c) basins
 d) superclines
 e) batholiths

5. Compared to the elevation of a thin piece of continental crust, the highest elevation of a thick piece in isostatic balance will be __________.
 a) higher
 b) the same
 c) lower
 d) overturned
 e) older

6. Orogenesis refers to those processes that collectively produce __________.
 a) a mountain system
 b) earthquakes
 c) oceanic plates
 d) trenches
 e) subduction zones

7. A fracture with horizontal displacement parallel to its surface trend is called a(n) __________ fault.
 a) dip-slip
 b) joint
 c) strike-slip
 d) oblique-slip
 e) overturned

8. The rock immediately above a fault surface is commonly called the __________.
 a) anticline
 b) foot-wall
 c) syncline
 d) hanging wall
 e) dip-slip

9. The removal of material by erosion will cause the crust to __________.
 a) subduct
 b) rise
 c) fold
 d) subside
 e) thicken

10. Folding is usually the result of __________.
 a) tensional forces
 b) shear forces
 c) faulting
 d) jointing
 e) compressional forces

11. As heat and pressure increase, ductile deformation __________.
 a) stops occurring
 b) becomes less likely
 c) is replaced by elastic deformation
 d) can cause erosion
 e) becomes more likely

12. Which one of the following pairs is properly matched?
 a) normal fault-tensional forces
 b) thrust fault-tensional forces
 c) reverse fault-tensional forces
 d) normal fault-compressional forces

13. The total accumulated displacement from earthquakes and creep along the San Andreas fault system is approximately __________ kilometers.
   a) 120
   b) 255
   c) 380
   d) 415
   e) 560

14. Where two oceanic plates converge, __________ subduction zones often occur.
   a) Andean-type
   b) Himalayan-type
   c) American-type
   d) Aleutian-type
   e) Arctic-type

15. Which one of the following features is formed by crustal upwarping?
   a) syncline
   b) basin
   c) anticline
   d) graben
   e) accretionary wedge

16. The most important difference between faults and joints is that joints __________.
   a) occur along folds
   b) are often parallel
   c) have no displacement
   d) are usually vertical
   e) are very rare

17. The final orogeny that produced the Appalachians occurred about __________ million years ago.
   a) 20
   b) 80
   c) 170
   d) 250
   e) 400

18. It is assumed that many of the terranes found in the North American Cordillera were once crustal fragments scattered throughout the eastern __________ ocean basin.
   a) Pacific
   b) Atlantic
   c) Indian

19. The San Andreas fault system is a well-known example of a(n) __________ fault.
   a) normal
   b) transform
   c) thrust
   d) reverse
   e) overturned

20. Faults where the movement is primarily vertical are called __________ faults.
   a) transform
   b) oblique
   c) dip-slip
   d) random
   e) strike-slip

21. The collision and joining of crustal fragments to a continent is called continental __________.
   a) subduction
   b) isostasy
   c) accretion
   d) suturing
   e) aggradation

22. Where oceanic crust is being thrust beneath a continental mass, __________ subduction zones often occur.
   a) Andean-type
   b) Himalaya-type
   c) American-type
   d) Aleutian-type
   e) Arctic-type

23. Dip-slip faults are classified as __________ faults when the hanging wall moves up relative to the footwall.
   a) normal
   b) transform
   c) reverse
   d) tensional
   e) strike-slip

24. The __________ Mountains are primarily a volcanic arc produced by a subducting plate.
   a) Himalaya
   b) Andes
   c) Adirondacks
   d) Ural
   e) Appalachian

25. Which one of the following mountain ranges has formed where continental crusts have converged?
   a) Sierra Nevada
   b) Andes Mountains
   c) Himalaya Mountains
   d) Coast Ranges

True/false. For the following true/false questions, if a statement is not completely true, mark it false. For each false statement, change the **italicized** *word to correct the statement.*

1. _____ As erosion lowers the peaks of mountains, isostatic adjustment gradually *lowers* the mountains in response.
2. _____ At high temperatures and pressures, most rocks exhibit *ductile* deformation once their elastic limit is surpassed.
3. _____ The first encompassing explanation of orogenesis came as part of the *plate-tectonics* theory.
4. _____ *Joints* are fractures in rock along which appreciable displacement has occurred.
5. _____ The rock in a fault that is higher than the fault surface is referred to as the *hanging* wall.
6. _____ Strike-slip faults that are associated with plate boundaries are called *transform* faults.
7. _____ Under surface conditions, rocks that exceed their elastic limit exhibit *brittle* failure.
8. _____ Most mountain building occurs in *tensional* environments.
9. _____ The crustal thickness for some mountain chains is greater than *twice* the average thickness of the continental crust.
10. _____ Where the axis of an anticline descends to the ground, the fold is said to be *plunging*.
11. _____ In large basins that contain sedimentary rock sloping at low angles, the *oldest* rocks are found near the center of the structure.
12. _____ When stress is applied, rocks first respond by deforming *plastically*.
13. _____ Faults in which the movement is primarily *horizontal* are called dip-slip faults.
14. _____ In a plunging *anticline,* the outcrop pattern “points” in the direction of the plunge.
15. _____ Wavelike undulations of sedimentary and volcanic rocks are called *grabens*.
16. _____ *Aleutian-type* subduction zones occur where two oceanic plates converge.
17. _____ In a normal fault, the hanging wall moves *downward* relative to the footwall.
18. _____ Most folds result from *compressional* stresses in the crust.
19. _____ The Tetons of Wyoming and the Sierra Nevada of California are *fault-block* mountains.
20. _____ The two most common types of folds are anticlines and *haloclines*.
21. _____ In a reverse fault, the footwall moves *upward* relative to the hanging wall.
22. _____ The *Appalachians* resulted from collisions between North America, Europe, and northern Africa.
23. _____ *Normal* faulting is often the type that occurs at spreading centers, where plates are diverging.
24. _____ Most major episodes of mountain building have occurred along *divergent* plate boundaries.
25. _____ *Fault-block* mountains are common in the Basin and Range Province of the southwestern United States.
26. _____ *Terrane* refers to any crustal fragment whose geologic history is distinct from that of adjoining fragments.
27. _____ At its southern end, the San Andreas Fault connects with the Gulf of *Mexico*.

Word choice. Complete each of the following statements by selecting the most appropriate response.

1. When a rock's shape and size are permanently altered by folding and flowing, the rock has undergone [elastic/orogenetic/ductile] deformation.
2. Forces that pull apart Earth's crust are called [compressional/tensional] stresses.
3. Thousands of earthquakes have taken place along the San Andreas fault system throughout its [9/29]-million-year history.
4. The concept of a floating crust in gravitational balance is called [orogenesis/isostasy].
5. A fault with primarily vertical displacement is called a(n) [dip-slip/strike-slip/oblique-slip] fault.
6. Reverse faults are usually produced by [tensional/compressional] forces.
7. Changes in elevation caused by the addition or removal of weight from Earth's crust are called [isostatic/volcanic/orogenetic] adjustments.
8. Fault-block mountains are associated with [compressional/tensional] forces.
9. Volcanic arcs are associated with [diverging/subducting] tectonic plates.
10. Troughs or downfolds in rock are called [anticlines/synclines].
11. Most mountain building is associated with the [margins/central regions] of tectonic plates.
12. Grabens and horsts are associated with regions of tectonic-plate [convergence/divergence].
13. Continental crust is [more/less] dense than oceanic crust.
14. Over time, erosion and isostatic adjustments will [bury/expose] a mountain's core.
15. The central regions of upwarped mountains are usually composed of metamorphic and [sedimentary/igneous] rocks.
16. The accretion of terranes [is/is not] associated with the development of the West Coast of North America.

Written questions

1. Why don't anticlines always appear as hills, even though the rocks beneath the surface are folded upward?

2. Briefly describe the composition and formation of an accretionary wedge.

3. Describe the process responsible for the formation of Earth's major mountain systems.

For other interesting and pertinent information, be sure to visit the *Earth Science* companion website at

http://www.prenhall.com/tarbuck

11

Geologic Time

It has only been as recently as the late eighteenth and early nineteenth centuries that scientists began to recognize the immensity of Earth history and the importance of time as a component in all geologic processes. Although they knew that the Earth was very old, these pioneering geologists had no way of knowing its true age. Was it tens of millions, hundreds of millions, or even billions of years old? Using what information they had, these early scientists developed a geologic time scale that showed the sequence of events based on relative dating principles, not on the actual ages of the events. As a result of the discovery of radioactivity and the development of radiometric dating techniques, geologists now can assign fairly accurate numerical dates to many events in Earth history.

Chapter eleven examines the fundamental principles used in determining the sequence of events that have taken place on Earth as well as how fossils are used to correlate rock units in different localities. Following an explanation of radioactivity, the essentials and significance of radiometric dating are investigated. The chapter concludes with a general overview of the geologic time scale.

Learning Objectives

After reading, studying, and discussing this chapter, you should be able to:

Describe the doctrine of uniformitarianism.

During the seventeenth and eighteenth centuries, *catastrophism* influenced the formulation of explanations about Earth. Catastrophism states that Earth's landscapes have been developed primarily by great catastrophes. By contrast, *uniformitarianism,* one of the fundamental principles of modern geology advanced by *James Hutton* in the late 1700s, states that the physical, chemical, and biological laws that operate today have also operated in the geologic past. The idea is often summarized as "the present is the key to the past." Hutton argued that processes that appear to be slow-acting could, over long spans of time, produce effects that were just as great as those resulting from sudden catastrophic events.

Explain the difference between numerical and relative dating.

The two types of dates used by geologists to interpret Earth history are (1) *relative dates,* which put events in their *proper sequence of formation,* and (2) *numerical dates,* which pinpoint the *time in years* when an event took place.

List the laws and principles used in relative dating.

Relative dates can be established using the *law of superposition, principle of original horizontality, principle of cross-cutting relationships, inclusions,* and *unconformities.*

Discuss correlation and its use.

Correlation, the matching up of two or more geologic phenomena in different areas, is used to develop a geologic time scale that applies to the whole Earth.

Describe fossils and fossilization.

Fossils are the remains or traces of prehistoric life. Some fossils form by *petrification, replacement,* or *carbonization.* Others are *impressions, casts,* or have been *preserved in amber.* Fossils can also be indirect evidence of prehistoric life such as *tracks* or *burrows*. The special conditions that favor preservation are *rapid burial* and the possession of *hard parts* such as shells, bones, or teeth.

Explain the principle of fossil succession.

Fossils are used to *correlate* sedimentary rocks that are from different regions by using the rocks' distinctive fossil content and applying the *principle of fossil succession*. It states that fossil organisms succeed one another in a definite and determinable order, and therefore any time period can be recognized by its fossil content.

Discuss the basic structure of atoms.

Each atom has a nucleus containing *protons* (positively charged particles) and *neutrons* (neutral particles). Orbiting the nucleus are negatively charged *electrons*. The *atomic number* of an atom is the number of protons in the nucleus. The *mass number* is the number of protons plus the number of neutrons in an atom's nucleus. *Isotopes* are variants of the same atom but with a different number of neutrons and hence a different mass number.

Describe radioactivity and the common types of radioactive decay.

Radioactivity is the spontaneous breaking apart (decay) of certain unstable atomic nuclei. Three common types of radioactive decay are (1) emission of alpha particles from the nucleus, (2) emission of beta particles (or electrons) from the nucleus, and (3) capture of electrons by the nucleus.

Discuss the process of radioactive decay.

An unstable *radioactive isotope,* called the *parent,* will decay and form stable *daughter products*. The length of time for one half of the nuclei of a radioactive isotope to decay is called the *half-life* of the isotope. If the half-life of the isotope is known, and the parent/daughter ratio can be measured, the age of a sample can be calculated.

Describe the geologic time scale.

The *geologic time scale* divides Earth's history into units of varying magnitude. It is commonly presented in chart form, with the oldest time and event at the bottom and the youngest at the top. The principal subdivisions of the geologic time scale, called *eons,* include the *Archean, and Proterozoic* (together, these two eons are commonly referred to as the *Precambrian*), and, beginning about 542 million years ago, the *Phanerozoic*. The Phanerozoic (meaning "visible life") eon is divided into the following *eras: Paleozoic* ("ancient life"), *Mesozoic* ("middle life"), and *Cenozoic* ("recent life").

List some difficulties in dating the geologic time scale.

A significant problem in assigning numerical dates to units of time is that *not all rocks can be dated radiometrically*. A sedimentary rock may contain particles of many ages that have been weathered from different rocks that formed at various times. One way geologists assign numerical dates to sedimentary rocks is to relate them to datable igneous masses, such as dikes and volcanic ash beds.

Key Terms

Archean eon
angular unconformity
catastrophism
Cenozoic era
conformable
correlation
cross-cutting relationships
disconformity
eon
epoch
era
fossil
fossil succession, principle of
geologic time scale
half-life
inclusions
index fossil
Mesozoic era
nonconformity
numerical date
original horizontality, principle of
paleontology
Paleozoic era
period
Phanerozoic eon
Precambrian
Proterozoic eon
radioactivity
radiocarbon dating
radiometric dating
relative dating
superposition, law of
unconformity
uniformitarianism

Vocabulary Review

Choosing from the list of key terms, furnish the most appropriate response for the following statements.

1. Simply stated, the doctrine of __________ states that the physical, chemical, and biological laws that operate today have also operated in the geologic past.
2. The __________ is the age of "recent life" on Earth.
3. To state specifically the point in history when something took place, for example, the extinction of the dinosaurs about 65 million years ago, is to use a type of date called a(n) __________.
4. The "matching up" of rocks from similar ages but different regions is referred to as __________.
5. The time required for one-half of the nuclei in a radioactive sample to decay is called the __________ of the isotope.
6. The doctrine of __________ adheres to the idea that Earth's landscape was produced by sudden and often worldwide disasters of unknowable causes that no longer operate.
7. To place events in their proper sequence is to apply a type of dating technique called __________.
8. The Archean and the Proterozoic eons are commonly referred to as the __________.
9. To assume that rock layers that are inclined have been moved into that position by crustal disturbances is to apply a principle of relative dating known as __________.
10. To state that "in an undeformed sequence of sedimentary rocks, the oldest rock is at the bottom" is to use a basic principle of relative dating called the __________.
11. A(n) __________ is a break in the rock record during which deposition ceased, erosion removed previously formed rocks, and then deposition resumed.
12. A(n) __________ is the remains or trace of prehistoric life.
13. The spontaneous breaking apart of atomic nuclei is a process known as __________.
14. A(n) __________ is a subdivision of a geologic era.
15. Pieces of one rock unit contained within another are called __________.
16. The unit of geologic time known as the __________ is noted for its "ancient life."
17. A fossil of an organism that was widespread geographically but limited to a short span of geologic time is often referred to as a(n) __________.
18. The fact that fossils succeed one another in a definite and determinable order is known as the __________.
19. Using radioactive isotopes to calculate the ages of rocks and minerals is a procedure called __________.
20. The unit of geologic time called a(n) __________ represents the greatest expanse of time.
21. The scientific study of fossils is called __________.

Comprehensive Review

1. Write a brief statement that compares the doctrine of catastrophism with the doctrine of uniformitarianism.

2. Dinosaurs became extinct about 65 million years ago, while large coal swamps flourished in North America about 300 million years ago. Using these facts, write relative and numerical date statements.

    a) Relative date:

    b) Numerical date:

3. Briefly state the following:

    a) Law of superposition:

    b) Principle of original horizontality:

    c) Principle of fossil succession:

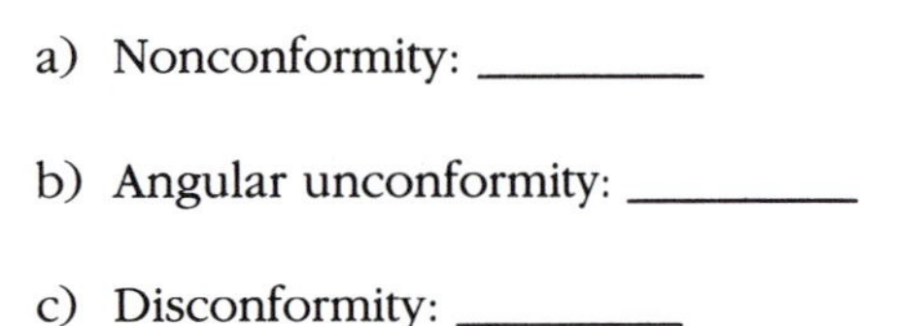

4. Using Figure 11.1, select the letter of the diagram that illustrates each of the following types of unconformities. (Unconformities are indicated with a dark wavy line and the small "v" symbol represents igneous rock.)

    a) Nonconformity: ________

    b) Angular unconformity: ________

    c) Disconformity: ________

Granite

A.

B.

C.

Figure 11.1

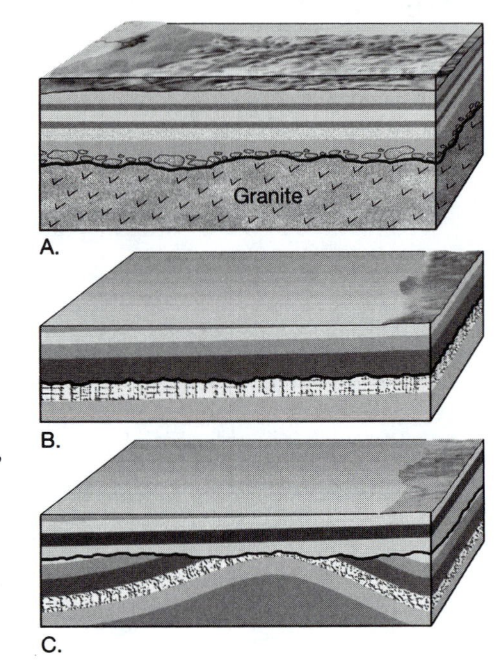

5. List the two special conditions that are necessary for an organism to be preserved as a fossil.

    1)

    2)

6. Examine the illustrations of the sedimentary rocks exposed along two separate cliffs in Figure 11.2. Different types of rocks are represented with different patterns. Write the number of the rock layer at Location 2 that correlates with the indicated rock layer at Location 1.

    Location 1

    a) layer b   correlates with

    b) layer e   correlates with

    Location 2

    layer ________

    layer ________

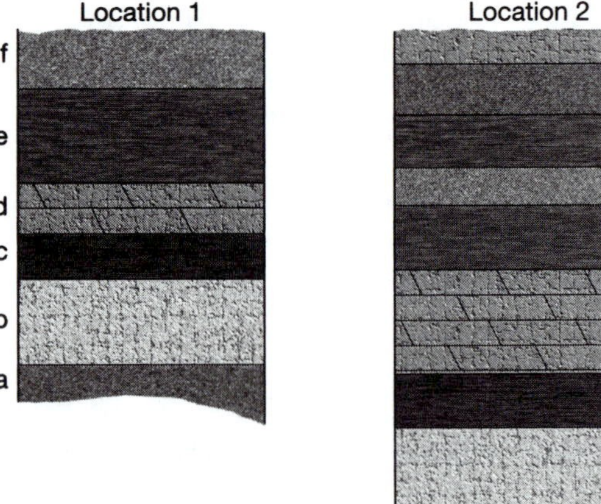

Figure 11.2

7. Figure 11.3 illustrates two types of fossilization. Write the letter of the photograph that is representative of each of the following types.

   a) Cast: _________

   b) Impression: _________

8. Explain the difference between an atom's atomic number and its mass number.

A.

B.

Figure 11.3

9. List three common types of radioactive decay.

   1)

   2)

   3)

10. Why can radiometric dating be considered reliable?

11. Briefly explain how carbon-14 is produced in the upper atmosphere.

12. What is the primary problem in assigning numerical dates to the units of time of the geologic time scale?

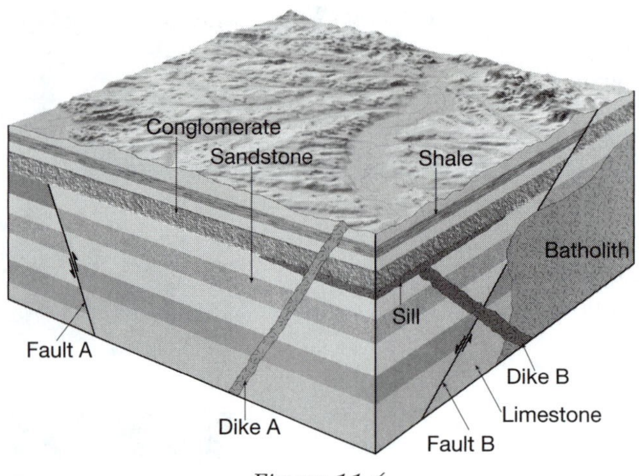

Figure 11.4

13. Using Figure 11.4, answer the following questions concerning the relative ages of the features illustrated. Also, when indicated, list the law or principle of relative dating you used to arrive at your answer.

   a) Dike B is __________ [older, younger] than fault B.

      Law or principle:

   b) The shale is __________ [older, younger] than the sandstone.

      Law or principle:

   c) Dike B is __________ [older, younger] than the batholith.

   d) The sandstone is __________ [older, younger] than Dike A.

   e) The conglomerate is __________ [older, younger] than the shale.

14. Assume a radioactive isotope has a half-life of 50,000 years. If the ratio of radioactive parent to stable daughter product is 1:3, how old is the rock containing the radioactive material?

Practice Test

Multiple choice. Choose the best answer for the following multiple-choice questions.

1. "The physical, chemical, and biological laws that operate today have also operated in the geologic past" is a statement of the doctrine of __________.
   a) uniformitarianism
   b) universal truth
   c) catastrophism
   d) unity
   e) paleontology

2. The task of matching up rocks of similar ages in different regions is known as __________.
   a) indexing
   b) correlation
   c) linking
   d) succession
   e) superposition

3. Fossils that are widespread geographically and are limited to a short span of time are referred to as __________ fossils.
   a) key
   b) matching
   c) succeeding
   d) index
   e) relative

4. A(n) __________ in the rock record represents a long period during which deposition ceased, erosion removed previously formed rocks, and then deposition resumed.
   a) numerical date
   b) superposition
   c) relative date
   d) inclusion
   e) unconformity

5. The process by which atomic nuclei spontaneously break apart (decay) is termed __________.
   a) ionization
   b) fusion
   c) reduction
   d) nucleation
   e) radioactivity

6. The doctrine of __________ implies that Earth's landscape has been developed by worldwide disasters over a short time span.
   a) uniformitarianism
   b) destruction
   c) catastrophism
   d) universal time
   e) suddenness

7. The greatest expanses of geologic time are referred to as __________.
   a) eras
   b) periods
   c) eons
   d) epochs
   e) systems

8. Nicolaus Steno proposed the most basic principle of relative dating, the law of __________.
   a) superposition
   b) secondary intrusion
   c) sequential stacking
   d) succession
   e) suddenness

9. Which eon of the geologic time scale means "visible life"?
   a) Phanerozoic
   b) Archean
   c) Hadean
   d) Proterozoic

10. The doctrine of __________ was put forth by James Hutton in his monumental work *Theory of the Earth*.
    a) destruction
    b) catastrophism
    c) suddenness
    d) uniformitarianism
    e) universal time

11. Rock layers that have been deposited essentially without interruption are said to be _____ strata.
    a) included
    b) conformable
    c) abnormal
    d) succeeding
    e) sequential

12. The major units of the geologic time scale were delineated during the __________.
    a) 1700s b) 1800s c) 1900s

13. The type of date that places events in proper sequence is referred to as a(n) __________ date.
    a) sequential
    b) secondary
    c) relative
    d) positional
    e) temporary

14. The remains or traces of prehistoric life are known as __________.
    a) indicators
    b) fossils
    c) replicas
    d) paleotites
    e) fissures

15. A(n) __________ consists of tilted rocks that are overlain by younger, more flat-lying strata.
    a) disconformity
    b) fault
    c) angular unconformity
    d) nonconformity
    e) contact unconformity

16. Atoms with the same atomic number but different mass numbers are called __________.
    a) protons
    b) isotopes
    c) ions
    d) variants
    e) nucleoids

17. A break that separates older metamorphic rocks from younger sedimentary rocks immediately above them is a type of unconformity called a(n) __________.
    a) disconformity
    b) nonconformity
    c) angular unconformity
    d) contact unconformity
    e) pseudoconformity

18. "Most layers of sediments are deposited in a horizontal position" is a statement of the principle of __________.
    a) original horizontality
    b) cross-cutting relationships
    c) fossil succession
    d) sediment
    e) cross-bedding

19. Which one of the following is NOT an era of the Phanerozoic eon?
    a) Proterozoic
    b) Mesozoic
    c) Paleozoic
    d) Cenozoic

20. Pieces of one rock unit contained within another are called __________.
    a) intrusions
    b) interbeds
    c) hosts
    d) conformities
    e) inclusions

21. Which one of the following is NOT a common type of radioactive decay?
    a) alpha particle emission
    b) nuclei ionization
    c) beta particle emission
    d) electron capture

22. Earth is about __________ years old.
    a) 4,000
    b) 4.5 million
    c) 5.8 million
    d) 4.5 billion
    e) 5.4 billion

23. A date that pinpoints the time in history when something took place is known as a(n) _____ date.
    a) relative
    b) exact
    c) known
    d) numerical
    e) factual

24. During radioactive decay, what will be the parent/daughter ratio after three half-lives?
    a) 1:3
    b) 1:4
    c) 1:5
    d) 1:6
    e) 1:7

25. The principle of fossil _____ states that fossil organisms succeed one another in a definite and determinable order.
    a) inclusions
    b) succession
    c) sequences
    d) superposition
    e) evolution

True/false. For the following true/false questions, if a statement is not completely true, mark it false. For each false statement, change the **italicized** *word to correct the statement.*

1. _____ The entire geologic time scale was created using *numerical* dates.
2. _____ Acceptance of the concept of uniformitarianism means the acceptance of a very *short* history for Earth.
3. _____ *Index* fossils are useful for correlating the rocks of one area with those of another.
4. _____ To state that a rock is 120 million years old is an example of applying a(n) *numerical* date.
5. _____ Rapid burial and the possession of *soft* parts are necessary conditions for fossilization.
6. _____ The rates of geologic processes have undoubtedly *varied* through time.
7. _____ All rocks *can* be dated radiometrically.
8. _____ Every element has a different number of *electrons* in the nucleus.
9. _____ *Catastrophists* believed that Earth was only a few thousand years old.
10. _____ When magma works its way into a rock and crystallizes, we can assume that the intrusion is *older* than the rock that has been intruded.
11. _____ During radioactive decay, as the percentage of radioactive parent atoms declines, the proportion of stable daughter atoms *decreases*.
12. _____ In addition to being important tools for correlation, fossils are also useful *environmental* indicators.
13. _____ Most layers of sediment are deposited in a *horizontal* position.
14. _____ Widespread occurrence of abundant, visible life marks the beginning of the *Phanerozoic* eon.
15. _____ The fossil record of those organisms with *hard* body parts that lived in areas of sedimentation is quite abundant.
16. _____ No place on Earth has a complete set of *conformable* strata.
17. _____ Among the major contributions of geology is the concept that Earth history is exceedingly *long*.
18. _____ The "present is the key to the past" summarizes the basic idea of *catastrophism*.
19. _____ *Index* fossils are important time indicators.
20. _____ An accurate radiometric date can only be obtained if the material remains in a *closed* system during the entire period since its formation.
21. _____ The scientific study of fossils is called *paleontology*.

Word choice. Complete each of the following statements by selecting the most appropriate response.

1. The most useful fossils for geologic dating are called [index/parent] fossils.
2. In the geologic time scale, [eons/eras/periods] represent the greatest expanses of time.
3. Isotopes of an element have different numbers of [electrons/neutrons/protons].
4. [James Ussher/James Hutton] put forth the fundamental principle of geology known as uniformitarianism.
5. Carbon-14 dating is used with [igneous/organic] materials.
6. Catastrophism states that the forces that shaped the major features of Earth were [the same as/different from] the current causes.
7. With most isotopes used in radiometric dating, the ratio of the original isotope to its [emitted particles/daughter product] is measured.
8. When rocks are deposited sequentially without interruption they are said to be [sequential/conformable].
9. The best index fossils formed from organisms that had [narrow/wide] geographic ranges and existed for [long/short] spans of time.
10. The rates of natural processes [do/do not] vary at different times and different places.
11. The entire geologic time scale was originally created using methods of [relative/numerical] dating.
12. Fossils would be best preserved in [coarse/fine] sediments.
13. Isotopes produced by the decay of radioactive elements are called the [parent/daughter] products.
14. Uniformitarianism requires that Earth be [older/younger] than the age required by catastrophism.
15. One of the radioactive decay products of uranium-238 is [nitrogen/carbon/radon], a colorless, odorless, invisible gas.
16. Inclusions are [younger/older] than the rock they are found in.
17. Rocks from several localities have been dated at more than [3/5] billion years.
18. Placing events in proper sequence is known as [relative/numerical] dating.
19. Angular unconformities are most likely to follow periods of [volcanism/folding and erosion].
20. [Igneous/Sedimentary] rocks are the least suitable for radiometric dating.

Written questions

1. Explain the difference between radiometric and relative dating.

2. List three laws or principles that are used to determine relative dates.

3. How are fossils helpful in geologic investigations?

For other interesting and pertinent
information, be sure to visit
the *Earth Science* companion website at

http://www.prenhall.com/tarbuck

12

Earth's Evolution Through Geologic Time

Earth has a long and complex history. Time and again, the splitting and colliding of continents has resulted in the formation of new ocean basins and the creation of great mountain ranges. Not only has the nature of life on our planet experienced dramatic changes through time, but the atmosphere, hydrosphere, and geosphere have also evolved to become what we see today.

The previous chapter investigated the tools that geologists have at their disposal for interpreting Earth's past. Using these tools, and the clues that are contained in the rock record, Earth scientists have been able to decipher many of the complex events of the geological past. The goal of Chapter twelve is to provide a brief overview of the history of Earth and its life forms by examining how our physical environment assumed its present form and how our planet's inhabitants changed through time.

Learning Objectives

After reading, studying, and discussing this chapter, you should be able to:

Discuss the early origin of Earth.

The history of Earth began about 13.7 billion years ago when the first elements were created during the *Big Bang*. It was from this material, plus other elements ejected into interstellar space by now defunct stars, that Earth along with the rest of the solar system formed. As material collected, high velocity impacts of chunks of matter called *planetesimals* and the decay of radioactive elements caused the temperature of our planet to steadily increase. Iron and nickel melted and sank to form the metallic core, while rocky material rose to form the mantle and Earth's initial crust.

Discuss the evolution of Earth's atmosphere and oceans.

Earth's primitive atmosphere, which consisted mostly of water vapor and carbon dioxide, formed by a process called *outgassing*, which resembles the steam eruptions of modern volcanoes. About 3.5 billion years ago, photosynthesizing bacteria began to release oxygen, first into the oceans and then into the atmosphere. This began the evolution of our modern atmosphere. The oceans formed early in Earth's history, as water vapor condensed to form clouds and torrential rains filled low-lying areas. The salinity in seawater came from volcanic outgassing, and from elements weathered and eroded from Earth's primitive crust.

List the events of Precambrian history.

The Precambrian, which is divided into the Archean and Proterozoic eons, spans nearly 90 percent of Earth's history, beginning with the formation of Earth about 4.5 billion years ago. During this time, much of Earth's stable continental crust was created through a multi-stage process. First, partial melting of the mantle generated magma that rose to form volcanic island arcs and oceanic plateaus. These thin crustal fragments collided and accreted to form larger crustal blocks called *cratons*. Cratons, which form the core of modern continents, were created mainly during the Precambrian.

Describe supercontinents and the supercontinent cycle.

Supercontinents are large landmasses that consist of all, or nearly all, existing continents. *Pangaea* was the most recent supercontinent, but a massive southern continent called *Gondwana*, and perhaps an even larger one, *Rodinia*, preceded it. The splitting and reassembling of supercontinents have generated most of Earth's major mountain belts. In addition, the movement of these crustal blocks have profoundly affected Earth's climate, and have caused sea level to rise and fall.

Discuss the geologic history of the Phanerozoic Eon.

The time span following the close of the Precambrian, called the *Phanerozoic eon*, encompasses 542 million years and is divided into three eras: *Paleozoic, Mesozoic*, and *Cenozoic*. The Paleozoic era was dominated by continental collisions as the supercontinent of Pangaea assembled, forming the Caledonian, Appalachian, and Ural Mountains. Early in the Mesozoic, much of the land was above sea level. However, by the middle Mesozoic, seas invaded western North America. As Pangaea began to break apart, the westward-moving North American plate began to override the Pacific plate, causing crustal deformation along the entire western margin of North America. Most of North America was above sea level throughout the Cenozoic. Owing to their different relations with plate boundaries, the eastern and western margins of the continent experienced contrasting events. The stable eastern margin was the site of abundant sedimentation as isostatic adjustment raised the Appalachians, causing streams to erode with renewed vigor and deposit their sediment along the continental margin. In the West, building of the Rocky Mountains (the *Laramide Orogeny*) was coming to an end, the Basin and Range Province was forming, and volcanic activity was extensive.

Describe Earth's earliest life forms in the Precambrian.

The first known organisms were single-celled bacteria, *prokaryotes*, which lack a nucleus. One group of these organisms, called cyanobacteria, that used solar energy to synthesize organic compounds (sugars) evolved. For the first time, organisms had the ability to produce their own food. Fossil evidence for the existence of these bacteria includes layered mounds of calcium carbonate called *stromatolites*.

Discuss the life of the Paleozoic.

The beginning of the Paleozoic is marked by the *appearance of the first life-forms with hard parts* such as shells. Therefore, abundant Paleozoic fossils occur, and a far more detailed record of Paleozoic events can be constructed. Life in the early Paleozoic was restricted to the seas and consisted of several invertebrate groups, including trilobites, cephalopods, sponges, and corals. During the Paleozoic, organisms diversified dramatically. Insects and plants moved onto land, and lobe-finned fishes that adapted to land became the first amphibians. By the Pennsylvanian period, large tropical swamps, which became the major coal deposits of today, extended across North America, Europe, and Siberia. At the close of the Paleozoic, a mass extinction destroyed 70 percent of all vertebrate species on land and 90 percent of all marine organisms.

Discuss the life of the Mesozoic.

The Mesozoic era, literally the era of middle life, is often called the "*Age of Reptiles*." Organisms that survived the extinction at the end of the Paleozoic began to diversify in spectacular ways. *Gymnosperms* (cycads, conifers, and ginkgoes) became the dominant trees of the Mesozoic because they could adapt to the drier climates. Reptiles became the dominant land animals. The most awesome of the Mesozoic reptiles were the *dinosaurs*. At the close of the Mesozoic, many large reptiles, including the dinosaurs, became extinct.

Discuss the life of the Cenozoic.

The Cenozoic is often called the "*Age of Mammals*" because these animals replaced the reptiles as the dominant vertebrate life forms on land. Two groups of mammals, the marsupials and the placentals, evolved and expanded during this era. One tendency was for some mammal groups to become very large. However, a wave of late *Pleistocene* extinctions rapidly eliminated these animals from the landscape. Some scientists suggest that early humans hastened their decline by selectively hunting the larger animals. The Cenozoic could also be called the "*Age of Flowering Plants*." As a source of food, flowering plants (angiosperms) strongly influenced the evolution of both birds and herbivorous (plant-eating) mammals throughout the Cenozoic era.

Key Terms

banded iron formations
cratons
eukaryotes
outgassing
planetesimals
prokaryotes
protoplanets
solar nebula
stromatolites
supercontinent
supercontinent cycle
supernova

Vocabulary Review

Choosing from the list of key terms, furnish the most appropriate response for the following statements.

1. The most common Precambrian fossils are __________, layered mounds of calcium carbonate.
2. __________ means "planets in the making."
3. The cells of these primitive organisms lack organized nuclei and they reproduce asexually.
4. The explosive death of a massive star, perhaps 10 to 20 times more massive than the Sun, is called a __________.
5. The Sun is thought to have formed from a large rotating cloud of interstellar dust and gas called the __________.
6. As our solar system was forming, matter gradually came together by collisions and the resulting asteroid-sized objects known as __________ would eventually become planets.
7. Earth's initial atmosphere formed by a process called __________, through which gases trapped in the planet's interior are released.
8. Iron oxide deposits in the Precambrian that consist of alternating layers of iron-rich rocks and chert are called __________.
9. __________ are large landmasses that contain all, or nearly all, of the existing continents.
10. __________ were primitive organisms in the Precambrian that exhibited a nucleus in their cellular structure.

Comprehensive Review

1. What was the source of the gases that formed Earth's original atmosphere? What were the gases that most likely made up the original atmosphere?

2. Where did the significant percentage of oxygen in Earth's present atmosphere come from?

3. Write the name of the geologic era during which each of the following events occurred.

   a) Humans develop:

   b) First one-celled organisms:

   c) First birds:

   d) First fishes:

   e) Formation of Pangaea:

   f) Dinosaurs dominant:

   g) First shelled organisms:

   h) Mammals dominant:

   i) First reptiles:

   j) Breakup of Pangaea:

4. Briefly summarize the geological development of the North American Continent.

5. What important event in animal evolution marks the beginning of the Paleozoic era?

6. What are two hypotheses for the extinction of the dinosaurs, along with many other plant and animal groups, at the end of the Mesozoic era?

   1)

   2)

7. Compare the geologic events that occurred along the eastern and western margins of North America during the Cenozoic era.

8. Following the demise of most Mesozoic reptiles, what were the four principal directions that mammals followed in their subsequent development and specialization?

   1)

   2)

   3)

   4)

9. List the ages and briefly summarize the events of the following geologic eras.

   a) Cenozoic era:

   b) Mesozoic era:

   c) Paleozoic era:

Practice Test

Multiple choice. Choose the best answer for the following multiple-choice questions.

1. Life on Earth began during the __________ era.
   a) Cenozoic
   b) Mesozoic
   c) Paleozoic
   d) Precambrian
   e) Evolutionary

2. Earth's original atmosphere consisted of gases expelled from within the planet during a process known as __________.
   a) exhalation
   b) expulsion
   c) venting
   d) intrusion
   e) outgassing

3. Large "core areas" of Precambrian rocks, called __________, exist on each continent.
   a) plateaus
   b) shields
   c) roots
   d) complexes
   e) rifts

4. Once the available metal __________ was oxidized, substantial quantities of oxygen began to accumulate in Earth's atmosphere.
   a) nickel
   b) copper
   c) magnesium
   d) lead
   e) iron

5. The beginning of the Paleozoic era is marked by the appearance of the first life forms with __________.
   a) legs
   b) cells
   c) fins
   d) hard parts
   e) lungs

6. Some of the oldest microfossils are dated at about __________ years of age.
   a) 570 million
   b) 890 million
   c) 1.2 billion
   d) 2.8 billion
   e) 3.1 billion

7. Which era of the geologic time scale is often referred to as the "age of mammals"?
   a) Cenozoic
   b) Mesozoic
   c) Paleozoic
   d) Precambrian

8. Which one of the following was NOT a gas that geologists hypothesize existed in Earth's original atmosphere?
   a) water vapor
   b) carbon dioxide
   c) oxygen
   d) nitrogen

9. Which era of the geologic time scale is often referred to as the "age of reptiles"?
   a) Cenozoic
   b) Mesozoic
   c) Paleozoic
   d) Precambrian

10. The supercontinent of Pangaea formed during the late __________.
    a) Cenozoic
    b) Mesozoic
    c) Paleozoic
    d) Precambrian

11. The petroleum traps of the Gulf Coast occur in __________ strata.
    a) Cenozoic
    b) Mesozoic
    c) Paleozoic
    d) Precambrian

12. Which one of the following events is NOT associated with the late Paleozoic era?
    a) establishment of land plants
    b) evolution of amphibians
    c) evolution of bony fishes
    d) first flowering plants
    e) extensive coal swamps

13. As the first oceans formed on Earth, large amounts of the gas __________ became dissolved in the water.
    a) oxygen
    b) carbon dioxide
    c) sulfur oxide
    d) argon
    e) hydrogen

14. Which one of the following is notably absent in Precambrian rocks?
    a) oxidized iron
    b) metamorphic rocks
    c) silicate minerals
    d) fossil fuels
    e) igneous rocks

15. Which era of geologic time spans the largest percentage of Earth's history?
    a) Precambrian
    b) Mesozoic
    c) Paleozoic
    d) Cenozoic

16. Plants, employing a process called __________, are the major contributors of oxygen to the atmosphere.
    a) conversion
    b) oxidation
    c) hydrolysis
    d) photosynthesis
    e) calcification

17. The presence of elevated proportions of the element __________ in a thin layer of sediment deposited 65 million years ago supports the theory that a large asteroid or comet collided with Earth.
    a) iridium
    b) radon
    c) silicon
    d) chlorine
    e) carbon

18. Earth is about __________ years old.
    a) 4,000
    b) 4.5 million
    c) 5.8 million
    d) 4.5 billion
    e) 6.4 billion

19. One group of __________, exemplified by the fossil *Archaeopteryx,* led to the birds.
    a) reptiles
    b) large cats
    c) rodents
    d) fishes
    e) angiosperms

20. The late __________ extinction was the greatest of at least five mass extinctions to occur over the past 500 million years.
    a) Cenozoic
    b) Mesozoic
    c) Paleozoic
    d) Precambrian

21. About 65 million years ago, during a time of mass extinction called the __________ (or KT) boundary, more than half of all plant and animal species died out.
    a) Cretaceous-Tertiary
    b) Carboniferous-Triassic
    c) Cretaceous-Triassic
    d) Carboniferous-Tertiary

22. Toward the end of the __________, a mountain-building event formed the middle and southern ranges of the Rocky Mountains.
    a) Cenozoic
    b) Mesozoic
    c) Paleozoic
    d) Precambrian

23. Which one of the following was NOT a characteristic of primitive mammals?
    a) small size
    b) large brain
    c) short legs
    d) flat, five-toed feet

24. With the perfection of the shelled egg, reptiles quickly became the dominant animals during the __________ era.
    a) Cenozoic
    b) Mesozoic
    c) Paleozoic
    d) Precambrian

25. The beginning of the breakup of the supercontinent Pangaea occurred during the __________ era.
    a) Cenozoic
    b) Mesozoic
    c) Paleozoic
    d) Precambrian

26. Many of the most ancient fossils are preserved in __________, a hard, dense, chemical sedimentary rock.
    a) chert
    b) iron oxide
    c) limestone
    d) shale
    e) mudstone

27. __________ bacteria thrive in environments that lack free oxygen.
    a) Anaerobic
    b) Proaerobic
    c) Nucleated
    d) Placental
    e) Photobiotic

True/false. For the following true/false questions, if a statement is not completely true, mark it false. For each false statement, change the **italicized** *word to correct the statement.*

1. _____ *Plants* are responsible for dramatically altering the composition of the entire planet's atmosphere.
2. _____ The first life forms on Earth, probably *placentals,* did not need oxygen.
3. _____ The *Devonian* period is often called the "age of fishes."
4. _____ During the *Mesozoic,* the entire western margin of North America was subjected to a continuous wave of deformation as the North American plate began to override the Pacific plate.
5. _____ Many large mammals were rapidly eliminated during a wave of late *Pleistocene* extinctions.
6. _____ The climate of the Mesozoic era was *wetter* than that of the Paleozoic.
7. _____ The smallest division of geologic time is a(n) *period.*
8. _____ The dominant animals of the Mesozoic era were *amphibians.*
9. _____ By the *Pennsylvanian* period, large tropical swamps extended across North America.
10. _____ The ancestral Appalachian Mountains formed during the late Paleozoic as North America and *Africa* collided.
11. _____ The first appearance of mammals was during the *Mesozoic* era.
12. _____ In North America, the Canadian Shield encompasses slightly more than 7 million square kilometers.
13. _____ The beginning of the *Cenozoic* era is marked by the appearance of the first life forms with hard parts.
14. _____ Today, of the two groups of mammals, *marsupials* are found primarily in Australia.
15. _____ The *Mesozoic* era is the least-understood span of Earth's history.
16. _____ By late Devonian time, *lobe-finned* fish had evolved into true air-breathing amphibians.
17. _____ Evidence indicates that some dinosaurs, unlike their present-day reptile relatives, were *warm* blooded.
18. _____ The Basin and Range Province of western North America formed during the *Mesozoic* era.
19. _____ The most common Precambrian fossils are *cephalopods.*
20. _____ Most mammals, including humans, are *placental.*

Word choice. Complete each of the following statements by selecting the most appropriate response.

1. Most of the divisions of geologic time are based upon the [life forms/geologic formations] present at the time.
2. Rocks from the middle Precambrian contain most of Earth's [iron/gold] ore, mainly as the mineral hematite.
3. Life began to colonize the land during the [Paleozoic/Mesozoic] era.
4. Earth's early atmosphere was composed chiefly of gases expelled from [the oceans/molten rock].
5. The predominant plant life of the Cenozoic era are [angiosperms/gymnosperms].
6. In the early Paleozoic era, North America and Europe are thought to have been [near/far from] the equator.
7. The [Paleozoic/Precambrian] encompasses most of Earth's history.
8. Grasses and accompanying herbivores developed during the [Quaternary/Tertiary] period of the Cenozoic era.
9. In the Precambrian, plant fossils occur [before/after] animal fossils.
10. By the close of the [Paleozoic/Mesozoic] era, all the continents had fused into the supercontinent Pangaea.
11. The appearance of higher organisms with preservable hard parts occurred in the [Cambrian/Permian] period.
12. A [drier/wetter] climate at the end of the Paleozoic era is thought to have triggered the extinction of many species on land and sea.
13. Isolation caused by the separation of continents [increases/decreases] the diversity of animals in the world.
14. Most of North America was [above/below] sea level throughout the Cenozoic era.
15. Volcanic activity was common in [eastern/western] North America during much of Cenozoic time.

Written questions

1. List the two hypotheses most frequently used to explain the extinction of the dinosaurs.

2. Why is so little known about the Precambrian era?

3. Describe the evolutionary advancement that freed reptiles from needing to live part of their lives in water, as did the amphibians that preceded them.

4. List four traits that separate mammals from reptiles.

For other interesting and pertinent information, be sure to visit the *Earth Science* companion website at

http://www.prenhall.com/tarbuck

13

The Ocean Floor

The global ocean covers nearly 71 percent of Earth's surface; however, it has only been in the recent past that the sea has become an important focus of scientific study. Today we know that if all the water were drained from the ocean what we would see would be plains, mountains, canyons, and plateaus that would rival or exceed those found on the continents. How old are these features? How did they form? And what about the carpet of sediment that covers much of the seafloor? Where did it come from, and what can be learned by examining it? Chapter thirteen provides answers to these questions by investigating the extent of the world ocean, the features associated with passive and active continental margins and the ocean basin, as well as oceanic ridges and seafloor sediments.

Learning Objectives

After reading, studying, and discussing this chapter, you should be able to:

Define oceanography.

Oceanography is an interdisciplinary science that draws on the methods and knowledge of geology, chemistry, physics, and biology to study all aspects of the world ocean.

Describe the extent of the world ocean.

Earth is a planet dominated by oceans. Nearly 71 percent of Earth's surface area consists of oceans and marginal seas. In the Southern Hemisphere, often called the *water hemisphere*, about 81 percent of the surface is water. The world ocean can be divided into four main ocean basins: the *Pacific Ocean* (largest and deepest ocean), the *Atlantic Ocean* (about half the size of the Pacific), the *Indian Ocean* (slightly smaller than the Atlantic and mostly in the Southern Hemisphere), and the *Arctic Ocean* (the smallest and shallowest ocean). Oceanographers also recognize one additional ocean, the *Southern* or *Antarctic Ocean*, which occurs south of about 50 degrees south latitude. The average depth of the oceans is 3729 meters (12,234 feet).

Explain how ocean water depths are determined.

Ocean bathymetry is determined using echo sounders and multibeam sonars, which bounce sonic signals off the ocean floor. Delicate receivers intercept the reflected signals. Ship-based receivers record the reflected echoes and accurately measure the time interval of the signals. With this information, ocean depths are calculated and plotted to produce maps of ocean-floor topography. Recently *satellite measurements* of the shape of the ocean surface have added data for mapping ocean-floor features.

Describe the features of a passive continental margin.

The zones that collectively make up a *passive continental margin* include the *continental shelf* (a gently sloping, submerged surface extending from the shoreline toward the deep-ocean basin); the *continental slope* (the true edge of the continent, which has a steep slope that leads from the continental shelf into deep water); and in regions where trenches do not exist, the steep continental slope merges into a gradual incline known as the *continental rise* (which consists of sediments that have moved downslope from the continental shelf to the deep-ocean floor).

Discuss the origin of submarine canyons.

Submarine canyons are deep steep-sided valleys that originate on the continental slope and may extend to the deep-ocean basin. Many submarine canyons have been excavated by *turbidity currents*, which are downslope movements of dense, sediment-laden water.

Describe active continental margins.

Active continental margins are located primarily around the Pacific Ocean in areas where the leading edge of a continent is overrunning oceanic lithosphere. Here sediment scraped from the descending oceanic plate is plastered against the continent to form a collection of sediments called an *accretionary wedge*. An active continental margin generally has a narrow continental shelf, which grades into a deep-ocean trench.

List and describe the major topographic units of the ocean-basin floor.

The deep-ocean basin lies between the continental margin and the mid-oceanic ridge system. The features of the deep-ocean basin include *deep-ocean trenches* (the deepest parts of the ocean, where moving crustal plates descend back into the mantle), *abyssal plains* (among the most level places on Earth, consisting of thick accumulations of sediments that were deposited atop the low, rough portions of the ocean floor), *seamounts* and *guyots* (isolated volcanic peaks on the ocean floor that originate near the mid-ocean ridge or in association with volcanic hot spots), and *oceanic plateaus* (vast accumulation of basaltic lava flows).

Explain how atolls form.

Atolls form as corals and other organisms build a reef on the flanks of sinking volcanic islands. They gradually build the reef complex upward as the island slowly sinks.

Discuss the general structure of oceanic (mid-ocean) ridges.

The *oceanic (mid-ocean) ridge* winds through the middle of most ocean basins. Seafloor spreading occurs along this broad feature, which is characterized by an elevated position, extensive faulting, and volcanic structures that have developed on newly formed oceanic crust. Much of the geologic activity associated with ridges occurs along a narrow region on the ridge crest, called the *rift valley*, where magma moves upward to create new slivers of oceanic crust.

Describe how each of the three broad categories of seafloor sediments originates.

There are three broad categories of seafloor sediments. Terrigenous sediment consists primarily of mineral grains that were weathered from continental rocks and transported to the ocean; *biogenous sediment* consists of shells and skeletons of marine animals and plants; and *hydrogenous sediment* includes minerals that crystallize directly from seawater through various chemical reactions. The global distribution of marine sediments is affected by proximity to source areas and water temperatures that favor the growth of certain marine organisms.

Explain how seafloor sediments are used to study worldwide climatic changes.

Seafloor sediments are helpful when studying worldwide climatic changes because they often contain the remains of organisms that once lived near the sea surface. The numbers and types of these organisms change as the climate changes, and their remains in the sediments record these changes.

List several resources obtained from the seafloor.

Energy resources from the seafloor include *oil and natural gas* and large untapped deposits of *gas hydrates*. Other seafloor resources include *sand and gravel, evaporative salts*, and metals within *manganese nodules*.

Key Terms

abyssal plain	deep-ocean trench	oceanography
active continental margin	deep-sea fan	passive continental margin
bathymetry	echo sounder	rift valley
biogenous sediment	gas hydrate	seamount
continental margin	graded bedding	submarine canyon
continental rise	guyot	tablemount
continental shelf	hydrogenous sediment	terrigenous sediment
continental slope	mid-ocean ridge	turbidite
continental volcanic arc	oceanic plateau	turbidity current
deep-ocean basin	oceanic ridge	volcanic island arc

Vocabulary Review

Choosing from the list of key terms, furnish the most appropriate response for the following statements.

1. __________ is the composite science that draws on the methods and knowledge of biology, chemistry, physics, and geology to study all aspects of the world ocean.
2. __________ is the measurement of ocean depths and the charting of the shape or topography of the ocean floor.
3. A long, narrow, deep ocean trough is known as a(n) __________.
4. The __________ is the gently sloping submerged surface of the continental margin that extends from the shoreline toward the deep-ocean basin.
5. The downslope movement of dense, sediment-laden water is known as a(n) __________.
6. Likely one of the most level places on Earth, a(n) __________ is an incredibly flat feature on the ocean floor.
7. The __________ is the site of seafloor spreading.
8. Sediment that consists primarily of mineral grains that were weathered from continental rocks and transported to the ocean is known as __________.
9. An isolated volcanic peak found on the ocean floor is known as a(n) __________.
10. Sediment consisting of minerals that crystallize directly from seawater through various chemical reactions is called __________.
11. Ocean sediment consisting of shells and skeletons of marine animals and plants is known as __________.
12. In regions where trenches do not exist, the steep continental slope merges into a more gradual incline known as the __________.
13. A(n) __________ is a submerged flat-topped seamount.
14. Much of the geologic activity on the ocean floor occurs along a relatively narrow zone on the mid-ocean ridge crest known as the __________.
15. The __________ marks the seaward edge of the continental shelf.
16. The instrument used to measure ocean depths by transmitting sound waves toward the ocean floor is the __________.
17. A(n) __________ is a turbidity current deposit characterized by graded bedding.
18. A rounded lump of hydrogenous sediment composed of a complex mixture of minerals that forms on the floor of the ocean basin is referred to as a(n) __________.
19. __________ is an unusually compact chemical structure made of water and natural gas.

20. Deep-ocean trenches are often paralleled by an arc-shaped row of active volcanoes called a __________.

21. A mantle plume can generate a large __________, which resembles a flood basalt province found on a continent.

Comprehensive Review

1. Briefly contrast the distribution of land and water in the Northern Hemisphere with that in the Southern Hemisphere.

2. Describe how ocean depths are measured.

3. Using Figure 13.1, select the letter that identifies each of the following features of the continental margin.

   a) Deep-sea fan: _____
   b) Submarine canyon: _____
   c) Continental slope: _____
   d) Continental shelf: _____
   e) Continental rise: _____

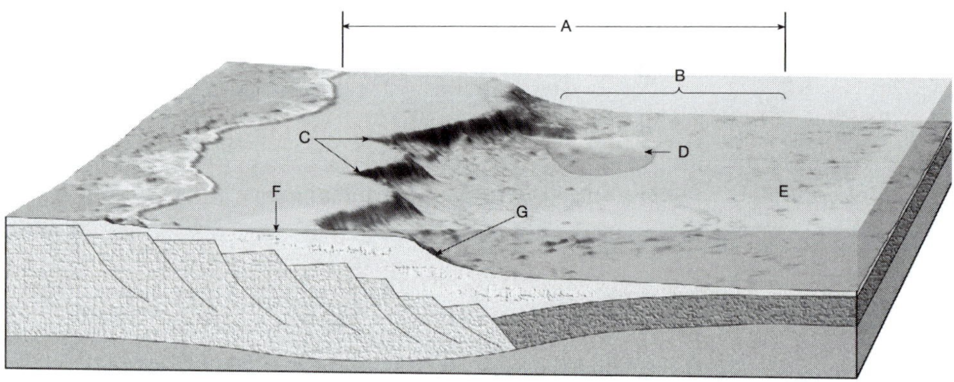

Figure 13.1

4. Describe abyssal plains and explain their origin. In Figure 13.1, letter _____ illustrates an abyssal plain.

5. What is the probable origin for submarine canyons?

6. List three water conditions that are necessary for the growth of coral.

   1)

   2)

   3)

7. Briefly describe the formation of an atoll.

8. List the three broad categories of seafloor sediments and briefly describe the origin of each.

   1)

   2)

   3)

9. Using Figure 13.2, select the letter that identifies each of the following ocean-basin features.

   a) Deep-ocean trench: _____

   b) Seamount: _____

   c) Mid-ocean ridge: _____

   d) Rift valley: _____

   e) Guyot: _____

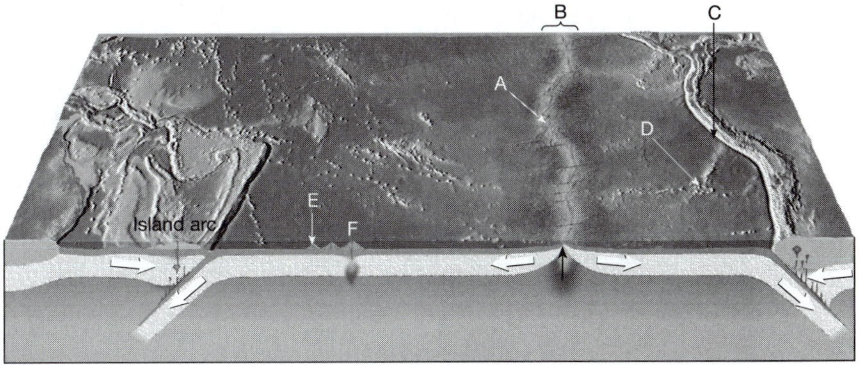

Figure 13.2

10. List three resources that are obtained from the seafloor.

Practice Test

Multiple choice. Choose the best answer for the following multiple-choice questions.

1. Approximately __________ percent of Earth's area is represented by oceans and marginal seas.
   a) 50
   b) 60
   c) 70
   d) 80
   e) 90

2. Important mineral deposits, including large reservoirs of petroleum and natural gas, are associated with __________.
   a) continental shelves
   b) ocean trenches
   c) oceanic ridges
   d) rift zones

3. The most prominent features in the oceans, forming an almost continuous mountain range, are the __________.
   a) deep-ocean trenches
   b) mid-ocean ridges
   c) seamounts
   d) abyssal plains
   e) guyots

4. The largest of Earth's water bodies is the __________.
   a) Pacific Ocean
   b) Mediterranean Sea
   c) Atlantic Ocean
   d) Gulf of Mexico
   e) Indian Ocean

5. As a consequence of the distribution of land and water on Earth, the Southern Hemisphere is referred to as the __________ hemisphere.
   a) water
   b) open
   c) land
   d) mixed
   e) dry

6. Which of the following is NOT one of the three major topographic units of the ocean basins?
   a) continental margins
   b) ocean-basin floor
   c) coastal plain
   d) oceanic (mid-ocean) ridges

7. The seaward edge of the continental shelf is marked by the __________.
   a) abyssal plain
   b) ocean trench
   c) continental slope
   d) mid-ocean ridge
   e) shoreline

8. Sediment that consists primarily of mineral grains that were weathered from continental rocks and transported to the ocean is known as __________ sediment.
   a) biogenous
   b) hydrogenous
   c) terrigenous

9. The __________ are likely the most level places on Earth.
   a) mid-ocean ridges
   b) abyssal plains
   c) deep-ocean trenches
   d) continental slopes
   e) guyots

10. Which one of the following is NOT a zone included in the continental margin?
    a) continental slope
    b) continental rise
    c) continental coastal plain
    d) continental shelf

11. Which of the following is NOT considered one of the four major oceans?
    a) Atlantic Ocean
    b) Equatorial Ocean
    c) Pacific Ocean
    d) Indian Ocean
    e) Arctic

12. Depths in the ocean are often measured using a(n) __________.
    a) altimeter
    b) laser
    c) submersible
    d) rope
    e) echo-sounder

13. Deep steep-sided valleys that originate on the continental slope and may extend to depths of 3 kilometers are referred to as __________.
    a) slope canyons
    b) abyssal plains
    c) rifts
    d) trenches
    e) submarine canyons

14. Where trenches do not exist, the steep continental slope merges into a more gradual incline known as the continental __________.
    a) abyss
    b) rise
    c) end
    d) coast
    e) run

15. The chaotic accumulation of deformed sediment and scraps of oceanic crust that can be found along active continental margins is called a(n) __________.
    a) continental rise
    b) accretionary wedge
    c) pelagic zone
    d) turbidite
    e) estuary

16. The deepest parts of the ocean are long, narrow features known as deep-ocean __________.
    a) ridges
    b) trenches
    c) scars
    d) holes
    e) rifts

17. Isolated volcanic peaks on the ocean floor are known as __________.
    a) ridges
    b) abyssals
    c) nodules
    d) seamounts
    e) rifts

18. Downslope movements of dense sediment-laden water are known as __________ currents.
    a) avalanche
    b) turbidity
    c) density
    d) sediment
    e) longshore

19. Which one of the following is NOT a water condition necessary for coral growth?
    a) warm
    b) sunlit
    c) deep
    d) clear

20. The speed of sound in water is about __________ meters per second.
    a) 500
    b) 1000
    c) 1500
    d) 2000
    e) 2500

True/false. For the following true/false questions, if a statement is not completely true, mark it false. For each false statement, change the **italicized** *word to correct the statement.*

1. _____ Deep-ocean trenches are associated with *active* continental margins.
2. _____ Because it has many shallow adjacent seas, the *Pacific* Ocean is the shallowest of the three major oceans.
3. _____ The true edge of the continent is the continental *slope*.
4. _____ The most economically important evaporative salt is *halite*.
5. _____ Turbidity currents originate along the continental *rise*.
6. _____ Deep-ocean *trenches* in the open ocean are often paralleled by volcanic island arcs.
7. _____ The most common sediment covering the deep-ocean floor is *sand*.
8. _____ Much of the geologic activity along mid-ocean ridges occurs in the *rift* zone.
9. _____ The continental *shelf* varies greatly in width from one continent to another.
10. _____ Deep-ocean trenches are the sites where moving crustal plates plunge back into the *mantle*.

11. _____ Oceanic (mid-ocean) ridges exist in *all* major oceans.
12. _____ The Northern Hemisphere can be referred to as the *land* hemisphere.
13. _____ Atolls owe their existence to the *sudden* sinking of oceanic crust.

Word choice. Complete each of the following statements by selecting the most appropriate response.

1. A portion of the [Mariana/Aleutian] trench is the deepest known part of the world ocean.
2. The continental shelf is underlaid by [oceanic/continental] crust.
3. The greatest number of seamounts have been identified in the [Pacific/Atlantic/Indian] Ocean.
4. The proportion of land to water is [larger/smaller] in the Northern Hemisphere than it is in the Southern Hemisphere.
5. A coral atoll forms during the [subsidence/uplift] of a seamount.
6. During the Pleistocene epoch (Ice Age) sea level was [higher/lower] than it is today.
7. Turbidity currents often form [deep-sea fans/guyots] on the flat ocean floor.
8. The largest ocean is the [Atlantic/Pacific/Indian].
9. [Biogenous/Terrigenous] sediments are most helpful in determining past climates.
10. The continental shelf and continental slope are both parts of [active/passive] continental margins.
11. Most of the geologic activity associated with a mid-ocean ridge occurs along the [rift/trench] zone.
12. Deep-ocean trenches are the sites where crustal plates plunge into Earth's [mantle/core].
13. The Atlantic Ocean has [more/less] extensive abyssal plains than the Pacific.
14. Coral reefs are built by [living organisms/chemical deposition].
15. Most seafloor sediment consists of [mud/sand].
16. Shells and skeletons of marine animals and plants form [biogenous/hydrogenous] sediment.
17. Oceanic (mid-ocean) ridges form along [divergent/convergent] plate boundaries.

Written questions

1. List and describe the major features of a typical ocean-basin floor.

2. What are the three major topographic units of a passive continental margin?

3. Relate deep-ocean trenches and oceanic (mid-ocean) ridges to seafloor spreading.

For other interesting and pertinent
information, be sure to visit
the *Earth Science* companion website at

http://www.prenhall.com/tarbuck

14

Ocean Water and Ocean Life

It is the salts, metals, and dissolved gases that give seawater its distinctly salty taste. In fact, every known naturally occurring element is found in at least trace amounts in ocean water. Like surface salinity, the surface temperature of ocean water also varies both with latitude and with depth. Together, temperature and salinity influence the density of seawater; and variations in density are responsible for large-scale vertical movements of ocean water.

Chapter fourteen investigates the salts of the sea and the factors responsible for variations in the salinity, temperature, and density of ocean water. Also examined are ocean life and ocean feeding relationships, including food chains and webs.

Learning Objectives

After reading, studying, and discussing this chapter, you should be able to:

Discuss the chemical composition of seawater.

Salinity is the amount of dissolved substances in water, usually expressed in parts per thousand (‰). Seawater salinity in the open ocean averages 35‰. The principal elements that contribute to the ocean's salinity are *chlorine* (55 percent) and *sodium* (31 percent), which combine to produce table salt. The primary *sources of the elements in sea salt* in the ocean are *chemical weathering* of rocks on the continents and *volcanic outgassing*.

Explain what causes variations in seawater salinity.

Variations in seawater salinity are primarily caused by changing the *water content*. Natural processes that add large amounts of fresh water to seawater and *decrease salinity* include *precipitation, runoff from land, icebergs melting*, and *sea ice melting*. Processes that remove large amounts of fresh water from seawater and *increase salinity* include *formation of sea ice* and *evaporation*. Seawater salinity in the open ocean *ranges from* 33‰ *to* 38‰ with some marginal seas experiencing considerably more variation.

Describe how ocean temperatures vary with latitude and depth.

The ocean's *surface temperature* is related to the amount of solar energy received and varies as a *function of latitude*. Low-latitude regions have distinctly colder water at depth, creating a *thermocline*, which is a layer of rapidly changing temperature. No thermocline exists in high-latitude regions, because the water column is *isothermal*.

Describe how ocean temperatures have changed through time.

Water's unique thermal properties have caused the *ocean's temperature to remain stable* for long periods of time, facilitating the development of life on Earth. Experiments have been conducted that *send sound through the ocean* to determine if the ocean's temperature is increasing as a result of global warming.

List the factors that affect seawater density.

Seawater density is mostly affected by water temperature but also by salinity. Low-latitude regions have distinctly denser (colder) water at depth, creating a *pycnocline*, which is a layer of rapidly changing density. No pycnocline exists in high-latitude regions, because the water column is *isopycnal.*

Explain the ocean's layered temperature and density structures.

Most open-ocean regions exhibit a *three-layered structure based on water density*. The shallow *surface mixed zone* has warm and nearly uniform temperatures. The *transition zone* includes a prominent thermocline and associated pycnocline. The *deep zone* is continually dark and cold and accounts for 80 percent of the water in the ocean. In high latitudes the three-layered structure does not exist.

Discuss the effect of increasing atmospheric carbon dioxide on the ocean.

The *increasing amount of atmospheric carbon dioxide released by the burning of fossil fuels is causing the ocean to become more acidic*, with serious implications for marine chemistry and marine life.

Describe the classification of marine organisms based on habitat and mobility.

Marine life is superbly adapted to the oceans. Marine organisms can be classified into one of three groups based on habitat and mobility. *Plankton* are free-floating forms with little power of locomotion, *nekton* are swimmers, and *benthos* are bottom dwellers. *Most of the ocean's biomass is planktonic.*

List the criteria used to establish marine life zones.

Three criteria are frequently used to establish *marine life zones*. Based on *availability to sunlight*, the ocean can be divided into the *photic zone* (which includes the *euphotic zone*) and the *aphotic zone*. Based on *distance from the shore*, the ocean can be divided into the *intertidal zone*, the *neritic zone*, and the *oceanic zone*. Based on *water depth*, the ocean can be divided into the *pelagic zone* and the *benthic zone* (which includes the *abyssal zone*).

Discuss primary productivity in the ocean.

Primary productivity is the amount of carbon fixed by organisms through the synthesis of organic matter using energy derived from solar radiation (*photosynthesis*) or chemical reactions (*chemosynthesis*). Chemosynthesis is much less significant than photosynthesis in worldwide oceanic productivity. Photosynthetic productivity in the ocean varies due to the *availability of nutrients and amount of solar radiation*.

Describe how oceanic photosynthetic productivity varies with latitude.

Oceanic photosynthetic productivity varies at different latitudes because of *seasonal changes* and the *development of a thermocline*. In *polar oceans*, the availability of *solar radiation limits productivity*, even though nutrient levels are high. In *tropical oceans*, a strong *thermocline* exists year-round, so the *lack of nutrients generally limits productivity*. In *temperate oceans*, productivity peaks in the spring and fall and is *limited by the lack of solar radiation in winter* and by the *lack of nutrients in summer*.

Summarize oceanic feeding relationships, food chains, and food webs.

The Sun's energy is utilized by *phytoplankton* and converted to *chemical energy*, which is passed through different *trophic levels*. On average, only about *10 percent* of the mass taken in at one trophic level is passed on to the next. As a result, the *size of individuals increases* but the *number of individuals decreases* with each trophic level of the *food chain* or *food web*. Overall, the total biomass of populations decreases at successive trophic levels.

Key Terms

abyssal zone	food web	phytoplankton
aphotic zone	intertidal zone	plankton
benthic zone	nekton	primary productivity
benthos	neritic zone	pycnocline
biomass	oceanic zone	salinity
density	pelagic zone	thermocline
euphotic zone	photic zone	trophic level
food chain	photosynthesis	zooplankton

Vocabulary Review

Choosing from the list of key terms, furnish the most appropriate response for the following statements.

1. The term __________ describes organisms living on or in the ocean bottom.
2. The total amount of solid material dissolved in water, often expressed in parts-per-thousand, is referred to as __________.
3. The upper part of the ocean into which sunlight penetrates is called the __________.
4. __________ include all organisms (algae, animals, and bacteria) that drift with ocean currents.
5. The area where the land and ocean meet and overlap is called the __________.
6. The amount of carbon fixed by organisms through the synthesis of organic matter using energy derived from solar radiation or chemical reactions is called __________.
7. A __________ is a sequence of organisms through which energy is transferred, starting with an organism that is a primary producer.
8. The __________ is the ocean layer where there is a rapid change of density with depth.
9. __________ include all animals capable of moving independently of the ocean currents, by swimming or other means of propulsion.
10. Seaward from the low-tide line is the __________, which covers the gently sloping continental shelf out to the shelf break.
11. Open ocean of any depth, where animals swim or float freely, is called the __________.
12. __________ is defined as mass-per-unit volume.
13. In the deep __________ there is no sunlight.
14. The marine life zone beyond the continental shelf is referred to as the __________.
15. Among plankton, algae are called __________.
16. The layer of the ocean where there is a rapid change in temperature with depth is known as the __________.
17. Animals that feed through a __________ rather than a food chain are more likely to survive because they have alternate foods to eat.
18. In surface waters of the ocean, strong sunlight supports the process of __________ by marine algae, which either directly or indirectly provide food for the vast majority of marine organisms.
19. Among plankton, animals are called __________.
20. The __________ is a subdivision of the benthic zone that includes the deep-ocean floor.
21. Each of the feeding stages of a food chain is called a __________.
22. Most of Earth's __________, the mass of all living organisms, consists of plankton adrift in the oceans.
23. The __________ is the portion of the photic zone near the surface, where light is strong enough for photosynthesis to occur.

Comprehensive Review

1. List the two most abundant salts in seawater. What are the two primary sources for most dissolved substances in the ocean?

2. List two processes that increase and two processes that decrease seawater salinity.

3. Briefly describe why ocean-surface salinity changes with latitude.

4. What are the factors that influence seawater density? Which has the greatest influence on surface seawater density?

5. Compare the vertical changes in ocean water temperature and density in low latitudes to those in high latitudes.

6. Describe the ocean's three-layer structure.

7. How are marine organisms classified according to where they live and how they move?

8. Using Figure 14.1, select the letter that identifies each of the following marine life zones.

a) Euphotic zone: _____ c) Aphotic zone: _____ e) Photic zone: _____

b) Neritic zone: _____ d) Abyssal zone: _____ f) Pelagic zone: _____

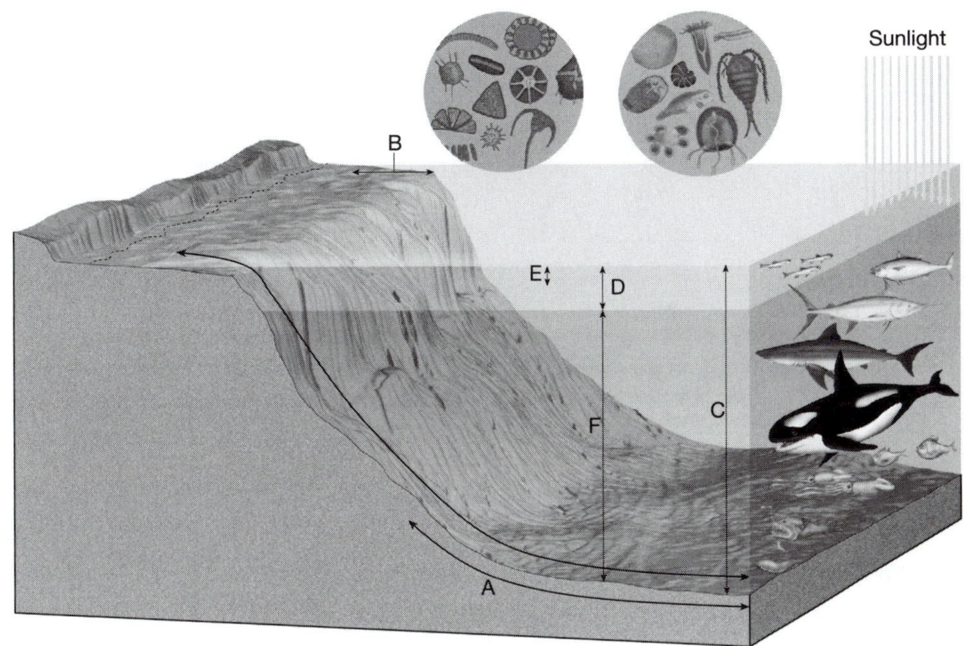

Figure 14.1

9. List the two factors that influence a region's photosynthetic productivity.

10. Briefly compare the oceanic productivity in polar oceans to that in tropical oceans.

11. What is the difference between a food chain and a food web?

Practice Test

Multiple choice. Choose the best answer for the following multiple-choice questions.

1. The principal source of water in the oceans is a process called __________.
   a) weathering c) outgassing e) tectonism
   b) infiltration d) evaporation

2. Which one of the following is NOT a process that decreases the salinity of seawater?
   a) runoff from land c) icebergs melting e) sea ice melting
   b) precipitation d) evaporation

3. Seawater consists of about __________ percent (by weight) of dissolved mineral substances.
   a) 1.5 c) 3.5 e) 5.5
   b) 2.5 d) 4.5

4. The layer of rapid temperature change with depth in the ocean is known as the __________.
   a) halocline c) thermocline e) tempegrad
   b) centicline d) faricline

5. Which one of the following is NOT a zone in the three-layered structure of the ocean according to density?
   a) surface mixed zone c) transition zone
   b) deep zone d) intertidal zone

6. __________ include all organisms that drift with ocean currents.
   a) Nekton c) Benthos e) Neritic
   b) Plankton d) Photic

7. The upper part of the ocean into which sunlight penetrates is called the __________ zone.
   a) neritic c) pelagic e) abyssal
   b) aphotic d) photic

8. In __________ oceans, the zooplankton biomass peaks in June and continues at a relatively high level until winter darkness begins in October.
   a) tropical c) subtropical
   b) polar d) temperate

9. The term __________ describes organisms living on or in the ocean bottom.
   a) benthos c) nekton e) euphotic
   b) pelagic d) plankton

10. The ocean's surface water temperature varies with the amount of __________ received.
    a) solar radiation c) freshwater e) magnesium chloride
    b) carbon dioxide d) salt

11. The layer of rapid density change with depth in the ocean is known as the __________.
    a) halograd c) thermocline
    b) densigrad d) pycnocline

12. Highest overall oceanic productivity occurs in __________ regions.
    a) temperate c) polar
    b) tropical d) equatorial

13. __________ is defined as mass-per-unit volume.
    a) Density
    b) Pycnocline
    c) Specific gravity
    d) Salinity

14. Which one of the following is NOT a main oceanic producer of food?
    a) algae
    b) bacteria
    c) plants
    d) bacteria-like archaea
    e) sharks

15. The transfer of energy between trophic levels averages about __________ percent.
    a) 2
    b) 6
    c) 10
    d) 14
    e) 20

16. Salinity variations in the open ocean normally range from 33‰ to __________.
    a) 38‰
    b) 45‰
    c) 60‰
    d) 67‰
    e) 83‰

17. Which one of the following water bodies has the highest salinity?
    a) Lake Michigan
    b) Baltic Sea
    c) Red Sea
    d) Gulf of Mexico
    e) Hudson Bay

18. Which one of the following properties most determines the water's vertical position in the ocean?
    a) organic content
    b) density
    c) current velocity
    d) oxygen content

19. The second most abundant salt in the sea is __________.
    a) calcium chloride
    b) magnesium chloride
    c) potassium chloride
    d) sodium chloride
    e) sodium fluoride

20. Open ocean of any depth is called the __________ zone.
    a) neritic
    b) abyssal
    c) benthic
    d) planktonic
    e) pelagic

True/false. For the following true/false questions, if a statement is not completely true, mark it false. For each false statement, change the **italicized** *word to correct the statement.*

1. _____ Ocean temperature and density vary with depth and generally follow a *three-layered* structure in the open ocean.
2. _____ The overall makeup of seawater has remained relatively *constant* through time.
3. _____ A thermocline is not present in high latitudes; instead, the water column is *isothermal.*
4. _____ In the ocean, *high* salinities are found where evaporation is high.
5. _____ The area where land and ocean meet and overlap is called the *surf* zone.
6. _____ *Low* ocean-surface salinities occur where large quantities of fresh water are supplied by rivers.
7. _____ Energy transfer between trophic levels is *efficient.*
8. _____ In the ocean where heavy precipitation occurs, *higher* surface salinities prevail.
9. _____ When seawater freezes in winter, sea salts are *expelled* from the ice.
10. _____ Below the thermocline, temperatures *fall* only a few more degrees.
11. _____ Seawater density is influenced by salinity and *temperature.*
12. _____ A *thermocline* can act as a barrier that eliminates the supply of nutrients from deeper waters below.

13. _____ *Primary* productivity is the amount of carbon fixed by organisms through the synthesis of organic matter using energy derived from solar radiation or chemical reactions.
14. _____ A pycnocline is not present in high latitudes; instead, the water column is *isopycnal.*
15. _____ The addition of atmospheric carbon dioxide *raises* the pH of seawater.

Word choice. Complete each of the following statements by selecting the most appropriate response.

1. [Runoff/Evaporation] increases the salinity of seawater.
2. [Evaporation/Precipitation] contributes to the low salinities in the mid-latitudes and near the equator.
3. Seawater has many unique thermal properties that make it [resistant/susceptible] to changes in temperature.
4. High-latitude surface waters typically have [high/low] nutrient concentrations.
5. Oceanographers typically express salinity in parts per [thousand/million].
6. Photosynthesis occurs in the [aphotic/euphotic] zone.
7. In the ocean, fish are more abundant in [cold/warm] waters.
8. The shallow surface layer of the ocean is called the [mixed/transition] zone.
9. [Temperature/Salinity] has the greatest influence on surface seawater density.
10. In low- and middle-latitudes, the density of the ocean's surface layer is [greater/less] than the deep-water density.
11. Hydrothermal vents are a source of food in the [intertidal/abyssal] zone.
12. The surface mixed zone usually extends to about a depth of [300/800] meters.
13. A feeding stage in a food chain is called a [euphotic/trophic] level.
14. Variations in the salinity of ocean water are primarily the consequence of changes in the [salt/water] content of the solution.
15. Among plankton, the algae are called [phytoplankton/zooplankton].

Written questions

1. Describe the term *salinity.* What is the average salinity of the ocean? Describe the variation in ocean-surface salinity with latitude.

2. Briefly describe the ocean's three-layer structure.

3. What are the factors used to divide the ocean into distinct marine life zones? Name a few of these zones.

For other interesting and pertinent information, be sure to visit the *Earth Science* companion website at

http://www.prenhall.com/tarbuck

15

The Dynamic Ocean

Powered by many different forces, the restless waters of the ocean are in constant motion. For example, winds generate surface ocean currents, which influence coastal climate, and also produce waves that transmit energy from storms to distant shores, where their impact erodes the land. In some areas, density differences create deep-ocean circulation, which is important for ocean mixing and recycling nutrients. Furthermore, the Moon and Sun produce tides, which periodically raise and lower sea level. Chapter fifteen examines these movements of ocean waters and their effect upon coastal regions.

Learning Objectives

After reading, studying, and discussing this chapter, you should be able to:

List the factors that influence surface ocean currents.

The ocean's surface currents follow the general pattern of the world's major wind belts. Surface currents are parts of huge, slowly moving loops of water called *gyres* that are centered in the subtropics of each ocean basin. The *positions of the continents* and the *Coriolis effect* also influence the movement of ocean water within gyres. Because of the Coriolis effect, subtropical gyres move *clockwise in the Northern Hemisphere* and *counterclockwise in the Southern Hemisphere.* Generally, *four main currents* comprise each subtropical gyre.

Discuss the importance of surface ocean currents.

Ocean currents are important in navigation and for the effect they have on climates. Poleward-moving *warm ocean currents moderate winter temperatures* in the middle latitudes. Cold currents exert the greatest influence during summer in middle latitudes and year-round in the tropics. In addition to cooler temperatures, *cold currents are associated with greater fog frequency and drought.*

Describe upwelling.

Upwelling, the rising of colder water from deeper layers, is a wind-induced movement that brings cold, nutrient-rich water to the surface. *Coastal upwelling* is most characteristic along the west coasts of continents.

Describe deep-ocean circulation.

In contrast to surface currents, *deep-ocean circulation* is governed by *gravity* and driven by *density differences.* The two factors that are most significant in creating a dense mass of water are *temperature* and *salinity*, so the movement of deep-ocean water is often termed *thermohaline circulation.* Most water involved in thermohaline circulation begins in high latitudes at the surface when the salinity of the cold water increases due to sea-ice formation. This dense water sinks, initiating deep-ocean currents.

Describe the various components of the coastal zone.

The *shore* is the area extending between the lowest tide level and the highest elevation on land that is affected by storm waves. The *coast* extends inland from the shore as far as ocean-related features can be found. The shore is divided into the *foreshore* and *backshore*. Seaward of the foreshore are the *nearshore* and *offshore* zones.

Discuss the parts and composition of beaches.

A *beach* is an accumulation of sediment found along the landward margin of the ocean or a lake. Among its parts are one or more *berms* and the *beach face*. Beaches are composed of whatever material is locally abundant and should be thought of as material in transit along the shore.

List the factors that influence wave characteristics.

Waves are moving energy, and *most ocean waves are initiated by the wind*. The three factors that influence the *height*, *wavelength*, and *period* of a wave are (1) *wind speed*, (2) *length of time the wind has blown*, and (3) *fetch*, the distance that the wind has traveled across open water. Once waves leave a storm area, they are termed *swells*, which are symmetrical, longer-wavelength waves.

Discuss the motion of water particles in a wave.

As waves travel, *water particles transmit energy by circular orbital motion*, which extends to a depth equal to one half the wavelength. When a wave travels into shallow water, it causes the water particle motion to experience changes that can cause the wave to collapse, or *break*, and form *surf*.

Describe wave erosion and refraction.

Wave erosion is caused by *wave impact pressure*, and *abrasion* (the sawing and grinding action of water armed with rock fragments). The bending of waves is called *wave refraction*. Owing to refraction, wave impact is concentrated against the sides and ends of headlands.

Describe beach drift and longshore currents.

Most waves reach the shore at a slight angle. The uprush and backwash of water from each breaking wave moves the sediment in a zigzag pattern along the beach. This movement is called *beach drift*. Oblique waves also produce *longshore currents* within the surf zone that flow parallel to the shore and transport more sediment than beach drift. *Rip currents* occur perpendicular to the coast and move in the opposite direction of breaking waves.

Describe wave erosion and the features produced by wave erosion.

Erosion features include *wave-cut cliffs* (which originate from the cutting action of the surf against the base of coastal land), *wave-cut platforms* (relatively flat, benchlike surfaces left behind by receding cliffs), and *marine terraces* (uplifted wave-cut platforms). Erosional features also include *sea arches* (formed when a headland is eroded and two caves from opposite sides unite), and *sea stacks* (formed when the roof of a sea arch collapses).

List and describe the depositional features found along coasts.

Some of the depositional features that form when sediment is moved by beach drift and longshore currents are *spits* (elongated ridges of sand that project from the land into the mouth of an adjacent bay), *baymouth bars* (sandbars that completely cross a bay), and *tombolos* (ridges of sand that connect an island to the mainland or to another island). Along the Atlantic and Gulf coastal plains, the coastal region is characterized by offshore *barrier islands*, which are low ridges of sand that parallel the coast.

List the factors that influence shoreline erosion.

Local *factors that influence shoreline erosion* are (1) the proximity of a coast to sediment-laden rivers, (2) the degree of tectonic activity, (3) the topography and composition of the land, (4) prevailing winds and weather patterns, and (5) the configuration of the coastline and nearshore areas.

Describe some solutions to shoreline erosional problems.

Hard stabilization involves building hard, massive structures in an attempt to protect a coast from erosion or prevent the movement of sand along a beach. Hard stabilization includes *groins* (short walls built at a right angle to the shore to trap moving sand), *breakwaters* (structures built parallel to the shoreline to protect boats from the force of large breaking waves), and *seawalls* (armoring the coast to prevent waves from reaching the area behind the wall). *Alternatives to hard stabilization* include *beach nourishment*, which involves the addition of sand to replenish eroding beaches, and *relocation* of damaged or threatened buildings.

Discuss the differences between America's Pacific and Atlantic coasts.

Because of basic geological differences, the *nature of shoreline erosion problems along America's Pacific and Atlantic coasts is very different.* Much of the development along the Atlantic and Gulf coasts has occurred on barrier islands, which receive the full force of major storms. Much of the Pacific Coast is characterized by narrow beaches backed by steep cliffs and mountain ranges. A major problem facing the Pacific shoreline is a narrowing of beaches caused by irrigation and flood-control dams that interrupt the natural flow of sand to the coast.

Explain the differences between an emergent and submergent coast.

One commonly used classification of coasts is based upon changes that have occurred with respect to sea level. *Emergent coasts* often exhibit wave-cut cliffs and marine terraces and develop either because an area experiences uplift or as a result of a drop in sea level. Conversely, *submergent coasts* commonly display drowned river mouths called *estuaries* that are created when sea level rises or the land adjacent to the sea subsides.

Discuss the factors that influence tides.

Tides, the daily rise and fall in the elevation of the ocean surface at a specific location, are caused by the *gravitational attraction* of the Moon and, to a lesser extent, by the Sun. Near the times of new and full moons, the Sun and Moon are aligned and their gravitational forces are added high and low tides. These are called the *spring tides*. Conversely, at about the times of the first and third quarters of the Moon, when the gravitational forces of the Moon and Sun are at right angles, the daily *tidal range* is less. These are called *neap tides*.

Describe the three main tidal patterns.

Three main tidal patterns exist worldwide. A *diurnal tidal pattern* exhibits one high and low tide daily; a *semidiurnal tidal pattern* exhibits two high and low tides daily of about the same height; and a *mixed tidal pattern* usually has two high and low tides daily of different heights.

Discuss tidal currents.

Tidal currents are horizontal movements of water that accompany the rise and fall of tides. *Tidal flats* are the areas that are affected by the advancing and retreating tidal currents. When tidal currents slow after emerging from narrow inlets, they deposit sediment that may eventually create *tidal deltas*.

Key Terms

abrasion
backshore
barrier island
baymouth bar
beach
beach drift
beach face
beach nourishment
berm
breakwater
coast
coastline
Coriolis effect
diurnal tidal pattern
emergent coast
estuary
foreshore
groin
gyre
hard stabilization
longshore current
marine terrace
mixed tidal pattern
neap tide
nearshore zone
offshore zone
rip current
sea arch
sea stack
seawall
semidiurnal tidal pattern
shore
shoreline
spit
spring tide
submergent coast
surf
thermohaline circulation
tidal current
tidal delta
tidal flat
tombolo
upwelling
wave-cut cliff
wave-cut platform
wave height
wavelength
wave period
wave refraction

Vocabulary Review

Choosing from the list of key terms, furnish the most appropriate response for the following statements.

1. An elongated ridge of sand that projects from the land into the mouth of an adjacent bay is called a(n) __________.
2. __________ is the movement of sediment in a zigzag pattern along a beach.
3. The large, circular surface-current pattern found in each ocean is called a(n) __________.
4. The __________ is the vertical distance between the trough and crest of a wave.
5. Deep-ocean circulation is called __________.
6. Turbulent water created by breaking waves is called __________.
7. The deflective force of Earth's rotation on all free-moving objects is known as the __________.
8. __________ and relocation are two alternatives to hard stabilization of eroding beaches.
9. The horizontal distance separating successive wave crests is the __________.
10. A(n) __________ is a low ridge of sand that parallels a coast at a distance from 3 to 30 kilometers offshore.
11. The horizontal flow of water accompanying the rise and fall of a tide is called a(n) __________.
12. A drowned river mouth is called a(n) __________.
13. The rising of colder water from deeper ocean layers to replace warmer surface water is called __________.
14. A(n) __________ is a ridge of sand that connects an island either to the mainland or to another island.
15. The time interval between the successive passage of wave crests is known as the __________.
16. When a sea arch collapses, it leaves an isolated remnant called a(n) __________.
17. A coast that develops either because the area experiences uplift or a drop in sea level is a(n) __________.
18. A(n) __________ is a sandbar that completely crosses a bay.

19. The area affected by alternating tidal currents is known as a(n) __________.
20. The bending of a wave is called __________.
21. A(n) __________ is a movement of water perpendicular to the coast in the opposite direction of breaking waves.
22. A nearshore current that flows parallel to the shore is called a(n) __________.
23. A(n) __________ is a relatively flat, benchlike surface left behind by a receding wave-cut cliff.
24. A(n) __________ is often highly irregular because the sea typically floods the lower reaches of river valleys.
25. A __________ is an accumulation of sediment found along the landward margin of the ocean or a lake.
26. A(n) __________ is a barrier built at a right angle to the beach to trap sand that is moving parallel to the shore.
27. The __________ is the wet, sloping surface that extends from the berm to the shoreline.
28. The highest tidal variation that occurs during the times of new and full moons is called a(n) __________.
29. A(n) __________ forms when a rapidly moving tidal current emerges from a narrow inlet and slows, depositing its load of sediment.
30. Along a shoreline, the __________ is landward of the high-tide shoreline.

Comprehensive Review

1. Briefly describe the influence that the Coriolis effect has on the movement of ocean waters.

2. Using Figure 15.1, select the letter that identifies each of the following surface currents.

   a) Gulf Stream: _____ d) West Wind Drift: _____ g) North Pacific: _____

   b) Peru: _____ e) Benguela: _____ h) Equatorial (North and South): _____

   c) Canary: _____ f) Labrador: _____

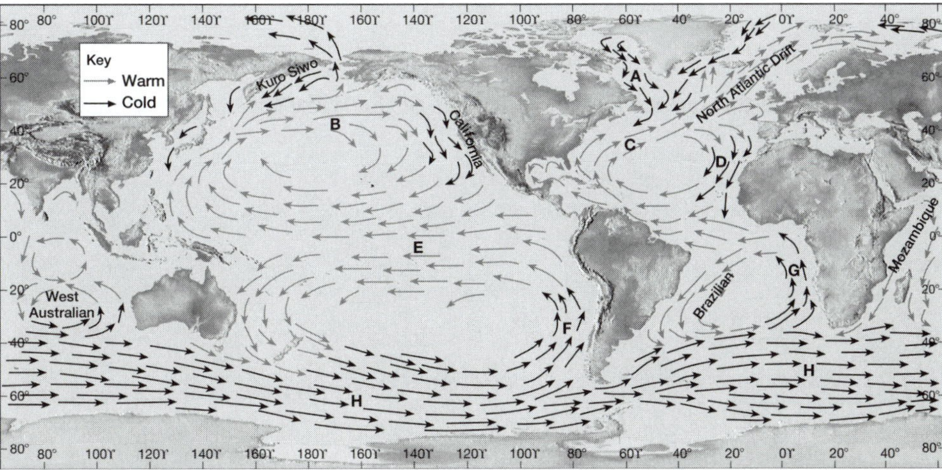

Figure 15.1

3. Describe the effect cold ocean currents have on climates.

4. In addition to the gravitational influence of the Sun and Moon, what are two other factors that influence the tide?

 1)

 2)

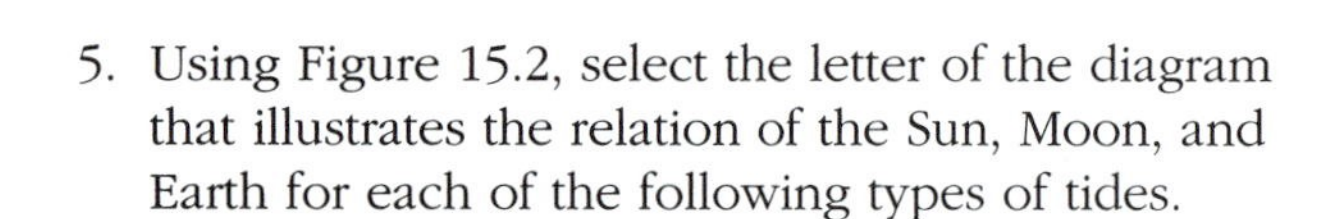

5. Using Figure 15.2, select the letter of the diagram that illustrates the relation of the Sun, Moon, and Earth for each of the following types of tides.

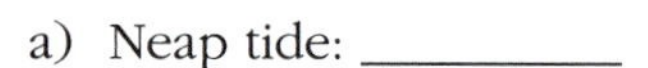

 a) Neap tide: __________

 b) Spring tide: __________

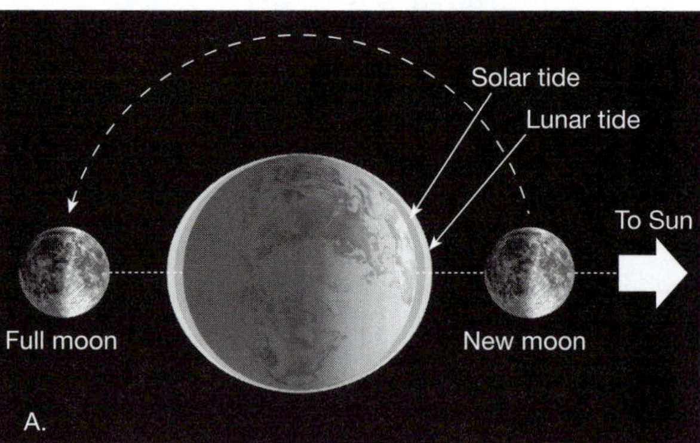

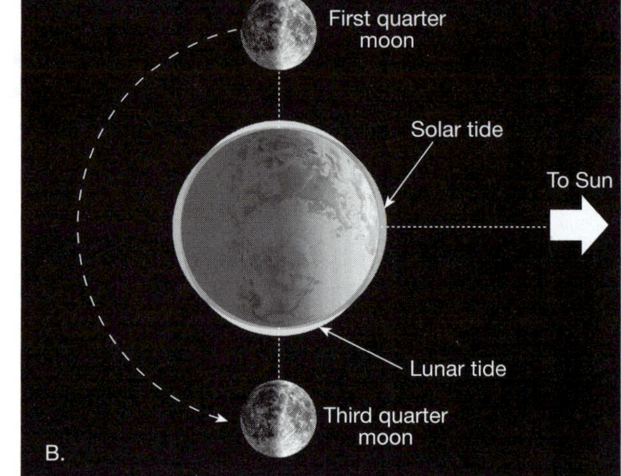

Figure 15.2

6. Using Figure 15.3, write the proper term for the part of the wave illustrated by each of the following letters.

 a) Letter A: __________

 b) Letter B: __________

 c) Letter C: __________

 d) Letter D: __________

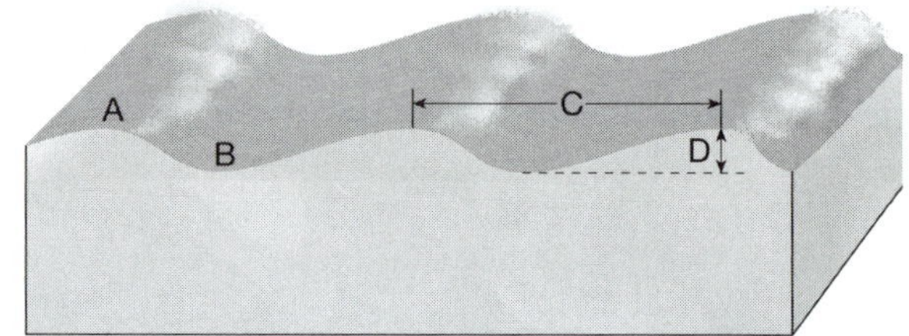

Figure 15.3

7. List the three factors that influence the height, length, and period of open water waves.

    1)

    2)

    3)

8. What are two types of hard stabilization used to protect eroding beaches?

9. What two factors are most significant in creating a dense mass of ocean water?

    1)

    2)

10. Where are the two major regions where dense water masses are created?

    1)

    2)

11. Describe the effect that wave refraction has on the energy of waves in a bay.

12. List two mechanisms that are responsible for transporting sediment along a coast.

    1)

    2)

13. Figure 15.4 illustrates the changes that take place through time along an initially irregular, relatively stable coastline. Using Figure 15.4, select the letter that illustrates each of the following features.

    a) Baymouth bar: _____

    b) Spit: _____

    c) Sea stack: _____

    d) Estuary: _____

    e) Wave-cut cliff: _____

    f) Tombolo: _____

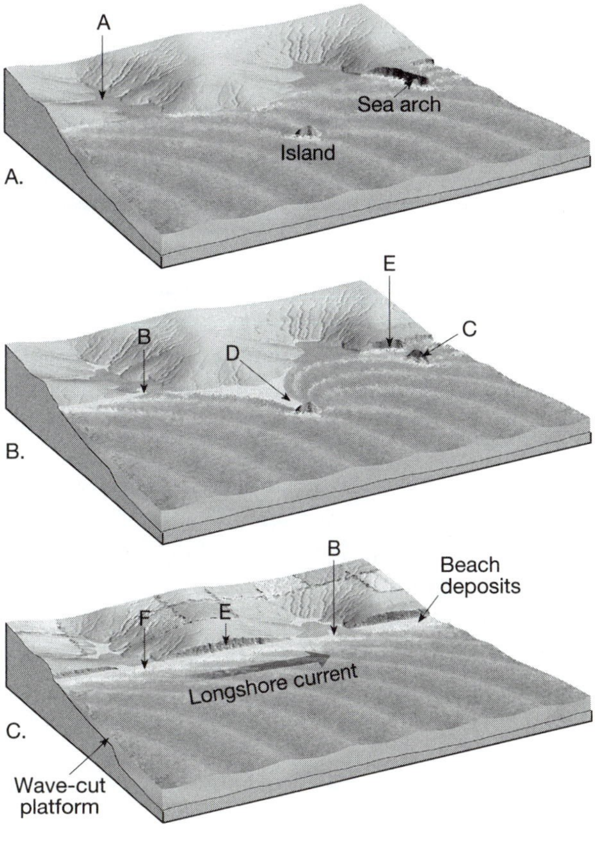

Figure 15.4

14. What are two ways that barrier islands may originate?

    1)

    2)

15. What are three factors that influence the type and degree of shoreline erosion?

    1)

    2)

    3)

16. Briefly describe the conditions that could result in the formation of the following types of coasts. Also list one feature you would expect to find associated with each type.

    a) Emergent coast:

       Feature: __________

    b) Submergent coast:

       Feature: __________

17. Briefly describe the high- and low-water characteristics of each of the three types of tides.

    a) Semidiurnal tidal pattern:

    b) Diurnal tidal pattern:

    c) Mixed tidal pattern:

Practice Test

Multiple choice. Choose the best answer for the following multiple-choice questions.

1. The movement of particles of sediment in a zigzag pattern along a beach is known as __________.
   a) longshore transportation c) turbidity current e) fetch
   b) beach drift d) beach abrasion

2. A barrier built at a right angle to the beach to trap sand that is moving parallel to the shore is known as a __________.
   a) seawall c) pier e) headland
   b) groin d) stack

3. A ridge of sand that connects an island to the mainland or to another island is called a __________.
   a) baymouth bar c) sea stack e) spit
   b) sea arch d) tombolo

4. Where the atmosphere and ocean are in contact, energy is passed from the moving air to the water through __________.
   a) friction c) oscillation e) translation
   b) radiation d) collision

5. The rising of cold water from deeper layers to replace warmer surface water is called __________.
   a) translation
   b) gyration
   c) modulation
   d) combination
   e) upwelling

6. Which one of the following is NOT a west coast desert influenced by cold offshore waters?
   a) Namib
   b) Gobi
   c) Atacama

7. Exposed wave-cut cliffs and platforms are frequently found along __________ coasts.
   a) submergent
   b) emergent

8. The densest water in all of the oceans is formed in __________ waters.
   a) Equatorial
   b) Antarctic
   c) North Pacific
   d) North Atlantic
   e) Indian Ocean

9. The energy and motion of most waves is derived from __________.
   a) currents
   b) tides
   c) gravity
   d) wind
   e) earthquakes

10. Other than wind, the most significant factor that influences the movement of ocean water is the __________.
    a) translation force
    b) Coriolis effect
    c) force of gravity
    d) density effect
    e) ocean slope

11. Which one of the following features occurs more frequently along the Atlantic and Gulf coasts than along the Pacific Coast?
    a) wave-cut cliffs
    b) barrier islands
    c) sea stacks
    d) wave-cut platforms

12. Tidal currents that advance into the coastal zone as the tide rises are called __________ currents.
    a) density
    b) turbidity
    c) flow
    d) longshore
    e) ebb

13. The Indian Ocean exists mostly in the __________ Hemisphere.
    a) Northern
    b) Southern

14. In the open sea, the movement of water particles in a wave becomes negligible at a depth equal to about __________ the distance from crest to trough.
    a) one-fourth
    b) one-third
    c) one-half
    d) two-thirds
    e) three-fourths

15. Deep-ocean circulation is referred to as __________ circulation.
    a) density
    b) subsurface
    c) ocean-floor
    d) thermohaline
    e) abyssal

16. The turbulent water created by breaking waves is called __________.
    a) surf
    b) backwash
    c) modulation
    d) refraction
    e) oscillation

17. The areas affected by alternating tidal currents are known as tidal __________.
    a) swamps
    b) marshes
    c) flats
    d) platforms
    e) estuaries

18. Daily changes in the elevation of the ocean surface are called __________.
    a) currents
    b) waves
    c) density currents
    d) tides
    e) estuaries

19. The large central area of the North Atlantic that has no well-defined currents is known as the __________ Sea.
    a) Red
    b) Baltic
    c) Sargasso
    d) Central Atlantic
    e) Central

20. Highly irregular coasts with numerous estuaries frequently characterize a(n) __________ coast.
    a) submergent
    b) emergent

21. The horizontal distance separating wave crests is known as the __________.
    a) wavelength
    b) fetch
    c) wave height
    d) wave period
    e) oscillation

22. The largest daily tidal ranges are associated with __________ tides.
    a) spring
    b) ebb
    c) neap
    d) flow
    e) lunar

23. The time interval between the passage of successive wave crests is called the __________.
    a) wave height
    b) wavelength
    c) wave period
    d) fetch
    e) oscillation

24. A wave begins to "feel bottom" at a water depth equal to about __________ its wavelength.
    a) one-fourth
    b) one-third
    c) one-half
    d) two-thirds
    e) three-fourths

25. Massive barriers built to prevent waves from reaching the shore are called __________.
    a) groins
    b) seawalls
    c) stacks
    d) barrier islands
    e) piers

26. The greatest difference between high- and low-water heights occurs during a __________ tide.
    a) spring
    b) flow
    c) neap
    d) back
    e) ebb

27. Which one of the following types of tidal patterns is most common along the Atlantic Coast of the United States?
    a) mixed tidal pattern
    b) ebb tidal pattern
    c) semidiurnal tidal pattern
    d) flow tidal pattern
    e) diurnal tidal pattern

True/false. For the following true/false questions, if a statement is not completely true, mark it false. For each false statement, change the **italicized** *word to correct the statement.*

1. _____ Turbulent water currents within the surf zone that flow parallel to the shore are called *longshore* currents.
2. _____ Evaporation and/or freezing tend to make the water *less* salty.
3. _____ *Reach* is the distance that wind has traveled across the open water.
4. _____ The two most significant factors in creating a dense mass of water are temperature and *salinity*.
5. _____ In the open sea, it is the wave *form* that moves forward, not the water itself.
6. _____ In the Southern Hemisphere, ocean currents form *counterclockwise* gyres.

7. _____ Groins have proved to be *unsatisfactory* solutions to shoreline erosion problems.
8. _____ The tide-generating potential of the Sun is slightly less than *half* that of the Moon.
9. _____ Two caves on opposite sides of a headland can unite to form a sea *cliff*.
10. _____ *Cold* ocean currents often have a dramatic impact on the climate of tropical deserts along the west coasts of continents.
11. _____ Ocean currents transfer *excess* heat from the tropics to the heat-deficient polar regions.
12. _____ Wind speeds less than 3 kilometers per hour generally do not produce stable, progressive waves.
13. _____ In the Northern Hemisphere, the Coriolis effect causes ocean currents to be deflected to the *left*.
14. _____ The *shore* is the area that extends between the lowest tide level and the highest elevation on land that is affected by storm waves.
15. _____ Turbulent water created by breaking waves is called *drift*.
16. _____ The lowest daily tidal ranges are associated with *neap* tides.

Word choice. Complete each of the following statements by selecting the most appropriate response.

1. In deep water, the motion of a water particle is best described as [circular/zigzag/vertical].
2. Increasing the salinity of seawater will cause its density to [increase/decrease].
3. The greatest amount of shoreline erosion is caused by [currents/tides/waves].
4. The Coriolis effect causes ocean currents to be [clockwise/counterclockwise] in the Northern Hemisphere and [clockwise/counterclockwise] in the Southern Hemisphere.
5. Daily tidal changes are the result of Earth's [rotation/revolution].
6. The presence of marine terraces above sea level is one line of evidence that the area is a(n) [emergent/submergent] coast.
7. Recent research indicates that sea level is [falling/rising].
8. Ocean currents affect climate by storing and transporting [heat/plankton/water].
9. The Pacific Coast of North America is undergoing [uplift/subsidence].
10. The largest thermohaline currents begin in [equatorial/polar] regions.
11. When waves break, they become [surf/gyres].
12. If Earth had an ocean with uniform depth covering its surface, the Moon would produce [one/two/three] tidal bulges.
13. Waves are created by [gravity/tides/wind].
14. A breakwater will often cause the beach behind it to [gain/lose] sand.
15. Damming rivers has caused West Coast beaches to [widen/narrow].
16. The [Canary Current/Gulf Stream] flows northward along the East Coast of the United States.
17. The zigzag movement of sediment along a beach is called [beach/longshore] drift.
18. The horizontal flow of water associated with rising and falling tides is called a tidal [current/fetch].
19. Beaches downcurrent from groins will usually [gain/lose] sand.
20. [Barrier/Tombolo] islands consist of low ridges of sand that parallel coastlines.
21. A highly irregular coastline is indicative of a(n) [emergent/submergent] coast.
22. Narrowing of beaches results in the cliffs behind them being [more/less] vulnerable to erosion.

23. [Reach/Fetch/Period] is the distance that wind has blown over the water.
24. The bending of waves is called [oscillation/refraction].
25. Upwelling is caused by [gravity/tides/winds].
26. As a part of a beach, a [beach face/berm] is a relatively flat platform often composed of sand that is adjacent to a coastal dune or cliff and marked by a change in slope at the seaward edge.

Written questions

1. Describe coastal upwelling and the effect it has on fish populations.

2. Briefly describe thermohaline circulation.

3. What is the cause of ocean tides?

4. Describe the motion of water particles in deep-water waves and waves approaching a shallow shore.

For other interesting and pertinent information, be sure to visit the *Earth Science* companion website at

http://www.prenhall.com/tarbuck

THE ATMOSPHERE:

Composition, Structure, and Temperature

16

Earth's atmosphere is unique. No other planet in our solar system has an atmosphere with the exact mixture of gases or the heat and moisture conditions needed to sustain life as we know it. The gases that constitute Earth's atmosphere and the controls that act upon them to initiate changes in our weather are vital to our existence. Chapter sixteen examines this "ocean" of air in which we all must live. How does weather differ from climate? What is the extent and structure of the atmosphere? What causes the seasons? How is it heated? What factors control temperature variations over the globe? These are a few of the questions investigated as you begin this third unit of the textbook and continue your exploration of Earth's dynamic systems.

Learning Objectives

After reading, studying, and discussing this chapter, you should be able to:

Explain the difference between weather and climate.

Weather is the state of the atmosphere at a particular place for a short period of time. *Climate*, on the other hand, is a generalization of the weather conditions of a place over a long period of time.

List the most important elements of weather and climate.

The most important *elements*, those quantities or properties that are measured regularly, of weather and climate are (1) air *temperature*, (2) *humidity*, (3) type and amount of *cloudiness*, (4) type and amount of *precipitation*, (5) air *pressure*, and (6) the speed and direction of the *wind*.

List the major and variable components of air.

If water vapor, dust, ozone, and other variable components of the atmosphere were removed, clean, dry air would be composed almost entirely of *nitrogen* (N), about 78 percent of the atmosphere by volume, and *oxygen* (O_2), about 21 percent. *Carbon dioxide* (CO_2), although present only in minute amounts (0.036 percent), is important because it has the ability to absorb heat radiated by Earth and thus helps keep the atmosphere warm. Among the variable components of air, *water vapor* is very important because it is the source of all clouds and precipitation, and, like carbon dioxide, it is also a heat absorber.

Explain the importance of ozone in the atmosphere.

Ozone (O_3), the triatomic form of oxygen, is concentrated in the 10- to 50-kilometer altitude range of the atmosphere; it is important to life because of its ability to absorb potentially harmful ultraviolet radiation from the Sun.

Describe the extent and structure of the atmosphere.

Because the atmosphere gradually thins with increasing altitude, it has no sharp upper boundary but simply blends into outer space. Based on temperature, the atmosphere is divided vertically into four layers. The *troposphere* is the lowermost layer. In the troposphere, temperature usually decreases with increasing altitude. This *environmental lapse rate* is variable but averages about 6.5°C per kilometer

(3.5°F per 1000 feet). Essentially all important weather phenomena occur in the troposphere. Above the troposphere is the *stratosphere*, which exhibits warming because of absorption of ultraviolet radiation by ozone. In the *mesosphere* temperatures again decrease. Upward from the mesosphere is the *thermosphere*, a layer with only a minute fraction of the atmosphere's mass and no well-defined upper limit.

List the principal motions of Earth.

The two principal motions of Earth are (1) *rotation*, the spinning of Earth about its axis, which produces the daily cycle of daylight and darkness, and (2) *revolution*, the movement of Earth in its orbit around the Sun.

Explain the causes of the seasons.

Several factors act together to cause the seasons. Earth's axis is inclined $23\frac{1}{2}°$ from the perpendicular to the plane of its orbit around the Sun *and remains pointed in the same direction* (toward the North Star) as Earth journeys around the Sun. As a consequence, Earth's orientation to the Sun continually changes. The yearly fluctuations in the angle of the Sun and length of daylight brought about by Earth's changing orientation to the Sun cause seasons.

List and describe the three mechanisms of heat transfer.

The three mechanisms of heat transfer are (1) *conduction*, the transfer of heat through matter by molecular activity; (2) *convection*, the transfer of heat by mass movement within a substance; and (3) *radiation*, the transfer of heat by electromagnetic waves.

Describe electromagnetic radiation.

Electromagnetic radiation is energy emitted in the form of rays, or waves, called electromagnetic waves. All radiation is capable of transmitting energy through the vacuum of space. One of the most important differences between electromagnetic waves is their *wavelengths*, which range from very long *radio waves* to very short *gamma rays*. *Visible light* is the only portion of the electromagnetic spectrum we can see. Some of the basic laws that govern radiation as it heats the atmosphere are (1) all objects emit radiant energy; (2) hotter objects radiate more total energy than do colder objects; (3) the hotter the radiating body, the shorter the wavelengths of maximum radiation; and (4) objects that are good absorbers of radiation are good emitters as well.

Describe how the atmosphere is heated.

The general drop in temperature with increasing altitude in the troposphere supports the fact that *the atmosphere is heated from the ground up*. The solar energy, primarily in the form of the shorter wavelengths, that penetrates through the atmosphere is ultimately absorbed at Earth's surface. Earth emits the absorbed radiation in the form of long-wave radiation. The atmospheric absorption of this long-wave *terrestrial radiation*, primarily by water vapor and carbon dioxide, is responsible for heating the atmosphere.

List the factors that cause temperature to vary from place to place.

The factors that cause temperature to vary from place to place, also called the *controls of temperature*, are (1) differences in the *receipt of solar radiation*—the greatest single cause; (2) the unequal *heating and cooling of land and water*, in which land heats more rapidly and to higher temperatures than water, and cools more rapidly and to lower temperatures than water; (3) *altitude*; (4) *geographic position*; (5) *cloud cover and albedo*; and (6) *ocean currents*.

Describe the general distribution of global surface temperatures.

Temperature distribution is shown on a map by using *isotherms*, which are lines that connect equal temperatures. Differences between January and July temperatures around the world can be explained in terms of the basic controls of temperature.

Key Terms

aerosols	greenhouse effect	spring equinox
air	heat	stratosphere
albedo	inclination of the axis	summer solstice
autumnal equinox	infrared	temperature
circle of illumination	isotherm	thermosphere
climate	mesosphere	Tropic of Cancer
conduction	ozone	Tropic of Capricorn
convection	radiation	troposphere
diffused light	reflection	ultraviolet
electromagnetic radiation	revolution	visible light
element (of weather and climate)	rotation	weather
environmental lapse rate	scattering	winter solstice

Vocabulary Review

Choosing from the list of key terms, furnish the most appropriate response for the following statements.

1. The bottom layer of the atmosphere, where temperature decreases with an increase in altitude, is the __________.
2. __________ is a word used to denote the state of the atmosphere at a particular place for a short period of time.
3. __________ is the spinning of Earth about its axis.
4. The outermost layer of the atmosphere is the __________.
5. __________ might be best described as an aggregate or composite of weather.
6. The total kinetic energy of all the atoms that make up a substance is called a(n) __________.
7. The __________ is located at $23\frac{1}{2}$ degrees north latitude.
8. A quantity or property of weather and climate that is measured regularly is termed a(n) __________.
9. __________ is the transfer of heat by the movement of a mass or substance from one place to another, usually vertically.
10. The fact that Earth's axis is not perpendicular to the plane of its orbit is referred to as the __________.
11. In the Northern Hemisphere, the time when the vertical rays of the Sun strike the equator moving southward is known as the __________.
12. The temperature decrease with increasing altitude in the troposphere is called the __________.
13. The __________ is located at $23\frac{1}{2}$ degrees south latitude.
14. The great circle that separates daylight from darkness is the __________.
15. In the atmospheric layer known as the __________, temperatures decrease with height until approximately 80 kilometers above the surface.
16. The transfer of energy through space by electromagnetic waves is known as __________.
17. The movement of Earth in its orbit around the Sun is referred to as __________.
18. __________ is the transfer of heat through matter by molecular activity.
19. The atmospheric layer immediately above the tropopause is the __________.
20. Although we cannot see it, we detect __________ radiation as heat.

21. In the Southern Hemisphere, the time when the vertical rays of the Sun are striking the Tropic of Cancer is known as the __________.
22. Tiny solid and liquid particles in the atmosphere are called __________.
23. The percentage of the total radiation that is reflected by a surface is called its __________.
24. The __________ refers to the mechanism responsible for heating the atmosphere.
25. In the Northern Hemisphere, the time when the vertical rays of the Sun are striking the Tropic of Cancer is known as the __________.
26. __________ is the process whereby light bounces back from an object at the same angle at which it encounters a surface.
27. An __________ is a line that connects points on a map that have the same temperature.

Comprehensive Review

1. Distinguish between the terms *weather* and *climate*.

2. Using Figure 16.1, select the letter that indicates each of the following parts of Earth's atmosphere.

   a) Mesosphere: _____

   b) Tropopause: _____

   c) Stratosphere: _____

   d) Troposphere: _____

   e) Stratopause: _____

   f) Thermosphere: _____

3. List and describe the two principal motions of Earth.

   1)

   2)

4. List the six most important elements of weather and climate.

   1) 4)

   2) 5)

   3) 6)

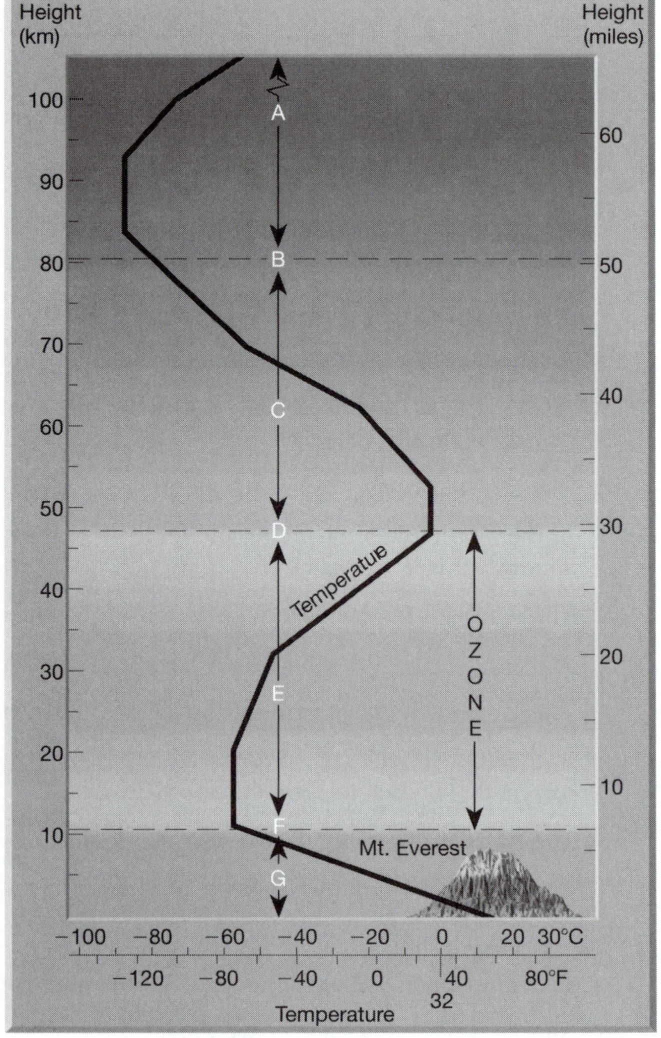

Figure 16.1

5. List two of the meteorological roles of atmospheric aerosols.

   1)

   2)

6. What are the two ways that the seasonal variation in the altitude of the Sun affects the amount of solar energy received at Earth's surface?

   1)

   2)

7. Using Figure 16.2, list the Northern Hemisphere season, date, and latitude of the vertical Sun for each of the four lettered positions.

   a) Position A:

   b) Position B:

   c) Position C:

   d) Position D:

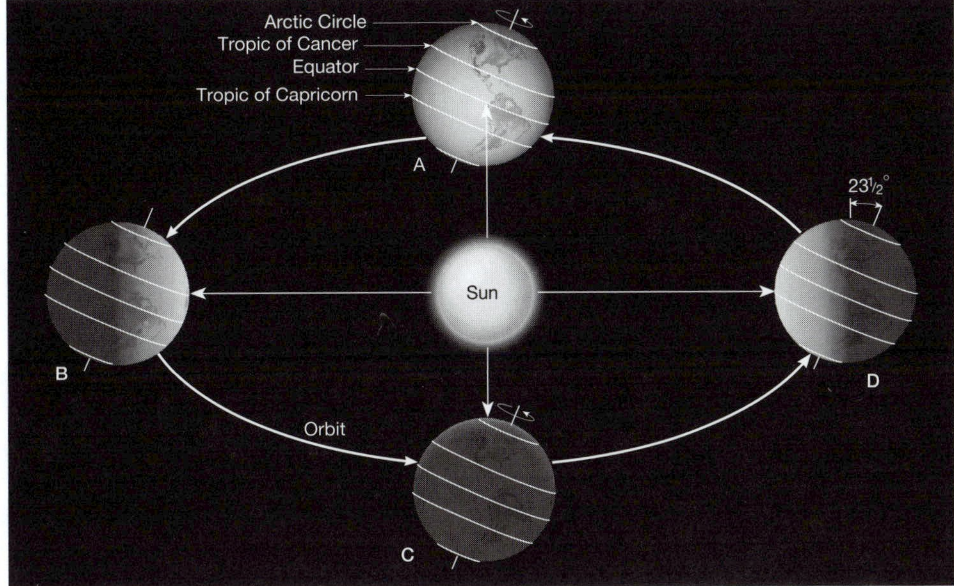

Figure 16.2

8. What are the two most abundant gases in clean, dry air? What are their percentages?

9. How does ozone differ from oxygen? What function does atmospheric ozone serve, especially for those of us on Earth?

10. In an average situation, approximately what percentage of incoming solar radiation is

    a) absorbed by the atmosphere and clouds: _____ percent

    b) absorbed by Earth's surface: _____ percent

    c) reflected by clouds: _____ percent

11. Beginning with the fact that more incoming solar radiation is absorbed by Earth's surface than by the atmosphere, describe the mechanism responsible for atmospheric heating.

12. Which two atmospheric gases are the principal absorbers of heat energy?

13. List and briefly describe the three mechanisms of heat transfer.

    1)

    2)

    3)

14. Briefly explain how each of the following temperatures are determined.

    a) Daily mean temperature:

    b) Annual mean temperature:

    c) Annual temperature range:

15. What is the effect of each of the following on temperature?

    a) Differential heating of land and water:

    b) Altitude:

    c) Geographic position:

    d) Cloud cover:

16. Why are the months of January and July often selected for the analysis of global temperature patterns?

Practice Test

Multiple choice. Choose the best answer for the following multiple-choice questions.

1. Which one of the following is NOT generally considered a major element of weather and climate?
   a) ocean currents
   b) humidity
   c) air temperature
   d) air pressure
   e) wind speed

2. A form of oxygen that combines three oxygen atoms into each molecule is called __________.
   a) trioxygen
   b) oxyurid
   c) oxulene
   d) ozone
   e) azide

3. Oxygen and ozone are efficient absorbers of incoming __________ radiation.
   a) ultraviolet
   b) infrared
   c) radio
   d) visible
   e) gamma

4. The generalization of atmospheric conditions over a long period of time is referred to as __________.
   a) meteorology
   b) climate
   c) global synthesis
   d) weather
   e) element analysis

5. On June 21 the vertical rays of the Sun strike a line of latitude known as the __________.
   a) equator
   b) Arctic circle
   c) Tropic of Capricorn
   d) Tropic of Cancer
   e) Antarctic circle

6. The red and orange colors of sunset and sunrise are the result of __________ in the atmosphere.
   a) moisture
   b) aerosols
   c) insects
   d) pollen
   e) argon

7. The bottom layer of the atmosphere in which we live is called the __________.
   a) mesosphere
   b) thermosphere
   c) troposphere
   d) hemisphere
   e) stratosphere

8. Which one of the following forms of radiation has the longest wavelength?
   a) radio waves
   b) ultraviolet
   c) blue light
   d) yellow light
   e) infrared

9. Ozone in the atmosphere is concentrated in a layer called the __________.
   a) troposphere
   b) stratosphere
   c) mesosphere
   d) thermosphere
   e) hemisphere

10. Which one of the following is NOT a mechanism of heat transfer?
    a) conduction
    b) condensation
    c) radiation
    d) convection

11. The temperature decrease with altitude in the troposphere is called the __________ lapse rate.
    a) environmental
    b) altitude
    c) ascending
    d) atmospheric
    e) incremental

12. On December 21 the vertical rays of the Sun strike a line of latitude known as the __________.
    a) equator
    b) Arctic circle
    c) Tropic of Capricorn
    d) Tropic of Cancer
    e) Antarctic circle

13. Which one of the following gases has the ability to absorb heat energy radiated by Earth?
    a) oxygen
    b) helium
    c) ozone
    d) nitrogen
    e) carbon dioxide

14. The chemicals contributing to the depletion of atmospheric ozone are referred to as __________.
    a) CCCs
    b) CFCs
    c) FCDs
    d) PCBs
    e) CFDs

15. Approximately what percentage of the solar energy that strikes the top of the atmosphere reaches Earth's surface?
    a) 30 percent
    b) 40 percent
    c) 50 percent
    d) 60 percent
    e) 70 percent

16. Which one of the following is NOT an important control of air temperature?
    a) heating of land and water
    b) altitude
    c) wind
    d) geographic position
    e) ocean currents

17. The two principal atmospheric absorbers of terrestrial radiation are carbon dioxide and __________.
    a) oxygen
    b) water vapor
    c) nitrogen
    d) ozone
    e) helium

18. When water changes from one state to another, __________ heat is stored or released.
    a) latent
    b) vapor
    c) fusion
    d) reduced
    e) elemental

19. Which one of the following yearly fluctuations is NOT responsible for the seasons?
    a) length of daylight
    b) Sun angle
    c) Earth-Sun distance

20. The value of the normal lapse rate is __________ C degrees per kilometer.
    a) 3.5
    b) 4.5
    c) 5.5
    d) 6.5
    e) 7.5

21. Ninety percent of the atmosphere lies below __________ miles.
    a) 5
    b) 10
    c) 15
    d) 20
    e) 25

22. In the Southern Hemisphere the greatest number of daylight hours occurs on __________.
    a) June 21
    b) December 21
    c) March 21
    d) April 21
    e) September 21

23. On the equinoxes the vertical rays of the Sun strike a line of latitude known as the __________.
    a) equator
    b) Arctic circle
    c) Tropic of Capricorn
    d) Tropic of Cancer
    e) Antarctic circle

24. The amount of energy received at Earth's surface is controlled by the __________ of the Sun.
    a) temperature
    b) altitude
    c) distance
    d) color
    e) rotation

25. Temperature distribution is shown on a map using lines called __________ that connect places of equal temperature.
    a) isotemps
    b) equagrads
    c) isobars
    d) isotherms

26. Which form of radiation do we detect as heat?
    a) radio waves
    b) ultraviolet
    c) red light
    d) yellow light
    e) infrared

27. Combustion of fossil fuels adds vast quantities of the gas __________ to the atmosphere.
    a) oxygen
    b) water vapor
    c) ozone
    d) nitrogen
    e) carbon dioxide

28. The annual temperature range will be greatest for a place located __________.
    a) at the pole
    b) on the equator
    c) in the interior of a continent
    d) at high altitude
    e) near the coast

29. Which one of the following surfaces has the highest albedo?
    a) concrete
    b) snow
    c) soil
    d) water
    e) forest

30. The most abundant gas in clean, dry air is __________.
    a) oxygen
    b) argon
    c) nitrogen
    d) carbon dioxide
    e) ozone

True/false. For the following true/false questions, if a statement is not completely true, mark it false. For each false statement, change the **italicized** *word to correct the statement.*

1. _____ In the Northern Hemisphere, December 21 is referred to as the *winter* solstice.
2. _____ *Climate* is a word used to denote the state of the atmosphere at a particular place over a short period of time.
3. _____ The *higher* the solar angle or altitude of the Sun, the more spread out and less intense is the solar radiation that reaches Earth's surface.
4. _____ On the *equinoxes*, all places on Earth receive 12 hours of daylight.
5. _____ Atmospheric ozone is an absorber of potentially harmful *ultraviolet* radiation from the Sun.
6. _____ The transfer of heat through matter by molecular activity is called *radiation*.
7. _____ An international agreement known as the *Montreal Protocol* represents a positive international response for eliminating ozone-depleting gases from general use.
8. _____ The most important difference among electromagnetic waves is their *wavelength*.
9. _____ If Earth's axis were not *inclined*, we would have no seasons.
10. _____ The atmosphere is divided into four layers on the basis of *composition*.
11. _____ The *higher* the solar angle or altitude of the Sun, the less distance solar radiation must penetrate through the atmosphere to Earth's surface.
12. _____ Essentially all important weather occurs in the *stratosphere*.
13. _____ The line separating the dark half of Earth from the lighted half is called the circle of *daylight*.
14. _____ In the Southern Hemisphere, June 21 is referred to as the *summer* solstice.
15. _____ There is a *sharp* boundary between the atmosphere and outer space.
16. _____ The hotter the radiating body, the *shorter* the wavelength of maximum radiation.
17. _____ Earth's two principal motions are rotation and *revolution*.

18. _____ In the United States, between 1980 and 2006 emissions of the five primary pollutants *increased* significantly.
19. _____ Heating an object causes its atoms to move *slowly.*
20. _____ The percentage of the total radiation that is reflected by an object is called its *albedo.*
21. _____ The principal source for atmospheric heat is *terrestrial* radiation.
22. _____ Land heats *more* rapidly than water.
23. _____ Compared to the Northern Hemisphere, the Southern Hemisphere has a *greater* annual temperature variation.
24. _____ One category of pollutants, the *primary* pollutants, are emitted directly from identifiable sources.
25. _____ The *visible* wavelengths of radiation from the Sun reach Earth's surface.

Word choice. Complete each of the following statements by selecting the most appropriate response.

1. The composition of air [is constant/varies].
2. On June 21 the vertical rays of the Sun strike the Tropic of [Cancer/Capricorn].
3. Temperature [increases/decreases] with an increase in altitude in the troposphere.
4. The changes in distance between Earth and the Sun throughout the year have [major/insignificant] influence on weather and climate.
5. Hotter objects radiate [more/less] total energy per unit area than colder objects.
6. The average value of the environmental lapse rate is [3.5/5.2]°F per 1000 feet.
7. Temperature variations are [greater/smaller] over large bodies of water than they are over land.
8. Carbon dioxide in the air has the ability to absorb [water/heat energy].
9. The seasonal north-south migration of isotherms is [greater/less] over the continents than over the oceans.
10. Air pressure [increases/decreases] as altitude increases.
11. Earth's axis is inclined [53.2/23.5] degrees from the perpendicular.
12. When water vapor changes from one state of matter to another it absorbs or releases [conduction/latent] heat.
13. Ozone in the atmosphere absorbs the potentially harmful [ultraviolet/infrared] radiation from the Sun.
14. On the date of an equinox, the vertical rays of the Sun strike the [equator/tropics].
15. The generalization of the weather conditions of a place over a long period of time is called [weather/climate].
16. All other factors being the same, the greatest daily temperature range would occur in a(n) [arid/humid] region.
17. The atmosphere is divided into four principal layers on the basis of [composition/temperature].
18. The greatest annual extremes of temperature occur over [land/water].
19. The upper boundary of the atmosphere [is/is not] sharply defined.
20. The hotter the object, the [longer/shorter] the wavelength of maximum radiation.
21. Annual temperature ranges are [greatest/least] near the equator.
22. The only mechanism of heat transfer that can transmit heat through the relative emptiness of space is [conduction/radiation].

23. Increased cloud cover will [increase/decrease] the rate of nighttime cooling.
24. Most of the radiation that heats the atmosphere directly is emitted by [the Sun/Earth].
25. The transfer of heat by mass movement within a substance is called [conduction/convection].
26. The specific heat for water is [greater/less] than the specific heat for land.

Written questions

1. What is the relation between CFCs and the ozone problem?

2. What causes the amount of solar energy reaching places on Earth's surface to vary with the seasons?

3. Briefly describe how Earth's atmosphere is heated.

For other interesting and pertinent
information, be sure to visit
the *Earth Science* companion website at

http://www.prenhall.com/tarbuck

17

Moisture, Clouds, and Precipitation

Water vapor is an odorless, colorless gas that mixes freely with the other gases of the atmosphere. In addition to being an important atmospheric heat absorber, it is also the source of clouds and all precipitation. Unlike the atmospheric gases oxygen and nitrogen, water can change from one state of matter (solid, liquid, or gas) to another at the temperatures and pressures experienced on Earth. Because of this unique property, water freely leaves the oceans as a gas and returns again as a liquid.

As you observe the day-to-day changes in weather, you might ask: How do we measure humidity? Why is it more humid in the summer than in winter? Why do clouds form on some days but not on others? What causes rain? Answers to these and other questions involving the role of water vapor in the atmosphere form the central theme of this chapter.

Learning Objectives

After reading, studying, and discussing this chapter, you should be able to:

List the processes that cause water to change from one state of matter to another.

Water vapor, an odorless, colorless gas, changes from one state of matter (solid, liquid, or gas) to another at the temperatures and pressures experienced near Earth's surface. The processes involved in *changing the state of matter* of water are *evaporation, condensation, melting, freezing, sublimation*, and *deposition*. During each change, *latent* (hidden) *heat* is either absorbed or released.

Explain mixing ratio, vapor pressure, relative humidity, and dew point.

Humidity is the general term to describe the amount of water vapor in the air. The methods used to express humidity quantitatively include (1) *mixing ratio*, the mass of water vapor in a unit of air compared to the remaining mass of dry air; (2) *vapor pressure*, that part of the total atmospheric pressure attributable to its water-vapor content; (3) *relative humidity*, the ratio of the air's actual water-vapor content compared to the amount of water vapor required for saturation at that temperature; and (4) *dew point*, that temperature to which a parcel of air would need to be cooled to reach saturation. When air is saturated, the pressure exerted by the water vapor, called the *saturation vapor pressure*, produces a balance between the number of water molecules leaving the surface of the water and the number returning. Because the saturation vapor pressure is temperature-dependent, at higher temperatures more water vapor is required for saturation to occur.

Describe how relative humidity can be changed.

Relative humidity can be changed in two ways. The first is by *adding or subtracting water vapor*. The second is by *changing the air's temperature*. When air is cooled, its relative humidity increases.

Explain the basic cloud-forming process.

The cooling of air as it rises and expands owing to successively lower pressure is the basic cloudforming process. Temperature changes in air brought about by expanding or compressing the air are called *adiabatic temperature changes*. Unsaturated air warms by compression and cools by expansion at the rather constant rate of 10°C per 1000 meters of altitude change, a figure called the *dry adiabatic rate*. If air rises high enough, it will cool sufficiently to cause condensation and form a cloud. From this point on, air that continues to rise will cool at the *wet adiabatic rate*, which varies from 5°C to 9°C per 1000 meters of ascent. The difference in the wet and dry adiabatic rates is caused by the condensing water vapor releasing *latent heat*, thereby reducing the rate at which the air cools.

List the processes that initiate the vertical movement of air.

Four mechanisms that can initiate the vertical movement of air are (1) *orographic lifting*, which occurs when elevated terrains, such as mountains, act as barriers to the flow of air; (2) *frontal wedging*, when cool air acts as a barrier over which warmer, less dense air rises; (3) *convergence*, which happens when air flows together and a general upward movement of air occurs; and (4) *localized convective lifting*, which occurs when unequal surface heating causes pockets of air to rise because of their buoyancy.

Describe stable and unstable air.

The *stability of air* is determined by examining the temperature of the atmosphere at various altitudes. Air is said to be *unstable* when the *environmental lapse rate* (the rate of temperature decrease with increasing altitude in the troposphere) is greater than the *dry adiabatic rate*. Stated differently, a column of air is unstable when the air near the bottom is significantly warmer (less dense) than the air aloft.

Discuss the conditions necessary for condensation.

For condensation to occur, air must be saturated. Saturation takes place either when air is cooled to its dew point, which most commonly happens, or when water vapor is added to the air. There must also be a surface on which the water vapor can condense. In cloud and fog formation, tiny particles called *condensation nuclei* serve this purpose.

List the criteria used to classify clouds.

Clouds are classified on the basis of their *appearance* and *height*. The three basic forms are *cirrus* (high, white, thin, wispy fibers), *cumulus* (globular, individual cloud masses), and *stratus* (sheets or layers that cover much or all of the sky). The four categories based on height are *high clouds* (bases normally above 6000 meters), *middle clouds* (from 2000 to 6000 meters), *low clouds* (below 2000 meters), and *clouds of vertical development*.

Describe the formation of fog.

Fog is defined as a cloud with its base at or very near the ground. Fogs form when air is cooled below its dew point or when enough water vapor is added to the air to bring about saturation. Various types of fog include *advection fog, radiation fog, upslope fog, steam fog*, and *frontal* (or *precipitation*) *fog*.

Discuss the formation of precipitation.

For *precipitation* to form, millions of cloud droplets must somehow join together into large drops. Two mechanisms for the formation of precipitation have been proposed. First, in clouds where the temperatures are below freezing, ice crystals form and fall as snowflakes. At lower altitudes the snowflakes melt and become raindrops before they reach the ground. In the second mechanism, large droplets form in warm clouds that contain large *hygroscopic* ("water seeking") *nuclei*, such as salt particles. As these big droplets descend, they collide and join with smaller water droplets. After many collisions the droplets are large enough to fall to the ground as rain.

List the forms of precipitation.

The forms of precipitation include *rain, snow, sleet, freezing rain (glaze), hail*, and *rime*.

Key Terms

absolute instability
absolute stability
adiabatic temperature change
advection fog
Bergeron process
calorie
cirrus
cloud
cloud of vertical development
collision-coalescence process
condensation
condensation nuclei
conditional instability
convergence
cumulus
deposition
dew point temperature
dry adiabatic rate
evaporation
fog
freezing
freezing nuclei
frontal fog
front
frontal wedging
glaze
hail
high cloud
humidity
hygrometer
hygroscopic nuclei
latent heat
localized convective lifting
low cloud
melting
middle cloud
mixing ratio
orographic lifting
parcel
precipitation fog
psychrometer
radiation fog
rain
rainshadow desert
relative humidity
rime
saturation
sleet
snow
stable air
steam fog
stratus
sublimation
supercooled
supersaturated
unstable air
upslope fog
vapor pressure
wet adiabatic rate

Vocabulary Review

Choosing from the list of key terms, furnish the most appropriate response for the following statements.

1. The general term for the amount of water vapor in air is __________.
2. The process whereby water vapor changes to the liquid state is called __________.
3. __________ is the ratio of the air's actual water-vapor content to its potential water-vapor capacity at a given temperature.
4. __________ results when elevated terrains, such as mountains, act as barriers to flowing air.
5. The part of total atmospheric pressure attributable to water-vapor content is called the __________.
6. A(n) __________ results when air is compressed or allowed to expand.
7. __________ refers to the energy stored or released during a change of state.
8. A(n) __________ is best described as visible aggregates of minute water droplets or tiny crystals of ice suspended in the air.
9. Frozen or semifrozen rain formed when raindrops freeze as they pass through a layer of cold air is called __________.
10. The term __________ refers to the maximum possible quantity of water vapor that the air can hold at any given temperature and pressure.
11. The rate of adiabatic temperature change in saturated air is the __________.
12. __________ is the conversion of a solid directly to a gas, without passing through the liquid state.
13. Sheets or layers of clouds that cover much or all of the sky are called __________ clouds.
14. The mass of water vapor in a unit of air compared to the remaining mass of dry air (e.g., grams per kilogram), is called __________.
15. __________ occurs when cool air acts as a barrier over which warmer, less dense air rises.
16. In the atmosphere, tiny bits of particulate matter, known as __________, serve as surfaces for water-vapor condensation.

17. The process by which a solid is changed to a liquid is referred to as __________.
18. In meteorology the term __________ is restricted to drops of water that fall from a cloud and have a diameter of at least 0.5 millimeter.
19. The temperature to which air would have to be cooled to reach saturation is the __________.
20. __________ is a deposit of ice crystals formed by the freezing of supercooled fog or cloud droplets on objects whose surface temperature is below freezing.
21. The conversion of a vapor directly to a solid is called __________.
22. Nearly spherical ice pellets that have concentric layers and are formed by the successive freezing of layers of water are referred to as __________.
23. The adiabatic rate of cooling or heating that applies only to unsaturated air is the __________.
24. Water-absorbing particles, such as salt, are termed __________.
25. The process of converting a liquid to a gas is called __________.
26. __________ is a type of fog that forms on cool, clear, calm nights, when Earth's surface cools rapidly by radiation.
27. A cloud with its base at or very near Earth's surface is referred to as __________.
28. A dry area on the lee side of a mountain that forms as air descends and is warmed by compression is called a(n) __________.
29. High, thin, delicate clouds are called __________ clouds.
30. Whenever air masses flow together, __________ is said to occur.
31. A(n) __________, the common unit of heat energy, is the amount of heat required to raise the temperature of 1 gram of water 1°C.
32. Water in the liquid state below 0°C is referred to as __________ water.
33. The __________ is an instrument used to measure relative humidity.

Comprehensive Review

1. Using Figure 17.1, select the letter that indicates each of the following processes.

   a) Freezing: _____ c) Deposition: _____ e) Sublimation: _____

   b) Evaporation: _____ d) Melting: _____ f) Condensation: _____

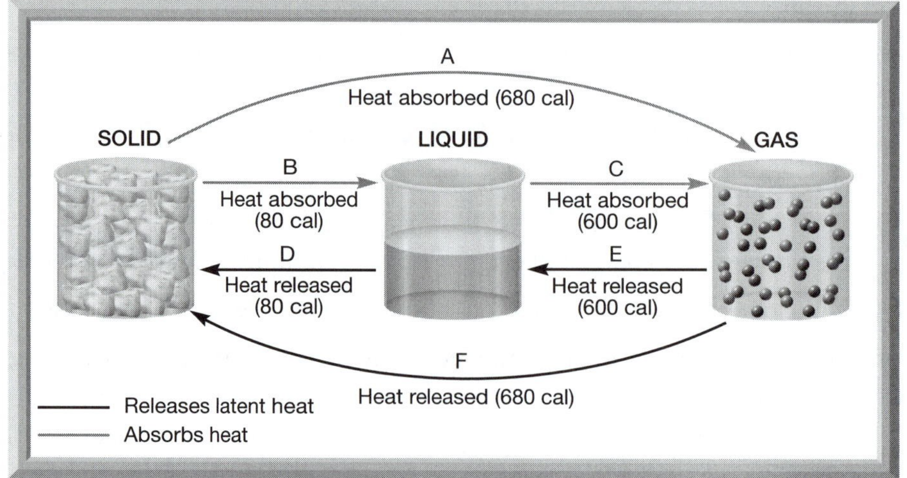

Figure 17.1

2. List two different ways that the relative humidity of a parcel of air can be increased.

   1)

   2)

3. Using Table 17.1 in the textbook, determine the relative humidity of a kilogram of air that has the following temperature and water-vapor content.

   a) Temperature: 25°C; Water-vapor content: 5 grams/kilogram

   b) Temperature: 41°F; Water-vapor content: 3 grams/kilogram

4. Using Table 17.1 in the textbook, determine the dew point of a kilogram of air that has the following temperature and relative humidity.

   a) Temperature: 25°C; Relative humidity: 50 percent

   b) Temperature: 59°F; Relative humidity: 20 percent

5. Explain the principle of the psychrometer for measuring relative humidity.

6. Using Table C.1 in the textbook's Appendix, determine the relative humidity of a parcel of air that yielded the following psychrometer readings:

   a) Dry-bulb temperature: 24°C; Wet-bulb temperature: 16°C

7. Why is the wet adiabatic rate of cooling of air less than the dry adiabatic rate?

8. What is the difference between stable and unstable air?

9. Briefly describe each of the following processes that lift air.

   a) Orographic lifting:

   b) Frontal wedging:

   c) Convergence:

   d) Localized convective lifting:

10. What two criteria are used as the basis for cloud classification?

    1)

    2)

11. Write the name of the cloud described by each of the following statements.

    a) Low, layered clouds with a uniform foglike appearance that frequently cover much of the sky:

    b) High, thin, delicate clouds that sometimes appear as hooked filaments called "mares' tails":

    c) Dense, fluffy, globular masses found between 2000 and 6000 meters above the surface:

12. Using Figure 17.2, select the letter of the photograph that illustrates the following cloud types.

    a) Cumulus: _____

    b) Cirrus: _____

A.

B.

Figure 17.2

13. Briefly describe the circumstances responsible for the formation of each of the following types of fog.

    a) Steam fog:

    b) Frontal fog:

14. List and briefly describe the two mechanisms that have been proposed to explain how precipitation forms.

    1)

    2)

15. Briefly describe the circumstances that result in the formation of each of the following forms of precipitation.

    a) Sleet:

    b) Hail:

Practice Test

Multiple choice. Choose the best answer for the following multiple-choice questions.

1. The ratio of the air's water-vapor content to its capacity at that same temperature is the _____.
   a) vapor pressure
   b) specific humidity
   c) wet adiabatic rate
   d) relative humidity
   e) dry adiabatic rate

2. The term _____ means "rainy cloud."
   a) ferrous
   b) nimbus
   c) stratus
   d) cirrus
   e) cumulus

3. Which one of the following changes from one state of matter to another at the temperatures and pressures experienced at Earth's surface?
   a) oxygen
   b) water
   c) carbon dioxide
   d) nitrogen
   e) methane

4. Which of the following is NOT a necessary condition for condensation?
   a) high altitude
   b) dew-point temperature reached
   c) surfaces
   d) saturation
   e) water vapor

5. Which one of the following processes involves the greatest quantity of heat energy?
   a) melting
   b) sublimation
   c) freezing
   d) evaporation
   e) condensation

6. Which one of the following is NOT a process that lifts air?
   a) convergence
   b) orographic lifting
   c) divergence
   d) frontal wedging
   e) localized convective lifting

7. Which form of precipitation is likely to occur when a layer of warm air with temperatures above freezing overlies a subfreezing layer near the ground?
   a) rain
   b) sleet
   c) snow
   d) hail
   e) drizzle

8. Air that has reached its water-vapor capacity is said to be __________.
   a) dry
   b) unstable
   c) soaked
   d) stable
   e) saturated

9. Which clouds are best described as sheets or layers that cover much or all of the sky?
   a) cirrus
   b) cumulus
   c) stratus
   d) altocumulus
   e) cumulonimbus

10. The temperature to which air would have to be cooled to reach saturation is called the __________ point.
    a) vapor
    b) dew
    c) adiabatic
    d) sublimation
    e) critical

11. When the environmental lapse rate is less than the dry adiabatic rate, a parcel of air will be __________.
    a) stable
    b) unstable

12. Which one of the following refers to the energy that is stored or released during a change of state of water?

    a) caloric heat
    b) ultraviolet heat
    c) latent heat
    d) geothermal heat
    e) evaporation heat

13. The process where cool air acts as a barrier over which warmer, less dense air rises is called __________.

    a) divergence
    b) frontal wedging
    c) orographic lifting
    d) collision
    e) subduction

14. Relative humidity is typically highest at __________.

    a) sunrise
    b) noon
    c) late afternoon
    d) early evening
    e) midnight

15. With which cloud type is hail most associated?

    a) cirrus
    b) stratus
    c) cumulonimbus
    d) nimbostratus
    e) cumulus

16. The wet adiabatic rate of cooling is less than the dry rate because __________.

    a) wet air is unsaturated
    b) of the release of latent heat
    c) dry air is less dense
    d) wet air is more common
    e) of the dew point

17. The conditions that favor the formation of radiation fog are __________.

    a) warm, cloudy, calm nights
    b) cool, cloudy, windy nights
    c) warm, clear, windy nights
    d) cool, clear, calm nights
    e) warm, cloudy, windy nights

18. The process responsible for deposits of white frost or hoar frost is __________.

    a) sublimation
    b) melting
    c) freezing
    d) evaporation
    e) deposition

19. When the environmental lapse rate is greater than the dry adiabatic rate, a parcel of air will be __________.

    a) stable
    b) unstable

20. Compared to clouds, fogs are __________.

    a) colder
    b) thicker
    c) of a different composition
    d) at lower altitudes
    e) drier

21. Which clouds are high, white, and thin?

    a) cirrus
    b) cumulus
    c) stratus
    d) nimbostratus
    e) cumulonimbus

22. When air expands or contracts, it will experience __________ temperature changes.

    a) volume
    b) radiation
    c) external
    d) adiabatic
    e) unnecessary

23. Which one of the following is NOT a process that has been proposed to explain the formation of precipitation?
    a) atmospheric subduction process
    b) collision-coalescence process
    c) Bergeron process

24. The conversion of a solid directly to a gas without passing through the liquid state is called __________.
    a) sublimation
    b) melting
    c) freezing
    d) evaporation
    e) deposition

25. Weather-producing fronts are parts of the storm systems called __________.
    a) mid-latitude cyclones
    b) hurricanes
    c) tornadoes
    d) tropical storms
    e) typhoons

26. Which one of the following is NOT produced by condensation?
    a) dew
    b) smog
    c) fog
    d) clouds

27. Which one of the following is a deposit of ice crystals formed by the freezing of supercooled fog or cloud droplets on objects whose surface temperature is below freezing?
    a) hail
    b) sleet
    c) snow
    d) rime
    e) glaze

28. The dry adiabatic rate is __________.
    a) 3.0°C/100 meters
    b) 0.5°C/meter
    c) 1°C/100 meters
    d) 1°C/1000 meters
    e) 0.5°C/1000 meters

29. That part of total atmospheric pressure that can be attributed to the water-vapor content is called __________ pressure.
    a) moisture
    b) isostatic
    c) vapor
    d) gas
    e) water

30. Which one of the following is NOT a type of fog?
    a) white fog
    b) radiation fog
    c) upslope fog
    d) steam fog
    e) frontal fog

True/false. For the following true/false questions, if a statement is not completely true, mark it false. For each false statement, change the **italicized** *word to correct the statement.*

1. _____ At the *dew-point* temperature, the air is both saturated and has a 100 percent relative humidity.
2. _____ A *dyne* is the amount of heat required to raise the temperature of 1 gram of water 1°C.
3. _____ Clouds are classified on the basis of their form and *color.*
4. _____ When water vapor condenses, it *releases* heat energy.
5. _____ When air expands, it will *warm.*
6. _____ When air masses flow together, *convergence* is said to occur.
7. _____ Assuming equal volumes, a *greater* quantity of water vapor is required to saturate warm air than cold air.
8. _____ Clouds associated with *unstable* air are towering and usually accompanied by heavy precipitation.

9. _____ The principle of the psychrometer relies on the fact that the amount of cooling of the *dry-bulb* thermometer is directly related to the dryness of the air.
10. _____ Above the ground, tiny bits of particulate matter serve as *condensation* nuclei.
11. _____ The water-vapor capacity of air is *temperature*-dependent.
12. _____ The *dry* adiabatic rate varies from about 5°C/1000 meters in moist air to 9°C/1000 meters in dry air.
13. _____ When the water-vapor content remains constant, a(n) *increase* in temperature results in an increase in relative humidity.
14. _____ A raindrop large enough to fall to the ground contains roughly 1 *million* times more water than a single cloud droplet.
15. _____ The energy absorbed by water molecules during evaporation is referred to as latent heat of *fusion*.
16. _____ When the surface temperature is about 4°C or higher, snowflakes usually melt before they reach the ground and continue their descent as rain.
17. _____ *Sleet* is defined as a cloud with its base at or very near the ground.
18. _____ Although highly variable, a general snow/water ratio of *ten (10)* units of snow to one unit of water is often used when exact information is not available.

Word choice. Complete each of the following statements by selecting the most appropriate response.

1. When water vapor condenses, it [absorbs/releases] heat.
2. The process of raindrop formation that relies on the fact that "giant" cloud droplets fall more rapidly than smaller droplets is referred to as the [collision-coalescence/Bergeron/evaporation] process.
3. The primary difference between a fog and a cloud is in the [droplet size/altitude/dew point].
4. The ratio of water vapor in the air to the air's water-vapor capacity at the same temperature is called the [specific/relative/absolute] humidity.
5. Freezing nuclei are [more/less] common in the atmosphere than condensation nuclei.
6. [Hail/Sleet/Glaze] is liquid raindrops that freeze as they fall through a layer of subfreezing air.
7. [More/Less] water vapor is required to saturate warm air than cold air.
8. Air that has a tendency to rise on its own is referred to as [stable/unstable] air.
9. As the relative humidity increases, the air's dew point [increases/decreases/remains the same].
10. Sinking air [compresses/expands] and therefore [warms/cools] adiabatically.
11. When the environmental lapse rate is greater than the dry adiabatic rate, the air exhibits [absolute instability/conditional instability/absolute stability].
12. When water vapor evaporates, it [absorbs/releases] heat.
13. [Cirrus/Stratus/Cumulus] clouds are best described as sheets or layers that cover much or all of the sky.
14. Cirrus clouds are most often composed of [ice crystals/water droplets].
15. The two ways of increasing relative humidity are by [increasing/decreasing] the temperature of the air and/or by [increasing/decreasing] the amount of water vapor in the air.
16. The wet adiabatic rate is [higher/lower] than the dry adiabatic rate due to the release of [residual/latent/kelvin] heat during [condensation/evaporation].

17. Ice crystals that grow by "absorbing" water vapor that is evaporating from liquid cloud droplets describes the [collision-coalescence/Bergeron/evaporation] process for the formation of precipitation.
18. When using a psychrometer, the greater the difference between the dry- and wet-bulb temperatures, the [higher/lower] the relative humidity.
19. Whenever air masses flow together, [convergence/divergence] occurs, and the result is a general [upward/downward] flow of air.
20. Orographic lifting [increases, decreases] the amount of rainfall on the windward sides of mountains.

Written questions

1. Describe how clouds are classified.

2. Explain why air cools as it rises through the atmosphere.

3. By comparing the environmental lapse rate to the adiabatic rate, describe when air will be stable and when it will be unstable.

4. Of the four mechanisms that lift air, which is most prevalent in your location? Explain how this mechanism is responsible for the precipitation of your area.

For other interesting and pertinent information, be sure to visit the *Earth Science* companion website at

http://www.prenhall.com/tarbuck

18

Air Pressure and Wind

Of all the elements of weather, air pressure is perhaps the least noticeable. When listening to a weather report, most people are interested in temperature, moisture conditions (humidity and precipitation), and perhaps wind. It is a rare person, however, who is concerned about air pressure. Although the hour-to-hour and day-to-day changes in air pressure are unnoticeable to human beings, they are very important to producing changes in our weather and forecasting changing conditions. It is, for example, the variations in air pressure from place to place that generate winds that in turn can bring changes in both temperature and humidity. Therefore, air pressure is one of the basic weather elements and a significant factor in weather forecasting. In this chapter you will not only examine the factors that affect pressure and wind, but also the various types of pressure systems and their associated winds, the global patterns of pressure and wind, and how pressure and wind are measured.

Learning Objectives

After reading, studying, and discussing this chapter, you should be able to:

Describe air pressure, how it is measured, and how it changes with altitude.

Air has weight: At sea level it exerts a pressure of 1 kilogram per square centimeter (14.7 pounds per square inch). *Air pressure* is the force exerted by the weight of air above. With increasing altitude, there is less air above to exert a force, and thus air pressure decreases with altitude—rapidly at first, then much more slowly. The unit used by meteorologists to measure atmospheric pressure is the *millibar*. *Standard sea-level pressure* is expressed as 1013.2 millibars. *Isobars* are lines on a weather map that connect places of equal air pressure.

Describe the instruments used to measure air pressure.

A *mercury barometer* measures air pressure using a column of mercury in a glass tube that is sealed at one end and inverted in a dish of mercury. As air pressure increases, the mercury in the tube rises; conversely, when air pressure decreases, so does the height of the column of mercury. A mercury barometer measures atmospheric pressure in "*inches of mercury*"—the height of the column of mercury in the barometer. Standard atmospheric pressure at sea level equals 29.92 inches of mercury. *Aneroid* ("without liquid") *barometers* consist of partially evacuated metal chambers that compress as air pressure increases and expand as pressure decreases.

Explain how the pressure gradient force, Coriolis effect, and friction influence wind.

Wind is the horizontal flow of air from areas of higher pressure to areas of lower pressure. Winds are controlled by the following combination of forces: (1) the *pressure gradient force* (amount of pressure change over a given distance); (2) *Coriolis effect* (deflective force of Earth's rotation—to the right in the Northern Hemisphere and to the left in the Southern Hemisphere); and (3) *friction* with Earth's surface (slows the movement of air and alters wind direction).

Discuss upper-air winds.

Upper-air winds, called *geostrophic winds*, blow parallel to the isobars and reflect a balance between the pressure gradient force and the Coriolis effect. Upper-air winds are faster than surface winds because friction is greatly reduced aloft. Friction slows surface winds, which in turn reduces the Coriolis effect. The result is air movement at an angle across the isobars toward the area of lower pressure.

Describe the movements of air associated with the two types of pressure centers.

The two types of pressure centers are (1) *cyclones*, or *lows* (centers of low pressure) and (2) *anticyclones*, or *highs* (high-pressure centers). In the Northern Hemisphere, winds around a low (cyclone) are counterclockwise and inward. Around a high (anticyclone), winds are clockwise and outward. In the Southern Hemisphere, the Coriolis effect causes winds to be clockwise around a low and counterclockwise around a high. Because air rises and cools adiabatically in a low-pressure system, cloudy conditions and precipitation are often associated with their passage. In a high-pressure system, descending air is compressed and warmed; therefore, cloud formation and precipitation are unlikely in an anticyclone, and "fair" weather is usually expected.

Describe the idealized global patterns of pressure and wind.

Earth's *global pressure zones* include the *equatorial low, subtropical high, subpolar low*, and *polar high*. The *global surface winds* associated with these pressure zones are the *trade winds, westerlies*, and *polar easterlies*.

Explain how the large seasonal temperature differences over continents disrupt the idealized global patterns of pressure and wind.

Particularly in the Northern Hemisphere, large seasonal temperature differences over continents disrupt the idealized, or zonal, global patterns of pressure and wind. In winter, large, cold landmasses develop a seasonal high-pressure system from which surface air flow is directed off the land. In summer, landmasses are heated and low pressure develops over them, which permits air to flow onto the land. These seasonal changes in wind direction are known as *monsoons*.

Discuss the general atmospheric circulation in the midlatitudes.

In the middle latitudes, between 30 and 60 degrees latitude, the general west-to-east flow of the westerlies is interrupted by migrating cyclones and anticyclones. The paths taken by these pressure systems are closely correlated to upper-level air flow and the polar *jet stream*. The average position of the polar jet stream, and hence the paths of cyclonic systems, migrates equatorward with the approach of winter and poleward as summer nears.

List the names and causes of the major local winds.

Local winds are small-scale winds produced by a locally generated pressure gradient. Local winds include *sea* and *land breezes* (formed along a coast because of daily pressure differences caused by the differential heating of land and water); *valley* and *mountain breezes* (daily wind similar to sea and land breezes except in a mountainous area where the air along slopes heats differently from the air at the same elevation over the valley floor); *chinook* and *Santa Ana winds* (warm, dry winds created when air descends the leeward side of a mountain and warms by compression); and the *country breeze*.

Discuss the two basic wind measurements.

The two basic wind measurements are *direction* and *speed*. Winds are always labeled by the direction *from* which they blow. Wind speed is measured using a *cup anemometer*. The instrument most commonly used to measure wind direction is the *wind vane*. When the wind consistently blows more often from one direction than from any other, it is termed a *prevailing wind*.

Describe El Niño.

El Niño is the name given to the periodic warming of the ocean that occurs in the central and eastern Pacific. It is associated with periods when a weakened pressure gradient causes the trade winds to diminish. A major El Niño event triggers extreme weather in many parts of the world.

List the factors that influence the global distribution of precipitation.

The global distribution of precipitation is strongly influenced by the global pattern of air pressure and wind, latitude, and distribution of land and water.

Key Terms

air pressure	cyclone	low	Santa Ana
aneroid barometer	divergence	mercury barometer	sea breeze
anticyclone	El Niño	monsoon	Southern Oscillation
barograph	equatorial low	mountain breeze	subpolar low
barometric tendency	geostrophic wind	polar easterlies	subtropical high
chinook	high	polar front	trade winds
convergence	isobar	polar high	valley breeze
Coriolis effect	jet stream	pressure gradient	westerlies
country breeze	land breeze	pressure tendency	wind
cup anemometer	La Niña	prevailing wind	wind vane

Vocabulary Review

Choosing from the list of key terms, furnish the most appropriate response for the following statements.

1. The stormy belt separating the westerlies from the polar easterlies is known as the __________.
2. A(n) __________ is a line on a weather map that connects places of equal air pressure.
3. A(n) __________ is a seasonal reversal of wind direction associated with large continents, especially Asia.
4. The deflective force of Earth's rotation on all free-moving objects is called the __________.
5. Air that flows horizontally with respect to Earth's surface is referred to as __________.
6. __________, a useful aid in short-range weather prediction, refers to the nature of the change in atmospheric pressure over the past several hours.
7. The __________ is an instrument used for measuring air pressure that consists of evacuated metal chambers that change shape as pressure changes.
8. When the wind consistently blows more often from one direction than from any other, it is termed a(n) __________.
9. The instrument most commonly used to determine wind direction is the __________.
10. A center of low atmospheric pressure is called a(n) __________.
11. __________ is the condition that exists when the distribution of winds within a given area results in a net horizontal inflow of air into the area.
12. A(n) __________ is a local wind blowing from land toward the water during the night in coastal areas.
13. The __________ is a belt of low pressure lying near the equator and between the subtropical highs.
14. A wind blowing down the leeward side of a mountain and warming by compression is called a(n) __________.
15. A(n) __________ is a local wind blowing from the sea during the afternoon in coastal areas.
16. The __________ is an instrument used to measure wind speed.
17. The amount of pressure change over a given distance is referred to as the __________.
18. A(n) __________ is a swift (120- to 240-km/hour) high-altitude wind.
19. A center of high atmospheric pressure is called a(n) __________.
20. The __________ is the pressure zone located at about the latitude of the Arctic and Antarctic circles.
21. The __________ are global winds that blow from the polar high toward the subpolar low.
22. A(n) __________ is an instrument that continuously records air-pressure changes.

23. __________ is the condition that exists when the distribution of winds within a given area results in a net horizontal outflow of air from the region.

24. The __________ is a region of several semipermanent anticyclonic centers characterized by subsidence and divergence located roughly between latitudes 25° and 35°.

25. The __________ are the dominant west-to-east winds that characterize the regions on the poleward sides of the subtropical highs.

26. The __________ are wind belts located on the equatorward sides of the subtropical highs.

27. The seesaw pattern of atmospheric pressure between the eastern and western Pacific is called the __________.

28. __________ is the name given to the periodic warming of the ocean that occurs in the central and eastern Pacific.

Comprehensive Review

1. Which element of weather is measured by each of the following instruments? Briefly describe the principle of each instrument.

   a) Mercury barometer:

   b) Wind vane:

   c) Aneroid barometer:

   d) Cup anemometer:

   e) Barograph:

2. List and describe the three forces that control the wind.

   1)

   2)

   3)

3. On Figure 18.1, complete each of the map views of the four pressure-center diagrams by labeling the isobars with appropriate pressures and drawing several wind arrows to indicate surface-air movement.

Northern Hemisphere

High (Anticyclone)

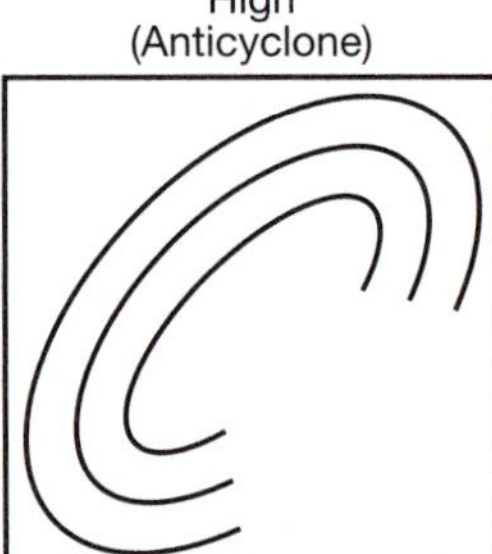

Low (Cyclone)

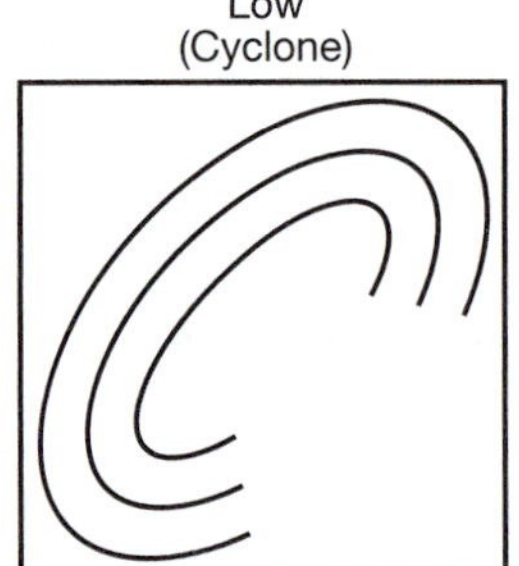

Southern Hemisphere

High (Anticyclone)

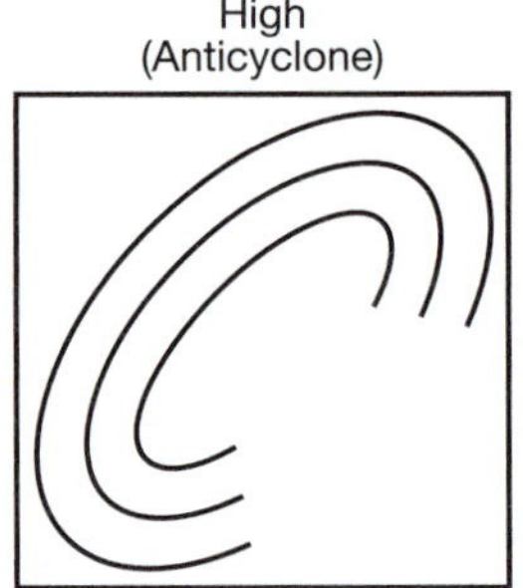

Low (Cyclone)

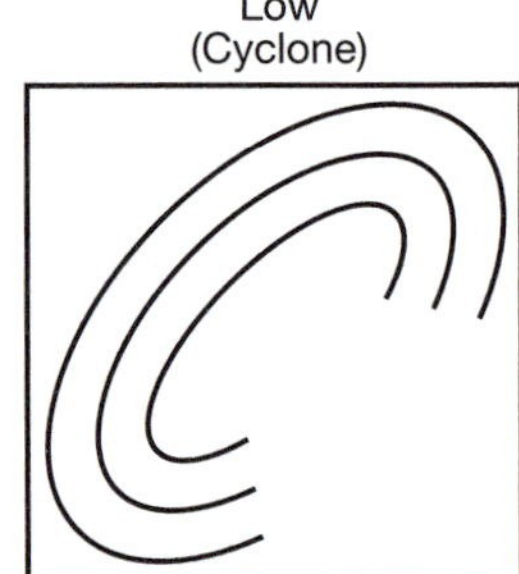

Figure 18.1

4. Referring to Figure 18.1, describe the general surface-air flow associated with each of the following pressure systems.

   a) Northern Hemisphere low (cyclone):

   b) Southern Hemisphere high (anticyclone):

5. Figure 18.2 illustrates side views of the air movements in a cyclone and an anticyclone. Using Figure 18.2, select the letter of the diagram that illustrates each of the following.

   a) Cyclone: 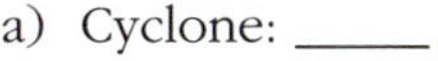

   b) Convergence aloft: _____

   c) Surface convergence: _____

   d) Anticyclone: _____

   e) Divergence aloft: _____

6. Why is "fair" weather usually associated with the approach of a high-pressure system (anticyclone)?

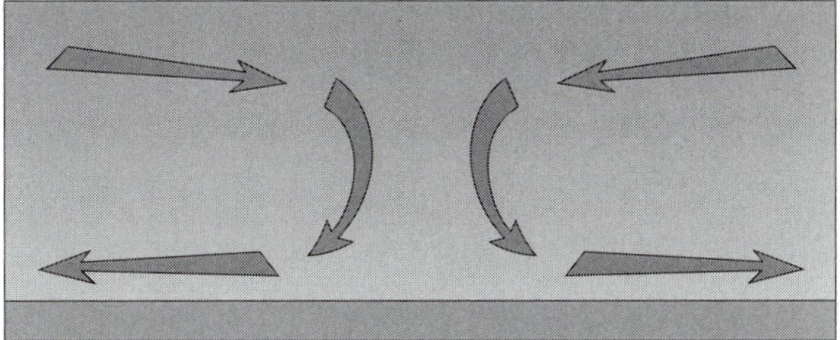

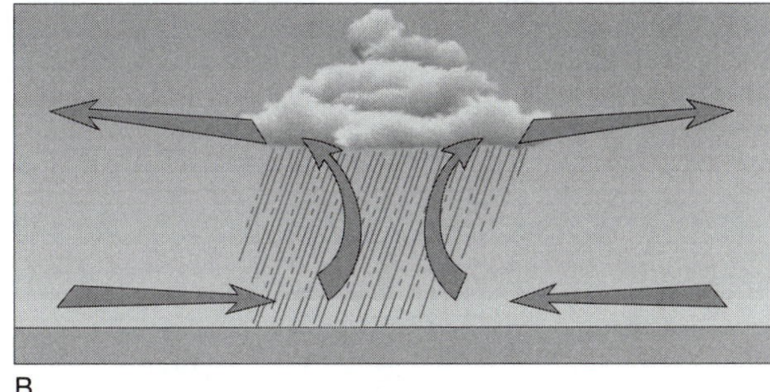

Figure 18.2

7. On Figure 18.3, write the names of the global pressure zones indicated by letters A through E, and the wind belts indicated by letters F through K.

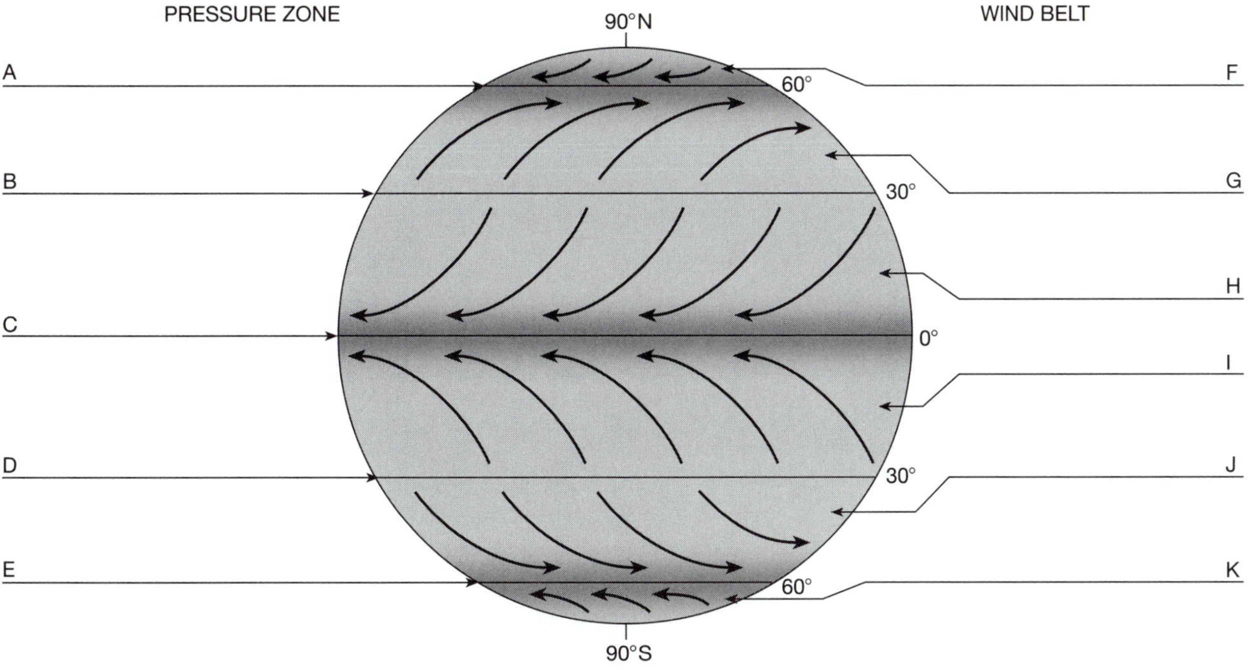

Figure 18.3

8. Write the name of the global pressure zone best described by each of the following statements.

    a) Warm and cold winds interacting to produce stormy weather: _____

    b) Subsiding dry air with extensive arid and semiarid regions: _____

    c) Warm, rising air marked by abundant precipitation: _____

    d) Cold, subsiding, polar air: _____

9. Briefly describe each of the following local winds.

    a) Chinook:

    b) Sea breeze:

    c) Country breeze:

10. If a wind is indicated with the following degree, from what compass direction is it blowing?

    a) 45 degrees: _____

    b) 270 degrees: _____

    c) 180 degrees: _____

Practice Test

Multiple choice. Choose the best answer for the following multiple-choice questions.

1. The only truly continuous pressure belt on Earth is the _____.
   a) Northern Hemisphere subtropical high
   b) equatorial low
   c) Southern Hemisphere subtropical high
   d) Southern Hemisphere subpolar low
   e) Northern Hemisphere subpolar low

2. The force exerted by the weight of the air above is called _____.
   a) air pressure
   b) convergence
   c) the Coriolis effect
   d) the pressure gradient
   e) divergence

3. Which one of the following is NOT a force that controls wind? _____
   a) magnetic force
   b) Coriolis effect
   c) pressure gradient force
   d) friction

4. Variations in air pressure from place to place are the principal cause of _____.
   a) snow
   b) rain
   c) wind
   d) clouds
   e) hail

5. Fair weather can usually be expected with the approach of a(n) _____.
   a) cyclone
   b) anticyclone

6. Centers of low pressure are called _____.
   a) anticyclones
   b) air masses
   c) cyclones
   d) jet streams
   e) domes

7. In the winter, large landmasses, particularly Asia, develop a seasonal _____.
   a) high-pressure system
   b) trade wind system
   c) low-pressure system
   d) chinook
   e) cyclonic circulation

8. Wind with a direction of 225 degrees would be a _____ wind.
   a) west
   b) northeast
   c) south
   d) southwest
   e) northwest

9. The general movement of low-pressure centers across the United States is from _____.
   a) north to south
   b) south to north
   c) east to west
   d) west to east
   e) northeast to southwest

10. Near the equator, rising air is associated with a pressure zone known as the _____.
    a) equatorial high
    b) tropical high
    c) equatorial low
    d) tropical low
    e) subtropical low

11. Standard sea-level pressure is _____.
    a) 1020.6 millibars
    b) 15.8 pounds per square inch
    c) 29.92 inches of mercury
    d) 3.7 kilograms per square centimeter
    e) 100 kilopascals

12. A sea breeze is most intense _____.
    a) during mid- to late afternoon
    b) in the late morning
    c) late in the evening
    d) at midnight
    e) at sunrise

13. High-altitude, high-velocity "rivers" of air are referred to as _____.
    a) cyclones
    b) tornadoes
    c) anticyclones
    d) air currents
    e) jet streams

14. The deflection of wind due to the Coriolis effect is strongest at _____.
    a) the equator
    b) the midlatitudes
    c) midnight
    d) the poles
    e) sunrise

15. Seasonal changes in wind direction associated with large landmasses and adjacent water bodies are called _____.
    a) chinooks
    b) geostrophic winds
    c) jet streams
    d) trade winds
    e) monsoons

16. Pressure data on a map are shown using lines that connect places of equal air pressure called _____.
    a) isotherms
    b) millibars
    c) aneroids
    d) kilograms
    e) isobars

17. Air is subsiding in the center of a(n) _____.
    a) low-pressure system
    b) jet stream
    c) surface convergence area
    d) high-pressure system
    e) chinook

18. A(n) _____ is an instrument that continuously records pressure changes.
    a) mercury barometer
    b) barograph
    c) aneroid barometer
    d) sling psychrometer
    e) cup anemometer

19. The surface winds between the subtropical high and equatorial low-pressure zones are the _____.
    a) polar easterlies
    b) sea breezes
    c) trade winds
    d) monsoon winds
    e) westerlies

20. Which one of the following does NOT describe the surface-air movement in a Northern Hemisphere low? _____
    a) inward
    b) counterclockwise
    c) net upward movement
    d) divergent

21. Winds that blow parallel to isobars are called _____.
    a) monsoon winds
    b) geostrophic winds
    c) sea breezes
    d) chinooks
    e) trade winds

22. Chinooks are warm, dry winds that commonly occur on the _____ slopes of the Rockies.
    a) northern
    b) eastern
    c) southern
    d) western

23. The westerlies and polar easterlies converge in a stormy region known as the _____.
    a) polar front
    b) equatorial low
    c) subtropical front
    d) subpolar high
    e) polar low

24. In the midlatitudes the movement of cyclonic and anticyclonic systems is steered by the _____.
    a) polar easterlies
    b) upper-level air flow
    c) trade winds
    d) pressure gradient
    e) ocean circulation

25. The surface winds between the subtropical high and subpolar low-pressure zones are the _____.
    a) polar easterlies
    b) sea breezes
    c) trade winds
    d) monsoon winds
    e) westerlies

True/false. For the following true/false questions, if a statement is not completely true, mark it false. For each false statement, change the **italicized** *word to correct the statement.*

1. _____ The ultimate driving force of wind is *solar* energy.
2. _____ The greater the pressure differences between two places, the *slower* the wind speed.
3. _____ Centers of high pressure are called *anticyclones*.
4. _____ In the Northern Hemisphere, the Coriolis effect causes the path of wind to be curved to the *left*.
5. _____ The *aneroid* barometer consists of metal chambers that change shape as air pressure changes.
6. _____ A *rising* barometric tendency often means that fair weather can be expected.
7. _____ Upper air flow is often nearly *parallel* to the isobars.
8. _____ Pressure decreases from the outer isobars toward the center in a(n) *anticyclone*.
9. _____ In India the winter monsoon is dominated by *wet* continental air.
10. _____ A steep pressure gradient indicates *strong* winds.
11. _____ A *low*-pressure system often brings cloudiness and precipitation.
12. _____ As wind speed increases, deflection by the Coriolis effect becomes *less*.
13. _____ The pressure zone of subsiding dry air, which encircles the globe near 30 degrees latitude north and south, is the *midlatitude* high.
14. _____ Wind direction is modified by both the Coriolis effect and *friction*.
15. _____ In both hemispheres an anticyclone is associated with *divergence* of the air aloft.
16. _____ In the Northern Hemisphere more cyclones are generated during the *warmer* months, when the temperature gradient is greatest.
17. _____ The rainiest regions of Earth are associated with *low*-pressure systems.
18. _____ El Niño refers to episodes of ocean warming in the equatorial *Atlantic* ocean.

Word choice. Complete each of the following statements by selecting the most appropriate response.

1. Wind flows from areas of [higher/lower] air pressure to regions of [higher/lower] air pressure.
2. Fair weather is predicted when the air pressure is [falling/rising].
3. During summers, continents will tend to have [higher/lower] air pressures than adjacent oceans.
4. The Coriolis force is [strongest/weakest] at the equator.
5. Divergence aloft [diminishes/maintains] a surface low-pressure center.
6. The amount of pressure change occurring over a given distance is indicated by the [angle/spacing] of isobars.
7. Cyclones usually move from [south to north/west to east] across the United States.
8. In the Southern Hemisphere, moving objects are deflected to the [right/left] by the Coriolis force.

9. Jet streams travel in a [west-to-east/east-to-west] direction.
10. Clouds and precipitation are most often associated with [high-/low-] pressure systems.
11. The Coriolis effect [does/does not] change the wind speed.
12. The prevailing winds in the midlatitudes are [easterlies/westerlies].
13. The greater the pressure gradient, the [higher/lower] the wind speed.
14. Sea breezes usually [raise/lower] temperatures along a coast.
15. Friction has the greatest effect on winds in the [upper/lower] atmosphere.
16. In general, high latitudes receive [more/less] precipitation than low latitudes.
17. Chinook and Santa Ana winds are warmed by [compression/expansion] of the air.
18. One of the [weakest/strongest] El Niño events on record occurred in 1997–98.

Written questions

1. How does the Coriolis effect influence the movement of air?

2. Briefly describe the weather that is usually associated with (1) a cyclone and (2) an anticyclone.

3. Describe the motions of air at the surface and aloft in a Northern Hemisphere cyclone.

For other interesting and pertinent information, be sure to visit the *Earth Science* companion website at

http://www.prenhall.com/tarbuck

19

Weather Patterns and Severe Storms

For most people living in the middle latitudes, including much of the United States, major changes in day-to-day weather are brought about by large bodies of moving air, called air masses, fronts, and traveling middle-latitude cyclones. It is often the boundary between two adjoining air masses with different temperature and moisture characteristics, called a front, that marks this change. In addition to investigating air masses, fronts, and the evolution of middle-latitude cyclones, Chapter nineteen also examines some of nature's most destructive forces, tornadoes and hurricanes. Thunderstorms, although less intense and far more common than tornadoes and hurricanes, are also discussed.

Learning Objectives

After reading, studying, and discussing this chapter, you should be able to:

Explain what an air mass is.

An *air mass* is a large body of air, usually 1600 kilometers (1000 miles) or more across, which is characterized by a *sameness of temperature and moisture* at any given altitude. When this air moves out of its region of origin, called the *source region*, it will carry these temperatures and moisture conditions elsewhere, perhaps eventually affecting a large portion of a continent.

Describe how air masses are classified and the general weather associated with each air mass type.

Air masses are classified according to (1) the nature of the surface in the source region and (2) the latitude of the source region. *Continental (c)* designates an air mass of land origin, with the air likely to be dry; whereas a *maritime (m)* air mass originates over water and therefore will be humid. *Polar (P)* air masses originate in high latitudes and are cold. *Tropical (T)* air masses form in low latitudes and are warm. According to this classification scheme, the *four basic types of air masses* are *continental polar (cP), continental tropical (cT), maritime polar (mP)*, and *maritime tropical (mT)*. Continental polar (cP) and maritime tropical (mT) air masses influence the weather of North America most, especially east of the Rocky Mountains. Maritime tropical air is the source of much, if not most, of the precipitation received in the eastern two-thirds of the United States.

Discuss the differences between warm fronts and cold fronts.

Fronts are boundaries that separate air masses of different densities, one warmer and often higher in moisture content than the other. A *warm front* occurs when the surface position of the front moves so that warm air occupies territory formerly covered by cooler air. Along a warm front, a warm air mass overrides a retreating mass of cooler air. As the warm air ascends, it cools adiabatically to produce clouds and frequently light-to-moderate precipitation over a large area. A *cold front* forms where cold air is actively advancing into a region occupied by warmer air. Cold fronts are about twice as steep and move more rapidly than do warm fronts. Because of these two differences, precipitation along a cold front is usually more intense and of shorter duration than is precipitation associated with a warm front.

Describe the primary midlatitude weather-producing systems.

The primary weather producers in the middle latitudes are *large centers of low pressure* that generally travel from *west to east*, called *middle-latitude cyclones*. These *bearers of stormy weather*, which last from a few days to a week, have a *counterclockwise circulation* pattern in the Northern Hemisphere and an *inward flow of air* toward their centers. Most middle-latitude cyclones have a *cold front and frequently a warm front* extending from the central area of low pressure. *Convergence and forceful lifting along the fronts* initiate cloud development and frequently cause precipitation. As a middle-latitude cyclone with its associated fronts passes over a region, it often brings with it abrupt changes in the weather. The particular weather experienced by an area depends on the path of the cyclone.

Discuss the cause of thunderstorms.

Thunderstorms are caused by the upward movement of warm, moist, unstable air, triggered by a number of different processes. They are associated with *cumulonimbus clouds* that generate heavy rainfall, thunder, lightning, and occasionally hail.

Describe tornadoes.

Tornadoes—destructive, local storms of short duration—are violent windstorms associated with severe thunderstorms that take the form of a rotating column of air that extends downward from a cumulonimbus cloud. Tornadoes are most often spawned along the cold front of a middle-latitude cyclone, most frequently during the spring months.

Describe the formation and demise of hurricanes.

Hurricanes, the greatest storms on Earth, are tropical cyclones with wind speeds in excess of 119 kilometers (74 miles) per hour. These complex tropical disturbances develop over tropical ocean waters and are fueled by the latent heat liberated when huge quantities of water vapor condense. Hurricanes form most often in late summer when ocean-surface temperatures reach 27°C (80°F) or higher and thus are able to provide the necessary heat and moisture to the air. Hurricanes diminish in intensity whenever they (1) move over cool ocean water that cannot supply adequate heat and moisture, (2) move onto land, or (3) reach a location where large-scale flow aloft is unfavorable.

Key Terms

air mass
air-mass weather
arctic (A) air mass
cold front
continental (c) air mass
Doppler radar
eye
eye wall
front
hurricane
lake-effect snow
maritime (m) air mass
middle-latitude cyclone
occluded front
occlusion
overrunning
polar (P) air mass
source region
stationary front
storm surge
thunderstorm
tornado
tornado warning
tornado watch
tropical (T) air mass
tropical depression
tropical storm
warm front

Vocabulary Review

Choosing from the list of key terms, furnish the most appropriate response for the following statements.

1. A(n) __________ is an immense body of air characterized by a similarity of temperature and moisture at any given altitude.
2. A tropical cyclonic storm having winds in excess of 119 kilometers (74 miles) per hour is called a(n) __________.
3. A(n) __________ is a boundary that separates different air masses, one warmer than the other and often higher in moisture content.
4. By international agreement, a(n) __________ is a tropical cyclone with winds between 61 and 119 kilometers (38 and 74 miles) per hour.

5. A(n) __________ is the area where an air mass acquires its characteristic properties of temperature and moisture.
6. A(n) __________ is a type of air mass that forms over land.
7. When the surface position of a front moves so that warm air occupies territory formerly covered by cooler air, it is called a(n) __________.
8. In the region between southern Florida and Alaska, the primary weather producer is the __________.
9. The __________ is the doughnut-shaped area of intense cumulonimbus development and very strong winds that surrounds the center of a hurricane.
10. An air mass that forms in low latitudes is called a(n) __________.
11. By international agreement, a(n) __________ is a tropical cyclone with maximum winds that do not exceed 61 kilometers (38 miles) per hour.
12. The __________ is a zone of scattered clouds and calm, averaging about 20 kilometers in diameter at the center of a hurricane.
13. Fairly constant weather that may take several days to traverse an area often represents a weather situation called __________.
14. When cold air is actively advancing into a region occupied by warmer air, the boundary is called a(n) __________.
15. A(n) __________ is a small, very intense cyclonic storm with exceedingly high winds, most often produced along cold fronts in conjunction with severe thunderstorms.
16. A(n) __________ is a type of air mass that originates over water.
17. When conditions appear favorable for tornado formation, a __________ is issued for areas covering about 65,000 square kilometers (25,000 miles).
18. The type of radar that can detect motion directly and hence greatly improve tornado and severe storm warnings is called __________.
19. A(n) __________ is a storm of relatively short duration produced by a cumulonimbus cloud and accompanied by strong wind gusts, heavy rain, lightning, thunder, and sometimes hail.
20. A(n) __________ is a dome of water that sweeps across the coast near the point where the eye of the hurricane makes landfall.
21. A(n) __________ is issued when a tornado funnel cloud has actually been sighted or is indicated by radar.
22. An air mass that originates in high latitudes is called a(n) __________.
23. When an active cold front overtakes a warm front, a new type of front, called a(n) __________ often forms.
24. During the winter a highly localized storm occurring along the leeward shores of the Great Lakes often creates what is known as a(n) __________.
25. When the surface position of a front does not move, the front is referred to as a(n) __________.
26. __________ is the process whereby a cold front catches and lifts a warm front.

Comprehensive Review

1. List and describe the two criteria used to classify air masses.

   1)

   2)

2. Using Figure 19.1, list the name and describe the temperature and moisture characteristics of the air mass identified on the figure by each of the following letters.

   A:

   B:

   F:

   G:

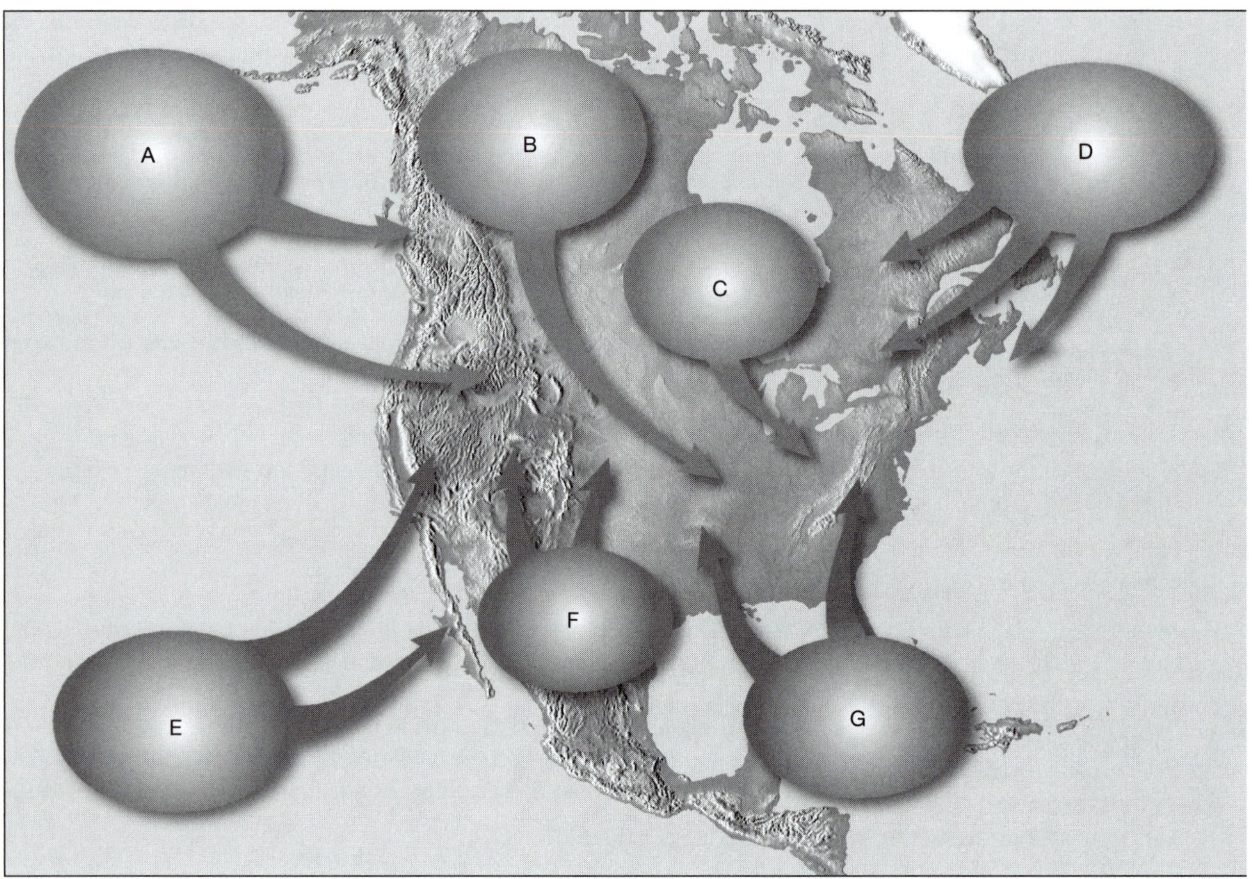

Figure 19.1

3. Which two air masses have the greatest influence on the weather of North America, especially east of the Rocky Mountains?

4. Sketch profiles (side views) of a typical warm front and a typical cold front and briefly describe each.

   a) Warm front:

   b) Cold front:

5. Of the two diagrams illustrated in Figure 19.2, which represents the more advanced stage in the development of a middle-latitude cyclone? What is the reason for your choice?

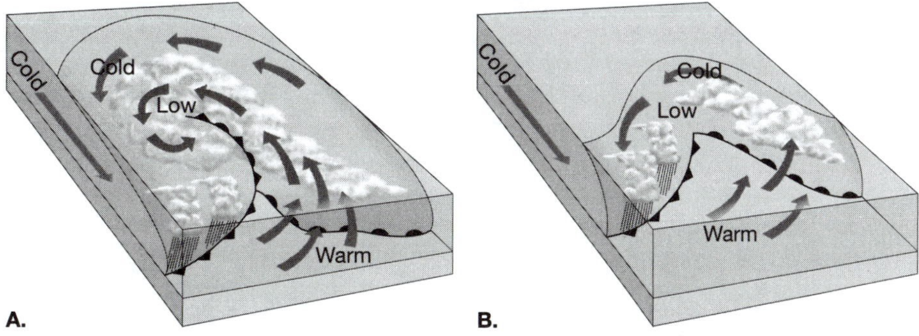

Figure 19.2

6. Using Figure 19.3, a diagram of a typical mature, middle-latitude cyclone, answer the following.

   a) Which type of front is shown at B?

   b) Which type of front is shown at D?

   c) Which air-mass type would most likely be found at C?

   d) Which air-mass type would most likely be found at G?

   e) The wind direction at C would most likely be from the __________.

   f) The wind direction at G would most likely be from the __________.

   g) As the center of the low moves to F, what would be the expected weather changes experienced by people living at C?

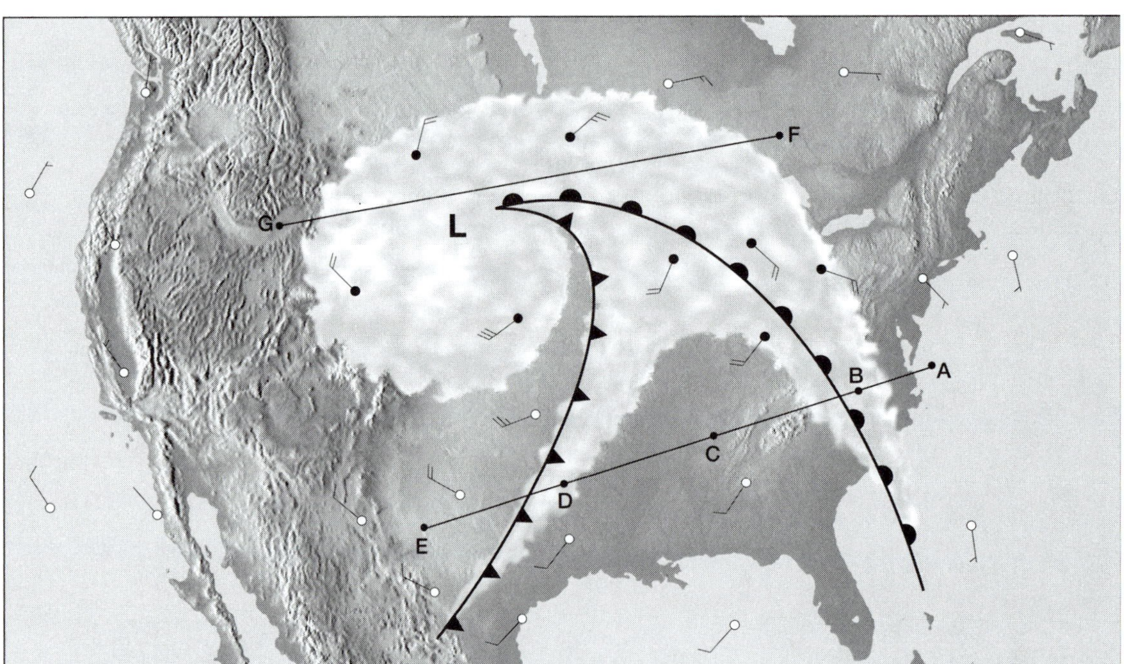

Figure 19.3

7. Describe the air flow at the surface and aloft for each of the following pressure systems.

   a) Cyclone:

   b) Anticyclone:

8. Compared to warm fronts, what are the two differences that largely account for the more violent nature of cold-front weather?

   1)

   2)

9. What are the atmospheric conditions associated with the formation of

    a) Thunderstorms?

    b) Tornadoes?

10. What are three conditions that can cause the intensity of hurricanes to diminish?

    1)

    2)

    3)

11. List the three categories of damage caused by hurricanes.

    1)

    2)

    3)

Practice Test

Multiple choice. Choose the best answer for the following multiple-choice questions.

1. An immense body of air characterized by a similarity of temperature and moisture at any given altitude is referred to as a(n) __________.
   a) cyclone
   b) air mass
   c) anticyclone
   d) air cell
   e) front

2. Hurricanes are classified according to intensity using the __________.
   a) Richter scale
   b) Saffir-Simpson scale
   c) Fujita intensity scale
   d) Doppler scale
   e) F-scale

3. Along which type of front is the intensity of precipitation greater but the duration shorter?
   a) warm front
   b) cold front

4. The boundary that separates different air masses is called a(n) __________.
   a) front
   b) cyclone
   c) storm
   d) contact slope
   e) anticyclone

5. Which air mass is most associated with lake-effect snows?
   a) mT
   b) cP
   c) cT
   d) mP

6. In the United States, thunderstorms typically form within __________ air masses.
   a) mT
   b) cP
   c) cT
   d) mP

7. The primary tornado warning system in use today involves both __________.
   a) observers and satellites
   b) barometers and wind vanes
   c) observers and barometers
   d) observers and radar
   e) satellites and radar

8. Which type of front has the steepest frontal surface?
   a) warm front
   b) cold front

9. The area in which an air mass acquires its characteristic properties of temperature and moisture is called its __________.
   a) area of origin
   b) location
   c) weather site
   d) source region
   e) classification region

10. Surface air flow in a Northern Hemisphere middle-latitude cyclone is __________.
    a) divergent and clockwise
    b) divergent and counterclockwise
    c) convergent and clockwise
    d) convergent and counterclockwise
    e) both convergent and divergent

11. The greatest tornado frequency in the United States is during the period __________.
    a) January through March
    b) April through June
    c) July through September
    d) October through December

12. Along a front, which air is always forced aloft?
    a) cooler, denser air
    b) the driest air
    c) the wettest air
    d) the fastest moving air
    e) warmer, less dense air

13. Following the passage of a cold front, winds often come from the __________.
    a) south to southeast
    b) east to southeast
    c) north to northwest
    d) west to southwest
    e) south to southwest

14. Which type of air mass originates in central Canada?
    a) mT
    b) cP
    c) mP
    d) cT

15. Pressure in a middle-latitude cyclone __________.
    a) decreases toward the center
    b) remains the same everywhere
    c) increases toward the center
    d) increases then decreases toward the center
    e) decreases then increases toward the center

16. Which one of the following is NOT an area where maritime tropical air masses that affect North America originate?
    a) Gulf of Mexico
    b) Hudson Bay
    c) Caribbean Sea
    d) Atlantic Ocean
    e) Pacific Ocean

17. The first sign of the approach of a warm front is the appearance of __________ clouds overhead.
    a) stratus
    b) cirrus
    c) cumulus
    d) cumulonimbus
    e) nimbostratus

18. The weather behind a cold front is dominated by a __________.
    a) subsiding, relatively cold air mass
    b) rising, relatively warm air mass
    c) mixed air mass
    d) rising, relatively cold air mass
    e) subsiding, relatively warm air mass

19. An mT air mass would be best described as __________.
    a) cold and dry
    b) warm and dry
    c) cold and wet
    d) warm and wet
    e) hot and dry

20. Hurricanes form in tropical waters between the latitudes of __________.
    a) 0 and 5 degrees
    b) 5 and 20 degrees
    c) 20 and 30 degrees
    d) 30 and 40 degrees

21. The source of the heat necessary to maintain the upward development of a thunderstorm is __________.
    a) surface heating
    b) friction
    c) latent heat
    d) electrical heat
    e) compressional heat

22. Which air mass is most associated with a "nor'easter" in New England?
    a) mT
    b) cP
    c) mP
    d) cT

23. Following the passage of a warm front, winds often come from the __________.
    a) north to northeast
    b) east to southeast
    c) north to northwest
    d) west to northwest
    e) south to southwest

24. Hurricanes develop most often in the __________.
    a) early winter
    b) late winter
    c) early summer
    d) late summer
    e) early spring

25. The center of a tornado is best characterized by its __________.
    a) very high pressure
    b) eye wall
    c) large calm area
    d) very low pressure
    e) subsidence

26. Along a(n) __________ front, the flow of air on both sides of the front is almost parallel.
    a) warm
    b) stationary
    c) cold
    d) occluded
    e) advancing

27. During the life cycle of a middle-latitude cyclone, the process of __________ occurs when the cold front overtakes the warm front and the warm sector is displaced aloft.
    a) frontalysis
    b) cyclogenesis
    c) differentiation
    d) cyclonics
    e) occlusion

28. Second only to floods, __________ is(are) responsible for the greatest number of storm-related deaths each year.
    a) lightning
    b) tornadoes
    c) landslides
    d) hurricanes
    e) strong winds

True/false. For the following true/false questions, if a statement is not completely true, mark it false. For each false statement, change the **italicized** *word to correct the statement.*

1. _____ A *front* usually marks a change in weather.
2. _____ Air masses are typically *5000* miles or more across.
3. _____ Cyclones are typically the bearers of *stormy* weather.
4. _____ A cP air mass originates over land and is likely to be cold and *dry*.
5. _____ Cold fronts advance *less* rapidly than warm fronts.
6. _____ Guided by the westerlies aloft, middle-latitude cyclones generally move *westward* across the United States.
7. _____ *Continental* tropical air is the source of much, if not most, of the precipitation received in the eastern two-thirds of the United States.
8. _____ Most severe weather occurs along *cold* fronts.
9. _____ Doppler radar has the ability to detect *motion* directly.
10. _____ On a weather map, the surface position of a *warm* front is shown by a line with semicircles extending into the cooler air.
11. _____ After the passage of a cold front, air pressure will most likely *rise*.
12. _____ Most severe thunderstorms in the midlatitudes form along or ahead of *warm* fronts.
13. _____ On the average, the slopes of cold fronts are about *half* as steep as warm-front slopes.

14. _____ In a middle-latitude cyclone, *convergence* and forceful lifting initiate cloud development.
15. _____ Cyclones and anticyclones are typically found *adjacent* to one another.
16. _____ More tornadoes are generated in the *western* United States than any other area of the country.
17. _____ In the western Pacific, hurricanes are called *typhoons*.
18. _____ A tornado *warning* is issued when a funnel cloud has actually been sighted or is indicated by radar.
19. _____ In a middle-latitude cyclone, *divergence* is occurring in the air aloft.
20. _____ A tornado with an F4 rating on the Fujita intensity scale produces *devastating* damage.
21. _____ The lowest pressures ever recorded in the Western Hemisphere are associated with *hurricanes*.
22. _____ Air flow aloft plays an *important* role in maintaining cyclonic and anticyclonic circulation.

Word choice. Complete each of the following statements by selecting the most appropriate response.

1. The term [tornado/cyclone] simply refers to the circulation around any low-pressure center, no matter how large or intense it is.
2. When a front advances, [cool/warm] air is forced aloft over the [cooler/warmer] air mass.
3. In 1900 a [hurricane/tornado] became the deadliest natural disaster in U.S. history.
4. On a weather map a cold front is shown by a line of [triangles/semicircles] extending into the warmer air.
5. The North [Atlantic/Pacific] Ocean produces the greatest number of hurricanes per year.
6. During the winter a North American cP air mass would most likely be cold and [moist/dry].
7. The greatest number of thunderstorms occur in association with [nimbostratus/cumulonimbus] clouds.
8. After a cold front passes, the weather will usually be [clear/stormy].
9. The center of a middle-latitude cyclone is a [high-/low-] pressure center.
10. A(n) [occluded/stationary] front often forms when an active cold front overtakes a warm front.
11. In a Northern Hemisphere middle-latitude cyclone, the circulation of the air around the center is [clockwise/counterclockwise].
12. The air within a hurricane's eye gradually [rises/descends] and [heats/cools] by [expansion/compression].
13. Due to lake-effect snows, snowfall is greatest on the [windward/leeward] shores of the Great Lakes.
14. Tornadoes are most frequent during the [spring/fall] and are most often spawned along the [cold/warm] front of a middle-latitude cyclone.
15. Prior to the development of a middle-latitude cyclone, the air flow in the adjacent air masses is [parallel/perpendicular] to the front.

Written questions

1. Describe the temperature and moisture characteristics of (1) a continental polar (cP) air mass and (2) a maritime tropical (mT) air mass.

2. Describe the weather conditions that an observer would experience as a middle-latitude cyclone passes with its center to the north.

3. What is the difference between a tornado watch and a tornado warning?

For other interesting and pertinent information, be sure to visit the *Earth Science* companion website at

http://www.prenhall.com/tarbuck

20

World Climates and Global Climate Change

To appreciate climate, it is important to understand that climate involves more than just the atmosphere. Rather, the atmosphere is the central component of a complex, connected, and interactive system upon which all life depends. Broadly defined, climate is the long-term behavior of this environmental system. To understand this climate system fully, one must also understand the Sun, oceans, ice sheets, solid Earth, and all forms of life. The focus of this chapter is to present the broad diversity of climates around the globe and to investigate how each climate group influences the nature of plant and animal life, the soil, and many external geological processes. Also examined is the significant impact climate has on people and the ways in which humans are changing global climate.

Learning Objectives

After reading, studying, and discussing this chapter, you should be able to:

Explain what is meant by Earth's climate system.

Climate is the aggregate of weather conditions for a place or region over a long period of time. Earth's *climate system* involves the exchanges of energy and moisture that occur among the atmosphere, hydrosphere, solid Earth, biosphere, and *cryosphere* (the ice and snow that exist at Earth's surface).

Describe the Köppen system of climate classification.

Climate classification brings order to large quantities of information, which aids comprehension and understanding and facilitates analysis and explanation. *Temperature and precipitation are the most important elements in a climatic description.* Many climate-classification schemes have been devised, with the value of each determined by its intended use. The *Köppen classification*, which uses mean monthly and annual values of temperature and precipitation, is a widely used system. The boundaries Köppen chose were largely based on the limits of certain plant associations. Five principal climate groups, each with subdivisions, were recognized. Each group is designated by a capital letter. Four of the climate groups (A, C, D, and E) are defined on the basis of temperature characteristics, and the fifth, the B group, has precipitation as its primary criterion.

Describe humid tropical (A) climates.

Humid tropical (A) climates are winterless, with all months having a mean temperature above 18°C. *Wet tropical climates* (Af and Am), which lie near the equator, have constantly high temperatures, year-round rainfall, and the most luxuriant vegetation (tropical rain forest) found in any climatic realm. *Tropical wet and dry climates* (Aw) are found poleward of the wet tropics and equatorward of the subtropical deserts where the rain forest gives way to the tropical grasslands and scattered drought-tolerant trees of the savanna. The most distinctive feature of this climate is the seasonal character of the rainfall.

Discuss dry (B) climates.

Dry (B) climates, in which the yearly precipitation is less than the potential loss of water by evaporation, are subdivided into two types: *arid* or *desert* (BW) and *semiarid* or *steppe* (BS). Their differences are primarily a matter of degree, with semiarid being a marginal and more humid variant of arid. Low-latitude deserts and steppes coincide with the clear skies caused by subsiding air beneath the subtropical high-pressure belts. Middle-latitude deserts and steppes exist principally because of their position in the deep interiors of large landmasses far removed from the oceans. Because many middle-latitude deserts occupy sites on the leeward sides of mountains, they can also be classified as *rainshadow deserts*.

Describe middle-latitude climates with mild winters (C climates).

Middle-latitude climates with mild winters (C climates) occur where the average temperature of the coldest month is below 18°C but above –3°C. Several C climate subgroups exist. *Humid subtropical climates* (Cfa) are located on the eastern sides of the continents, in the 25 to 40° latitude range. Summer weather is hot and sultry, and winters are mild. In North America the *marine west coast climate* (Cfb, Cfc) extends from near the United States–Canada border northward as a narrow belt into southern Alaska. The prevalence of maritime air masses means that mild winters and cool summers are the rule. *Dry-summer subtropical climates* (Csa, Csb) are typically located along the west sides of continents between latitudes 30 and 45 degrees. In summer the regions are dominated by stable, dry conditions associated with the oceanic subtropical highs. In winter they are within range of the cyclonic storms of the polar front.

Discuss humid middle-latitude climates with severe winters (D climates).

Humid middle-latitude climates with severe winters (D climates) are land-controlled climates that are absent in the Southern Hemisphere. The D climates have severe winters. The average temperature of the coldest month is –3°C, and the warmest monthly mean exceeds 10°C. *Humid continental climates* (Dfa, Dfb, Dwa, Dwb) are confined to the eastern portions of North America and Eurasia in the latitude range between approximately 40 and 50 degrees north latitude. Both winter and summer temperatures can be characterized as relatively severe. Precipitation is generally greater in summer than in winter. *subarctic climates* (Dfc, Dfd, Dwc, Dwd) are situated north of the humid continental climates and south of the polar tundras. The outstanding feature of subarctic climates is the dominance of winter. By contrast, summers in the subarctic are remarkably warm despite their short duration. The highest annual temperature ranges on Earth occur here.

Describe polar (E) climates.

Polar (E) climates are summerless, with the average temperature of the warmest month below 10°C. Two types of polar climates are recognized. The *tundra climate* (ET) is a treeless climate found almost exclusively in the Northern Hemisphere. The *ice-cap climate* (EF) does not have a single monthly mean above 0°C. As a consequence, the growth of vegetation is prohibited, and the landscape is one of permanent ice and snow.

Discuss highland climates.

Compared to nearby places of lower elevation, *highland climates* are cooler and usually wetter. Because atmospheric conditions fluctuate rapidly with changes in altitude and exposure, these climates are best described by their variety and changeability.

List several ways humans have modified the environment.

Humans have been modifying the environment for thousands of years. By altering ground cover with the use of fire and the overgrazing of land, people have modified such important climatological factors as surface albedo, evaporation rates, and surface winds.

Describe how humans have contributed to global warming.

By adding carbon dioxide and other trace gases (methane, nitrous oxide, and chlorofluorocarbons) to the atmosphere, humans may be significantly contributing to global warming.

Discuss climate-feedback mechanisms.

When any component of the climate system is altered, scientists must consider the many possible outcomes, called *climate-feedback mechanisms*. Changes that reinforce the initial change are called *positive-feedback mechanisms*. On the other hand, *negative-feedback mechanisms* produce results that are the opposite of the initial change and tend to offset it.

Discuss aerosols and their effect on climate.

Global climate is also affected by human activities that contribute to the atmosphere's *aerosol* (tiny, often microscopic, liquid and solid particles that are suspended in air) content. By reflecting sunlight back to space, aerosols have a net cooling efect. The effect of aerosols on today's climate is determined by the amount emitted during the preceding couple of weeks, while carbon dioxide remains for much longer spans and influences climate for many decades.

List some possible changes that may occur as the result of increased levels of carbon dioxide in the atmosphere.

Because the climate system is very complex, predicting specific regional changes that may occur as the result of increased levels of carbon dioxide in the atmosphere is highly speculative. However, some potential consequences of global warming include (1) altering the distribution of the world's water resources, (2) a probable rise in sea level, (3) a change in weather patterns, such as a greater intensity of tropical cyclones, and (4) changes in the extent of Arctic sea ice and permafrost.

Key Terms

arid climate
climate feedback mechanisms
climate system
desert climate
dry-summer subtropical climate
highland climate
humid continental climate
humid subtropical climate
ice cap climate
Köppen classification
marine west coast climate
polar climate
rainshadow desert
semiarid climate
steppe climate
subarctic climate
tropical rain forest
tropical wet and dry climate
tundra climate

Vocabulary Review

Choosing from the list of key terms, furnish the most appropriate response for the following statements.

1. Earth's __________ consists of the atmosphere, hydrosphere, solid Earth, biosphere, and cryosphere and involves the exchanges of energy and moisture that occur among the five parts.
2. Located on the eastern sides of the continents, in the 25 to 40° latitude range, is a climatic region known as the __________, which dominates the southeastern United States as well as other similarly situated areas around the world.
3. Nearly half of Australia is a(n) __________, and much of the remainder is a steppe climate.
4. One of the best-known and most-used systems of climate classification is the __________.
5. The __________ is a type of polar climate that does not have a single monthly mean above 0°C.
6. Situated north of the humid continental climate and south of the polar tundra is an extensive __________ region covering broad, uninterrupted expanses from western Alaska to Newfoundland in North America and from Norway to the Pacific coast of Russia in Eurasia.
7. The __________ is a climatic region situated on the western (windward) side of continents, from about 40 to 65° north and south latitude, that is dominated by the onshore flow of oceanic air.
8. The most luxuriant vegetation found in any climatic realm occurs in the __________.
9. A(n) __________ frequently occurs on the leeward side of a mountain where the conditions are often drier and more arid than the windward side.
10. The climatic region called the __________ is confined to the central and eastern portions of North America and Eurasia in the latitude range between approximately 40 and 50 degrees north latitude.
11. The transitional climatic region located poleward of the wet tropics and equatorward of the tropical deserts is called the __________.
12. The __________ is a treeless climate found almost exclusively in the Northern Hemisphere.

13. A(n) __________ is a marginal and more humid variant of the arid climate and represents a transition zone that surrounds the desert and separates it from the bordering humid climates.
14. Typically located along the west sides of continents between latitudes 35° and 45° and situated between the marine west coast climate on the poleward side and the subtropical steppes on the equatorward side, the climatic region known as the __________ is best described as transitional in character.
15. Compared to nearby places at lower elevations, sites with a(n) __________ are cooler and usually wetter.
16. The possible outcomes of altering the climate system are referred to as __________.

Comprehensive Review

1. What are the five parts of the climate system that constantly exchange energy and moisture?

2. What are the two most important elements in a climatic description? Why are these elements important?

3. What are the names, capital-letter designations, and characteristics of the five principal Köppen climate groups?

   1)

   2)

   3)

   4)

   5)

4. Using Figure 20.1, the climate diagram for St. Louis, Missouri, answer the following questions.

   a) During which month does the highest temperature occur? What is the highest temperature?

   b) During which month does the lowest temperature occur? What is the lowest temperature?

   c) What is the annual temperature range for St. Louis?

   d) Which month receives the greatest amount of precipitation?

   e) The precipitation is concentrated in which season?

   f) Using the Köppen system of climatic classification presented in Table 20.1 of the textbook and the data presented in the climate diagram, Figure 20.1, determine the Köppen climate classification for St. Louis, Missouri.

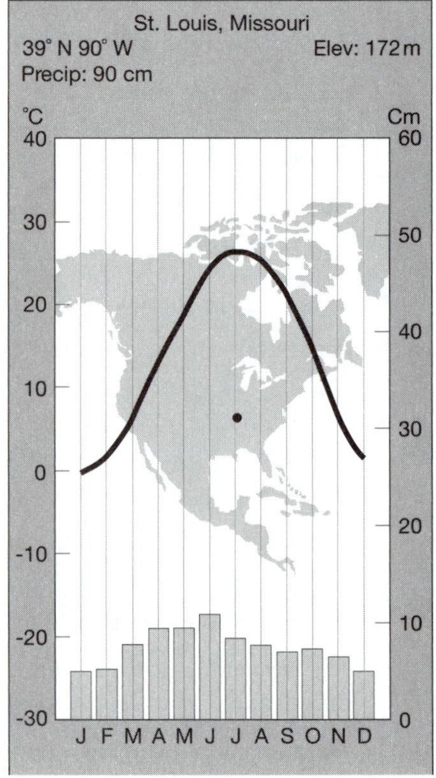

Figure 20.1

5. What are the three temperature/precipitation features that characterize the wet tropics? Where are the most continuous expanses of the wet tropics located?

6. What is the difference between these two climatic types: arid (desert) and semiarid (steppe)?

7. With what meteorological feature(s) do most low-latitude deserts and steppes coincide?

8. By referring to Figure 20.4 in the textbook, where in the United States is the location of the humid subtropical, warm summer climate region?

9. Examine the locations of the humid middle-latitude climates with severe winters (D climates) in Figure 20.4 of the textbook. Why are D climates absent in the Southern Hemisphere?

10. Indicate by name the climate group best described by each of the following statements.

    a) Potential evaporation exceeds precipitation:

    b) Average temperature of the warmest month is below 10°C:

    c) Very little change in monthly precipitation and temperature throughout the year:

    d) Situated on the windward side of continents, from 40 to 65° latitude:

    e) Location coincides with the subtropical high-pressure belts:

    f) Vast areas of northern coniferous forest:

    g) Tropical rain forest:

11. Why are the summer temperatures at Death Valley, California, consistently among the highest in the Western Hemisphere?

12. Compared with a rural environment, what impact does a city have on the following elements?

    a) Temperature:

    b) Fog and rain:

    c) Wind:

13. What are the two most important gases involved in the greenhouse effect of the atmosphere?

14. What are the two most prominent means by which humans contribute to the buildup of carbon dioxide in the atmosphere?

    1)

    2)

15. What are three potential weather changes that could occur as a result of the projected global warming?

    1)

    2)

    3)

Practice Test

Multiple choice. Choose the best answer for the following multiple-choice questions.

1. The best-known and most-used system for climate classification is the __________ classification.
   a) Smith
   b) Köppen
   c) Klaus
   d) Strahler
   e) Trewartha

2. As a result of broad continents in the middle latitudes, __________ climates are land-controlled climates.
   a) A
   b) B
   c) C
   d) D
   e) E

3. The description of the aggregate weather conditions of a place or region is termed __________.
   a) weather
   b) cyclogenesis
   c) climate
   d) synthesis
   e) averaging

4. Which one of the following is NOT a possible consequence of greenhouse warming?
   a) more frequent and intense hurricanes
   b) reduction in secondary pollutants
   c) rising sea levels
   d) increase of heat waves and droughts
   e) shifts in the paths of large-scale cyclonic storms

5. A __________ climate dominates the southeastern United States.
   a) marine west coast
   b) humid subtropical
   c) humid continental
   d) tropical wet and dry
   e) steppe

6. According to Köppen, the distribution of natural __________ was an excellent expression of the totality of climate.
   a) wind belts
   b) air masses
   c) cyclones
   d) vegetation
   e) pressure belts

7. The __________ refers to the snow and ice that exist at Earth's surface.
   a) crystalsphere
   b) cryosphere
   c) hydrosphere
   d) thermosphere
   e) exosphere

8. Many of the rainiest places in the world are located on __________ mountain slopes.
   a) windward
   b) gradual
   c) leeward
   d) steep
   e) barren

9. A type of polar climate, called the __________ climate, is a treeless climate found almost exclusively in the Northern Hemisphere.
   a) taiga
   b) tundra
   c) plains
   d) coastal
   e) highland

10. In the Köppen system the __________ climate group has precipitation as its primary criterion.
    a) A
    b) B
    c) C
    d) D
    e) E

11. Melting ice packs and reducing the albedo of a region, leading to increased warming, is an example of __________.
    a) a temperature inversion
    b) a climatic feedback mechanism
    c) frost control
    d) deforestation
    e) a climatic control

12. The highest temperature recorded in the Western Hemisphere (134°F) was set at __________.
    a) Celaya, Mexico
    b) Houston, TX
    c) Death Valley, CA
    d) Las Vegas, NM
    e) Tucson, AZ

13. Af and Am climates form a discontinuous belt astride the __________.
    a) Antarctic Circle
    b) Tropic of Cancer
    c) equator
    d) Tropic of Capricorn
    e) Atlantic Ocean

14. The consensus in the scientific community is that altering atmospheric composition by the addition of carbon dioxide and trace gases will eventually lead to a __________ planet with a different distribution of climatic regimes.
    a) cooler
    b) wetter
    c) warmer
    d) drier
    e) smaller

15. Poleward of the wet tropics and equatorward of the tropical deserts lies a transitional climatic region called the __________ climate.
    a) humid middle-latitude
    b) polar
    c) humid tropical
    d) tropical wet and dry
    e) highland

16. The two most important elements in a climatic description are temperature and __________.
    a) precipitation
    b) pressure
    c) wind speed
    d) altitude
    e) wind direction

17. The lowest seasonal temperature variations occur in the __________ climates.
    a) A
    b) B
    c) C
    d) D
    e) highland

18. The name __________ climate is often used as a synonym for the dry-summer subtropical climate.
    a) Canadian
    b) Hudson Bay
    c) African
    d) European
    e) Mediterranean

19. Which climates are absent in the Southern Hemisphere?
    a) A
    b) B
    c) C
    d) D
    e) E

20. The highest average annual rainfall in the world (some 486 inches) occurs in __________.
    a) California
    b) Brazil
    c) New Zealand
    d) Hawaii
    e) Japan

21. The best-known climatic effect of increased altitude is to cause lower __________.
    a) relative humidities
    b) wind velocities
    c) temperatures
    d) solar intensities
    e) rainfall amounts

22. If carbon dioxide levels reach projected levels, models predict that by the year 2100 mean global surface temperatures will increase by __________.
    a) 0.5 to 1.0°C
    b) 1.0 to 2.0°C
    c) 1.4 to 5.8°C
    d) 4.0 to 7.5°C

23. The subarctic climate region is often referred to as the __________ climate.
    a) taiga
    b) tundra
    c) plains
    d) coastal
    e) polar

24. The most studied and well-documented urban climate effect is the urban __________ island.
    a) pollution
    b) wind
    c) pressure
    d) precipitation
    e) heat

True/false. For the following true/false questions, if a statement is not completely true, mark it false. For each false statement, change the **italicized** *word to correct the statement.*

1. _____ The Köppen climate classification recognizes *three* principal groups, each designated by a capital letter.
2. _____ The *leeward* sides of mountains are often wet.
3. _____ Wet tropical climates are strongly influenced by the equatorial *high* pressures.
4. _____ Temperature and *pressure* are the most important elements in a climatic description.
5. _____ In a humid tropical (A) climate, all months have a mean temperature *above* 18°C.
6. _____ Models indicate that the temperature response in *polar* regions due to global warming triggered by carbon dioxide and trace gases could be as much as two to three times greater than the global average.
7. _____ Climatologists define a dry climate as one in which the yearly precipitation is not as great as the potential loss of water by *infiltration*.
8. _____ Wet tropical climates are restricted to elevations *below* 1000 meters.
9. _____ Köppen believed that the distribution of natural *vegetation* was an excellent expression of the totality of climate.
10. _____ In a polar (E) climate the average temperature of the warmest month is *above* 10°C.
11. _____ Much of California has a climate similar to that surrounding the *Mediterranean* Sea.
12. _____ A steppe climate is a marginal and *more* humid variant of a desert climate.
13. _____ Places with an Af or Am climate designation lie near the *equator.*
14. _____ Paralleling the rapid growth of *agriculture*, great quantities of carbon dioxide have been added to the atmosphere.
15. _____ Coinciding with the subtropical high-pressure belts are Earth's *low-latitude* deserts and steppes.

Word choice. Complete each of the following statements by selecting the most appropriate response.

1. The climate system involves the exchanges of energy and moisture among its [two/three/five] parts.
2. In the Köppen climate classification, precipitation is the primary criterion for the [B/C] group.
3. Greenhouse warming could cause a [rise/fall] in sea level.
4. The Köppen climate classification uses mean monthly and annual values of [pressure/temperature] and [precipitation/solar radiation].
5. The largest area of Cfb climate is found in [South America/Europe], for here there is no mountain barrier blocking the movement of cool maritime air.

6. The intense drought conditions of the tropical wet and dry climates are caused by the [equatorial low/subtropical high] pressure belts.
7. The taiga climate closely corresponds to the northern [deciduous/coniferous] forest region.
8. Total precipitation in the wet tropical climates often exceeds [200/800] centimeters per year.
9. Beneath the subtropical high-pressure belts, air is [rising/subsiding], compressed, and warmed.
10. During the twentieth century the global average temperature increased approximately [0.6/6.0]°C.
11. Many of the rainiest places in the world are located on the [windward/leeward] slopes of mountains.
12. Carbon dioxide represents about [0.036/3.60] percent of the gases that make up clean, dry air.
13. A climate diagram is a plot of the average monthly temperature and [precipitation/solar radiation] amounts that occur at a place throughout the year.
14. Along with carbon dioxide, [nitrogen/water vapor] is largely responsible for the greenhouse effect of the atmosphere.
15. Models indicate that global warming of the lower atmosphere triggered by carbon dioxide and trace gases [will/will not] be the same everywhere.

Written questions

1. Briefly describe the distribution of climates found in North America.

2. Why are wet tropical climates restricted to elevations below 1000 meters?

3. Briefly describe the two types of climate-feedback mechanisms.

For other interesting and pertinent information, be sure to visit the *Earth Science* companion website at

http://www.prenhall.com/tarbuck

21

Origins of Modern Astronomy

Today we realize that Earth is one of nine planets and numerous smaller bodies that orbit the Sun. Only recently, within the past century, have astronomers accepted the fact that the Sun is an average star in a much larger family of perhaps 100 billion stars that comprise the Milky Way galaxy and that the universe itself is populated with countless other galaxies. This view of Earth's position in space is considerably different than that held only a few hundred years ago, when Earth was thought to occupy a privileged position as the center of the universe. To help unravel the history of astronomy, this chapter begins with an examination of the observations and contributions of the individuals that led to modern astronomy. In addition, the primary motions of Earth are described in detail along with discussions of the phases of the Moon, lunar motions, and eclipses.

Learning Objectives

After reading, studying, and discussing this chapter, you should be able to:

Describe the geocentric theory of the universe held by many early Greeks.

Early Greeks held the *geocentric* ("Earth-centered") view of the universe, believing that Earth was a sphere that stayed motionless at the center of the universe. Orbiting Earth were the seven wanderers (*planetai* in Greek), which included the Moon, Sun, and the known planets—Mercury, Venus, Mars, Jupiter, and Saturn. To the early Greeks the stars traveled daily around Earth on a transparent, hollow sphere called the *celestial sphere.* In A.D. 141, *Claudius Ptolemy* presented the geocentric outlook of the Greeks in its most sophisticated form in a model that became known as the *Ptolemaic system.* The Ptolemaic model had the planets moving in circular orbits around a motionless Earth. To explain the *retrograde motion* of planets (the apparent westward, or opposite motion planets exhibit for a period of time as Earth overtakes and passes them), Ptolemy proposed that the planets orbited in small circles (*epicycles*), revolving along large circles (*deferents*).

List the astronomical contributions of the ancient Greek philosophers Aristotle, Anaxagoras, Aristarchus, Eratosthenes, and Hipparchus.

In the fifth century B.C. the Greek astronomer *Anaxagoras* reasoned that the Moon shines by reflected sunlight, and because it is a sphere, only half is illuminated at one time. *Aristotle* (384–322 B.C.) concluded that Earth is spherical. The first Greek to profess a Sun-centered, or *heliocentric*, universe was *Aristarchus* (312–230 B.C.). The first successful attempt to establish the size of Earth is credited to *Eratosthenes* (276–194 B.C.). The greatest of the early Greek astronomers was *Hipparchus* (second century B.C.), best known for his star catalog.

List the contributions to modern astronomy of Nicolaus Copernicus, Tycho Brahe, Johannes Kepler, Galileo Galilei, and Sir Isaac Newton.

Modern astronomy evolved through the work of many dedicated individuals during the 1500s and 1600s. *Nicolaus Copernicus* (1473–1543) reconstructed the solar system with the Sun at the center and the planets orbiting around it but erroneously continued to use circles to represent the orbits of planets. *Tycho Brahe's* (1546–1601) observations were far more precise than any made previously and are his legacy to astronomy. *Johannes Kepler* (1571–1630) ushered in the new astronomy with his three laws of planetary motion. After constructing his own telescope, *Galileo Galilei* (1564–1642) made many important discoveries that supported the Copernican view of a Sun-centered solar system. *Sir Isaac Newton* (1643–1727) was the first to formulate and test the law of universal gravitation, develop the laws of motion, and prove that the force of *gravity*, combined with the tendency of an object to move in a straight line (*inertia*), results in the elliptical orbits discovered by Kepler.

Discuss constellations.

As early as 5000 years ago people began naming the configurations of stars, called *constellations*, in honor of mythological characters or great heroes. Today 88 constellations are recognized that divide the sky into units, just as state boundaries divide the United States.

Describe the equatorial system for locating stars.

One method for locating stars, called the *equatorial system*, divides the celestial sphere into a coordinate system similar to the latitude-longitude system used for locations on Earth's surface. *Declination*, like latitude, is the angular distance north or south of the *celestial equator*. *Right ascension* is the angular distance measured eastward from the position of the *vernal equinox* (the point in the sky where the Sun crosses the celestial equator at the onset of spring).

List and describe the primary motions of Earth.

The two primary motions of Earth are *rotation* (the turning, or spinning, of a body on its axis) and *revolution* (the motion of a body, such as a planet or moon, along a path around some point in space). Another very slow motion of Earth is *precession* (the slow motion of Earth's axis that traces out a cone over a period of 26,000 years). Earth's rotation can be measured in two ways, making two kinds of days. The *mean solar day* is the time interval from one noon to the next, which averages about 24 hours. On the other hand, the *sidereal day* is the time it takes for Earth to make one complete rotation with respect to a star other than the Sun, a period of 23 hours, 56 minutes, and 4 seconds. Earth revolves around the Sun in an elliptical orbit at an average distance from the Sun of 150 million kilometers (93 million miles). At *perihelion* (closest to the Sun), which occurs in January, Earth is 147 million kilometers from the Sun. At *aphelion* (farthest from the Sun), which occurs in July, Earth is 152 million kilometers distant. The imaginary plane that connects Earth's orbit with the celestial sphere is called the *plane of the ecliptic*.

Discuss the cycle of the Moon.

One of the first astronomical phenomenon to be understood was the regular cycle of the phases of the Moon. The cycle of the Moon through its phases requires $29^1/_2$ days, a time span called the *synodic month*. However, the true period of the Moon's revolution around Earth is $27^1/_3$ days and is known as the *sidereal month*. The difference of nearly two days is due to the fact that as the Moon orbits Earth, the Earth-Moon system also moves in an orbit around the Sun.

Describe lunar and solar eclipses.

In addition to understanding the Moon's phases, the early Greeks also realized that eclipses are simply shadow effects. When the Moon moves in a line directly between Earth and the Sun, which can occur only during the new-Moon phase, it casts a dark shadow on Earth, producing a *solar eclipse*. A *lunar eclipse* takes place when the Moon moves within the shadow of Earth during the full-Moon phase. Because the Moon's orbit is inclined about 5 degrees to the plane that contains the Earth and Sun (the plane of the ecliptic), during most new- and full-Moon phases no eclipse occurs. Only if a new- or full-Moon phase occurs as the Moon crosses the plane of the ecliptic can an eclipse take place. The usual number of eclipses is four per year.

Key Terms

aphelion
astronomical unit (AU)
axial precession
celestial sphere
constellation
declination
ecliptic
equatorial system
geocentric
heliocentric
lunar eclipse
mean solar day
perihelion
perturbation
phases of the Moon
plane of the ecliptic
Ptolemaic system
retrograde motion
revolution
right ascension
rotation
sidereal day
sidereal month
solar eclipse
synodic month

Vocabulary Review

Choosing from the list of key terms, furnish the most appropriate response for the following statements.

1. The apparent westward drift of the planets with respect to the stars is called __________.
2. The __________, the time interval it takes for Earth to make one complete rotation (360°) with respect to a star other than the Sun, has a period of 23 hours, 56 minutes, and 4 seconds.
3. Aristarchus (312–230 B.C.) was the first Greek to profess a Sun-centered, or __________, universe.
4. The true period of the Moon's revolution around Earth, $27^1/_3$ days, is known as the __________.
5. Early Greeks held the __________ view of the universe, believing that Earth was a sphere that stayed motionless at its center.
6. __________ is the turning, or spinning, of a body on its axis.
7. The average distance from Earth to the Sun, about 150 million kilometers (93 million miles), is a unit of distance called the __________.
8. When the Moon moves in a line directly between Earth and the Sun, it casts a dark shadow on Earth, producing a __________.
9. The angular distance north or south of the celestial equator denoting the position of a stellar body is referred to as __________.
10. Any variance in the orbit of a body from its predicted path is called __________.
11. The point in the orbit of a planet where it is closest to the Sun is called __________.
12. The apparent annual path of the Sun against the backdrop of the celestial sphere is called the __________.
13. The __________ was an imaginary, transparent, hollow sphere on which the ancients believed the stars traveled daily around Earth.
14. A(n) __________ is an apparent group of stars originally named in honor of a mythological character or great hero.
15. The imaginary plane that connects Earth's orbit with the celestial sphere is called the __________.
16. The __________ is an Earth-centered model of the universe proposed in A.D. 141 that uses epicycles and deferents to describe a planet's motion.
17. __________ is the motion of a body, such as a planet or moon, along a path around some point in space.
18. The __________ refers to the progression of changes in the Moon's appearance during the month.
19. The angular distance of a stellar object measured eastward along the celestial equator from the position of the vernal equinox is called __________.

20. The cycle of the Moon through its phases requires $29^1/_2$ days, a time span called the __________.
21. During the full-Moon phase, when the Moon moves within the shadow of Earth, a __________ occurs.
22. The __________ is a method used to locate stellar objects that is very similar to the latitude-longitude system used on Earth's surface.
23. The point in the orbit of a planet where it is farthest from the Sun is referred to as __________.
24. The very slight movement of Earth's axis over a period of 26,000 years is known as __________.
25. The__________, the time interval from one noon to the next, averages about 24 hours.

Comprehensive Review

1. Describe the geocentric model of the universe held by the early Greeks.

2. What is retrograde motion? In Figure 21.1, a diagram of the orbit of Mars as observed from Earth over many weeks, between which two points is retrograde motion occurring? How did the geocentric model of Ptolemy explain retrograde motion?

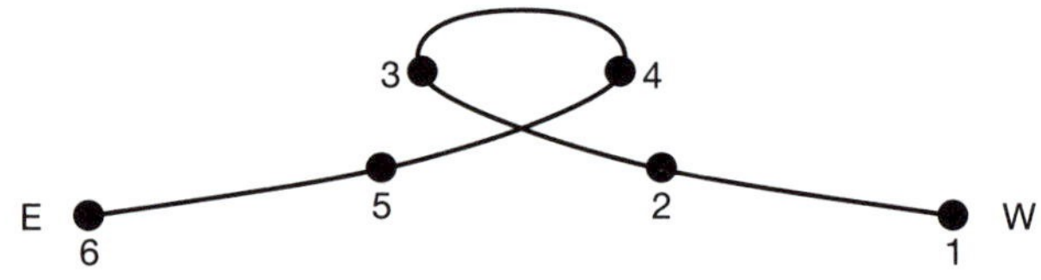

Figure 21.1

3. Why did the early Greeks reject the idea of a rotating Earth?

4. For each of the following statements, list the name of the scientist credited with the contribution.

   a) First to use the telescope for astronomy:

   b) Proposed circular orbits of the planets around the Sun in 1500s:

   c) Systematically measured the locations of heavenly bodies:

   d) Proposed three laws of planetary motion:

   e) Monumental work, *De Revolutionibus*:

   f) Proposed elliptical orbits for planets around the Sun:

   g) Discovered four satellites, or moons, orbiting Jupiter:

h) Wrote *Dialogue of the Great World Systems:*

i) Law of universal gravitation:

j) The first Greek to profess a heliocentric universe:

k) An early Greek astronomer best known for his star catalog:

l) Credited with the first successful attempt to establish the size of Earth:

5. Describe stellar parallax.

6. List each of Kepler's three laws of planetary motion.

 1)

 2)

 3)

7. List four of Galileo's astronomical discoveries.

 1)

 2)

 3)

 4)

8. Use Figure 21.2, a diagram of the orbit of a hypothetical planet, to answer the following questions.

 a) During which months is the planet moving fastest in its orbit?

 b) During which months is the planet moving slowest in its orbit?

 c) In addition to inertia, what force is keeping the planet in orbit?

 d) If the planet's mean distance from the Sun is 4 AUs, what would be its orbital period?

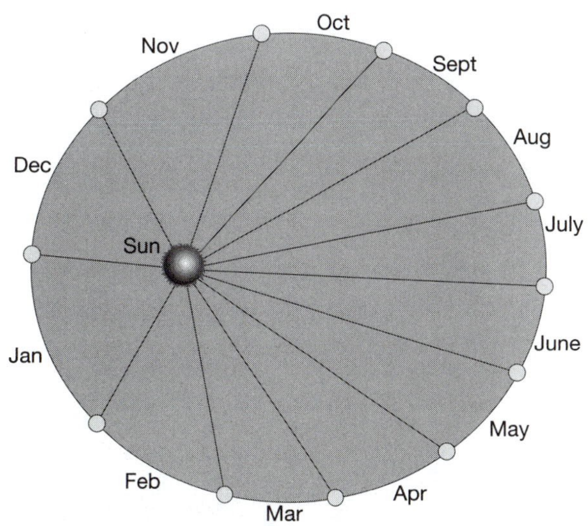

Figure 21.2

9. Briefly explain the equatorial system used for locating objects on the celestial sphere.

10. Using Figure 21.3, select the letter that identifies each of the following components of the celestial sphere.

    a) Celestial equator:

    b) North celestial pole:

    c) Declination of star G:

    d) Right ascension of star G:

11. List six motions that Earth is continuously experiencing.

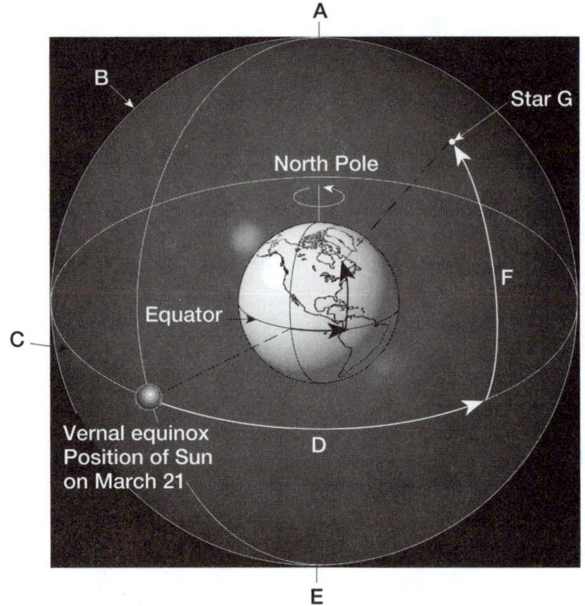

Figure 21.3

12. Describe the difference between the Moon's synodic and sidereal cycles.

13. Using Figure 21.4, select the number that illustrates the Moon's position in its orbit for each of the following phases.

    a) Full:

    b) Third quarter:

    c) Crescent (waxing):

    d) New:

    e) Gibbous (waning):

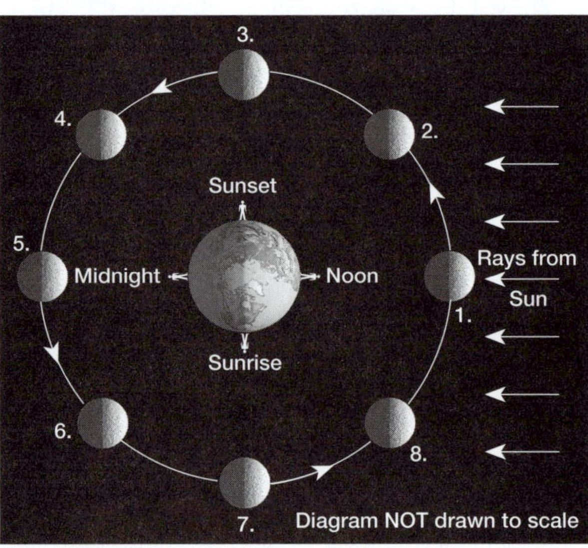

Figure 21.4

Practice Test

Multiple choice. Choose the best answer for the following multiple-choice questions.

1. Who was the ancient Greek that developed a geocentric model of the universe explaining the observable motions of the planets?
   a) Aristotle
   b) Ptolemy
   c) Copernicus
   d) Kepler
   e) Newton

2. During the period when the Moon's phases are changing from new to full, it is __________.
   a) waning
   b) approaching Earth
   c) waxing
   d) receding from Earth
   e) exhibiting retrograde motion

3. The apparent path of the Sun against the backdrop of the celestial sphere is called __________.
   a) declination
   b) perturbation
   c) the ecliptic
   d) right ascension
   e) the precession

4. One astronomical unit averages about __________.
   a) 93 million miles
   b) 210 million kilometers
   c) 39 million miles
   d) 93 million kilometers
   e) 150 million miles

5. At its closest position to the Sun, called __________, Earth is 147 million kilometers (91 million miles) distant.
   a) aphelion
   b) quadrature
   c) perihelion
   d) declination
   e) ascension

6. Ancient astronomers believed that __________.
   a) Earth revolved around the celestial sphere
   b) the Sun was the center of the universe
   c) the Sun was on the celestial sphere
   d) Earth was a "wanderer"
   e) Earth was the center of the universe

7. The __________ of an object is a measure of the total amount of matter it contains.
   a) density
   b) mass
   c) volume
   d) size
   e) velocity

8. It takes the Moon about __________ weeks to go from full-Moon phase to new-Moon phase.
   a) one
   b) two
   c) three
   d) four
   e) five

9. The apparent westward movement of a planet against the background of stars is called __________.
   a) retrograde motion
   b) reverse motion
   c) westward drift
   d) Ptolemaic motion
   e) deferent motion

10. The most visible stars in a constellation are generally named in order of their __________ by the letters of the Greek alphabet.
    a) color
    b) temperature
    c) size
    d) mass
    e) brightness

11. __________ refers to the fact that the direction in which Earth's axis points continuously changes.
    a) Perturbation
    b) Axial precession
    c) Percolation
    d) Progradation
    e) Progression

12. Which one of the following planets was unknown to the ancient Greeks?
    a) Earth
    b) Mars
    c) Uranus
    d) Mercury
    e) Venus

13. Earth moves forward in its orbit about __________ kilometers each second.
    a) 11
    b) 16
    c) 25
    d) 30
    e) 43

14. Using Tycho Brahe's data, which scientist proposed three laws of planetary motion?
    a) Newton
    b) Galileo
    c) Kepler
    d) Copernicus
    e) Ptolemy

15. A(n) __________ eclipse occurs when the Moon casts its shadow on Earth.
    a) lunar
    b) umbral
    c) sidereal
    d) solar
    e) synodic

16. The first accurate measurement of the size of Earth was made by __________.
    a) Aristotle
    b) Eratosthenes
    c) Aristarchus
    d) Ptolemy
    e) Newton

17. Another name for the North Star is __________.
    a) Vega
    b) Altair
    c) Andromeda
    d) Antares
    e) Polaris

18. The first modern astronomer to propose a Sun-centered solar system was __________.
    a) Galileo
    b) Newton
    c) Ptolemy
    d) Copernicus
    e) Brahe

19. The variance in the orbit of a planet from its predicted path is called __________.
    a) perturbation
    b) precession
    c) percolation
    d) progradation
    e) progression

20. Which scientist discovered that Venus has phases, just like the Moon?
    a) Ptolemy
    b) Copernicus
    c) Galileo
    d) Kepler
    e) Newton

21. Thirty degrees of right ascension is equivalent to __________ minutes.
    a) 15
    b) 30
    c) 45
    d) 60
    e) 120

22. The apparent shift in the position of a star caused by Earth's motion is called __________.
    a) stellar shift
    b) retrograde motion
    c) an epicycle
    d) stellar parallax
    e) revolution

23. The force that gravity exerts on an object is referred to as __________.
    a) mass
    b) pressure
    c) perturbation
    d) weight
    e) density

24. The ancient Greek, __________, reasoned that the Moon shines by reflected sunlight.
    a) Aristotle
    b) Anaxagoras
    c) Aristarchus
    d) Artemis
    e) Almagest

25. The shape of a planet's orbit is __________.
    a) circular
    b) irregular
    c) constantly changing
    d) elliptical
    e) parabolic

26. According to the ancients, the stars traveled around Earth on the transparent, hollow __________.
    a) deferent
    b) celestial sphere
    c) equatorial sphere
    d) solar orb
    e) stellar globe

27. Which of the following is correctly paired with its analogous measurement on Earth?
    a) right ascension-declination
    b) right ascension-latitude
    c) declination-latitude
    d) declination-right ascension
    e) declination-longitude

28. A day measured by using the stars rather than the Sun is called a(n) __________ day.
    a) sidereal
    b) mean solar
    c) synodic
    d) ecliptic
    e) average

29. According to Kepler's third law, the period of revolution of a planet is related to the planet's __________.
    a) distance to the Sun
    b) size
    c) mass
    d) acceleration curve
    e) gravitational attraction

30. The period of revolution of the Moon around Earth is $27^1/_3$ days, while the period of rotation of the Moon about its axis is __________ days.
    a) $27^1/_3$
    b) 28
    c) $29^1/_2$
    d) $30^1/_3$
    e) $33^1/_2$

True/false. For the following true/false questions, if a statement is not completely true, mark it false. For each false statement, change the **italicized** *word to correct the statement.*

1. _____ The orbits of the planets are *circular.*
2. _____ The early Greeks believed that there were *nine* wandering heavenly bodies.
3. _____ Probably the greatest of the early Greek astronomers was *Tycho Brahe*, best known for his star catalog.
4. _____ The farther away a star is, the *greater* will be its parallax.
5. _____ The cycle of the Moon through its phases as seen from Earth is called the *sidereal* month.
6. _____ The force of gravity is *inversely* proportional to the square of the distance between two objects.
7. _____ The mean solar day is *shorter* than the sidereal day.
8. _____ Total solar eclipses are only observed by people in the dark part of the Moon's shadow, called the *umbra.*
9. _____ A *geocentric* model holds that Earth is the center of the universe.
10. _____ Any variance in the orbit of a body from its predicted path is called *precession.*
11. _____ Gravity is directly proportional to *mass.*
12. _____ The orbital periods of the planets and their *distances* to the Sun are proportional.
13. _____ A *lunar* eclipse occurs when Earth's shadow is cast on the Moon.

14. _____ Our present calendar is accurate to within 1 *second* in 3000 years.
15. _____ Lunar eclipses occur during *full-Moon* phase.
16. _____ Earth is the center of the universe in a *heliocentric* model.
17. _____ During a new moon, the lunar hemisphere that faces Earth is the *opposite* hemisphere that faces Earth during full moon.
18. _____ In this model, Copernicus used *circles* to represent the orbits of the planets.
19. _____ The declination of a star is usually expressed in *hours*.
20. _____ Beginning in 1609, *Galileo Galilei* published his three laws of planetary motion.

Word choice. Complete each of the following statements by selecting the most appropriate response.

1. The actual time it takes for the Moon to complete one 360° revolution around Earth is called the [sidereal/synodic] month.
2. Most of the time planets move [eastward/westward] among the stars.
3. A [lunar/solar] eclipse occurs when the Moon casts its shadow on Earth.
4. The [nautical/astronomical] unit is equal to Earth's mean distance to the Sun, about 150 million kilometers (93 million miles).
5. The mass of an object is [constant/variable], but its weight [remains constant/varies] when gravitational forces change.
6. The synodic cycle of the Moon is [shorter/longer] than the sidereal cycle.
7. The calendar that we currently use is called the [Julian/Gregorian] calendar.
8. The belief that Earth is the center of the universe is a [geocentric/heliocentric] view.
9. Early Greeks believed that the Moon was illuminated by [reflected sunlight/internal fires].
10. The concept that once an object is in motion it will continue traveling at a uniform speed in a straight line is called [inertia/momentum].
11. A total [solar/lunar] eclipse is visible to anyone on the side of Earth facing the Moon.
12. Earth is closest to the Sun at [aphelion/perihelion].
13. The Ptolemaic model of the universe states that the planets move in [circular/elliptical] orbits around [Earth, the Sun].
14. [Aristotle/Aristarchus] believed that the universe was centered on the Sun.
15. The Moon's period of rotation is [more than/equal to/less than] its period of revolution around Earth.
16. All stars in a constellation [are/are not] located at the same distance from Earth.
17. The plane of the ecliptic is tilted relative to the celestial equator due to the inclination of Earth's [axis/orbit].
18. Solar eclipses occur during the [new-Moon/full-Moon] phase.
19. In the 1630s [Kepler/Galileo/Newton] was tried by the Inquisition and convicted of proclaiming doctrines contrary to religious beliefs.
20. A [quadrant/pendulum] can be used to prove that Earth rotates.

Written questions

1. Explain why planets appear to exhibit retrograde motion.

2. Briefly explain why eclipses don't occur every month.

3. List and describe four motions that Earth continuously experiences.

4. Describe the model of the universe proposed by Ptolemy in A.D. 141.

For other interesting and pertinent information, be sure to visit the *Earth Science* companion website at

http://www.prenhall.com/tarbuck

22

Touring Our Solar System

The Sun, with 99.85 percent of the mass of our solar system, is the dominant gravitational force within a rotating system of nine planets, their satellites, and numerous small asteroids, comets, and meteoroids. Although each planet has many distinguishing characteristics, collectively they fall nicely into two distinct groups, the Earth-like, or terrestrial planets (Mercury, Venus, Earth, and Mars) and the Jupiter-like (Jovian) planets (Jupiter, Saturn, Uranus, and Neptune). Following an overview of the planets and the Moon, Chapter twenty-two examines the unique features of the planets and lesser members of the solar system, emphasizing the new information obtained from recent planetary space probes.

Learning Objectives

After reading, studying, and discussing this chapter, you should be able to:

Describe the general characteristics of the two groups of planets in the solar system.

The planets can be arranged into two groups: the *terrestrial* (Earth-like) *planets* (Mercury, Venus, Earth, and Mars) and the *Jovian* (Jupiter-like) *planets* (Jupiter, Saturn, Uranus, and Neptune). When compared to the Jovian planets, the terrestrial planets are smaller, more dense, contain proportionally more rocky material, have slower rates of rotation, and meager atmospheres.

Describe the major features of the lunar surface and discuss the Moon's history.

The lunar surface exhibits several types of features. Most *craters* were produced by the impact of rapidly moving interplanetary debris (*meteoroids*). Bright, densely cratered *highlands* (*terrae*) make up most of the lunar surface. The dark, fairly smooth lowlands are called *maria* (singular, *mare*). Maria basins are enormous impact craters that have been flooded with layer upon layer of very fluid basaltic lava. All lunar terrains are mantled with a soil-like layer of gray, unconsolidated debris, called *lunar regolith*, which has been derived from a few billion years of meteoric bombardment. One hypothesis for the Moon's origin suggests that a Mars-sized object collided with Earth to produce the Moon. Scientists conclude that *the Moon evolved in four phases:* (1) *the original crust*, (2) *lunar highlands,* (3) *maria basins*, and (4) *youthful rayed craters*.

List the distinguishing features of each planet in the solar system.

Mercury is a small, dense planet that has no atmosphere and exhibits the greatest temperature extremes of any planet. *Venus*, the brightest planet in the sky, has a thick, heavy atmosphere composed of 97 percent carbon dioxide, a surface of relatively subdued plains and inactive volcanic features, a surface atmospheric pressure 90 times that of Earth's, and surface temperatures of 475°C (900°F). *Mars*, the "Red Planet," has a carbon dioxide atmosphere only 1 percent as dense as Earth's, extensive dust storms, numerous inactive volcanoes, many large canyons, and several valleys of debatable origin exhibiting drainage patterns similar to stream valleys on Earth. *Jupiter*, the largest planet, rotates rapidly, has a banded appearance caused by huge convection currents driven by the planet's interior heat, a *Great Red Spot* that varies in size, a thin ring system, and at least 16 moons (one of the moons, *Io*, is a volcanically active body). *Saturn* is best known for its system of rings. It also has a dynamic atmosphere with winds up to 930 miles per hour and "storms" similar to Jupiter's Great Red Spot. *Uranus* and *Neptune* are often called "the twins" because of similar structure and composition. A unique feature of Uranus is the fact that it rotates "on its side." Neptune has white cirruslike clouds above its main cloud deck and an Earth-size *Great Dark Spot*, assumed to be a large rotating storm similar to Jupiter's Great Red Spot.

List and describe the minor members of the solar system.

The minor members of the solar system include the *asteroids, comets, meteoroids,* and *dwarf planets*. Most asteroids lie between the orbits of Mars and Jupiter. No conclusive evidence has been found to explain their origin. Comets are made of frozen gases (water, ammonia, methane, carbon dioxide, and carbon monoxide) with small pieces of rocky and metallic material. Many travel in very elongated orbits that carry them beyond Pluto, and little is known about their origin. Meteoroids, small solid particles that travel through interplanetary space, become *meteors* when they enter Earth's atmosphere and vaporize with a flash of light. *Meteor showers* occur when Earth encounters a swarm of meteoroids, probably material lost by a comet. *Meteorites* are the remains of meteoroids found on Earth. Recently, Pluto was placed into a new class of solar system objects called dwarf planets.

Key Terms

asteroid
asteroid belt
coma
comet
cryo volcanism
dwarf planet
escape velocity
impact craters
inner planet
Jovian planet
Kuiper belt
lunar highlands
lunar regolith
maria
meteor
meteorite
meteoroid
meteor shower
Oört cloud
outer planets
planetesimals
protoplanets
solar nebula
terrae
terrestrial planet

Vocabulary Review

Choosing from the list of key terms, furnish the most appropriate response for the following statements.

1. The glowing head of a comet is called the __________.
2. A(n) __________ is a planet that has physical characteristics similar to those of Earth.
3. A(n) __________ is a small, planetlike body whose orbit lies mainly between Mars and Jupiter.
4. The dark regions on the Moon that resemble "seas" on Earth are __________.
5. A(n) __________ is a small solid particle that travels through interplanetary space.
6. Most short period comets originate in a region called the __________.
7. The physical characteristics of a(n) __________ are like those of Jupiter.
8. Long period comets form a spherical shell around the solar system, called the __________.
9. __________ is the soil-like layer of gray, unconsolidated debris that mantles all lunar terrains.
10. A(n) __________ is a luminous phenomenon observed when a small solid particle enters Earth's atmosphere and burns up.
11. The initial speed that an object needs before it can leave a planet and go into space is known as the __________.
12. In a nebula, the clumps of material that collide and gradually grow into asteroid-sized objects are called __________.
13. A small body made of frozen gases and small pieces of rocky and metallic materials that generally revolves about the Sun in an elongated orbit is a(n) __________.
14. The remains of a meteoroid, when found on Earth, is referred to as a(n) __________.

15. A spectacular display of numerous meteor sightings that occurs when Earth encounters a swarm of meteoroids is called a(n) __________.

16. The eruption of ices, such as that observed on Triton and other moons of the Jovian planets is called __________.

17. A meteorite that is a mixture of iron, nickel, and silicate minerals is called a(n) __________.

18. The bright areas on the Moon are called __________ (the latin word for lands).

Comprehensive Review

1. Which celestial objects constitute the solar system?

2. What are the gas, rock, and ice substances that compose the majority of the solar system?

   a) Gases:

   b) Rocks:

   c) Ices:

3. Describe the following two groups of planets, and list the planets that are included in each group.

   a) Terrestrial planets:

   b) Jovian planets:

4. For each of the following characteristics, write a brief statement that compares the terrestrial planets to the Jovian planets.

   a) Size:

   b) Density:

   c) Period of rotation:

   d) Mass:

   e) Number of known satellites:

   f) Period of revolution:

   g) General composition:

5. Why is it more difficult for gases to escape from the Jovian planets than from the terrestrial planets?

6. Using the photograph of the Moon, Figure 22.1, select the letter that identifies each of the following lunar features.

    a) Mare: _____

    b) Crater: _____

    c) Ray(s): _____

    d) Highland: _____

    e) Oldest feature: _____

Figure 22.1

7. Why would a 100-pound person weigh only 17 pounds on the Moon?

8. Describe the origin of lunar maria.

9. What is the most widely accepted model for the origin of the Moon?

10. By placing a number in front of the event, arrange the following lunar events in correct order from oldest (1) to most recent (5).

    a) _____ Crust solidifies

    b) _____ Debris gathers in space

    c) _____ Formation of rayed craters

    d) _____ Maria formation

    e) _____ Bombardment melts the outer shell and possibly the interior

11. List the name of the planet described by each of the following statements.

    a) Red planet:

    b) Thick clouds, dense carbon dioxide atmosphere, volcanic surface:

    c) Greatest diameter:

d) Has only satellite (Titan) with substantial atmosphere:

e) Hot surface temperature and no atmosphere:

f) Cassini's division:

g) Second only to the Moon in brilliance in the night sky:

h) Axis of rotation lies near the plane of its orbit:

i) Great Red Spot:

j) Moon (Triton) has lowest surface temperature in solar system:

k) Most prominent system of rings:

l) Carbon dioxide atmosphere 1 percent as dense as Earth's, volcanic surface:

m) Moons include Io, Callisto, Ganymede, and Europa:

n) Earth's "twin":

o) Great Dark Spot:

p) Revolves quickly but rotates slowly:

q) Longest period of revolution:

12. What is the most probable origin of the Martian moons?

13. Write the name of the planet to which each of the following moons belong.

a) Callisto:

b) Deimos:

c) Io:

d) Triton:

e) Miranda:

f) Charon:

g) Titan:

14. What is the theory for the origin of planetary ring particles?

15. What are the two hypotheses for the origin of the asteroids?

    1)

    2)

16. Using Figure 22.2, list the letter that identifies each of the following comet parts.

    a) Ionized gas tail: _____

    b) Coma: _____

    c) Nucleus: _____

    d) Dust tail: _____

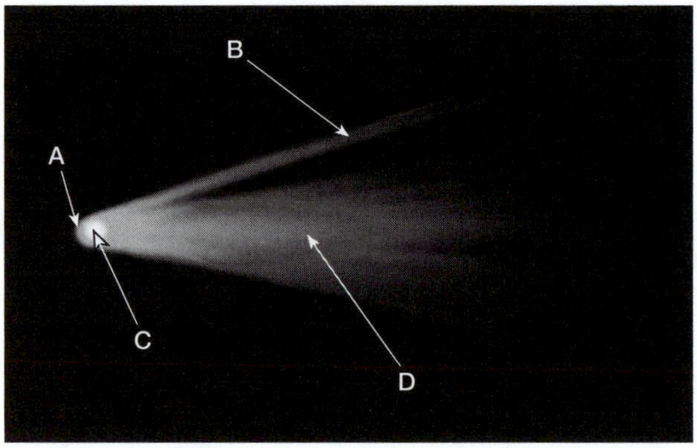

Figure 22.2

17. What are the two forces that contribute to the formation of a comet's tail?

    1)

    2)

18. What is the difference between a meteor and a meteorite?

19. List and briefly describe the three most common types of meteorites based upon composition.

    1)

    2)

    3)

Practice Test

Multiple choice. Choose the best answer for the following multiple-choice questions.

1. Which one of the following is proportionally more abundant on terrestrial planets than on Jovian planets?
   a) hydrogen
   b) silicate minerals
   c) methane
   d) ammonia ice
   e) helium

2. Which one of the following is NOT a terrestrial planet?
   a) Mercury
   b) Earth
   c) Mars
   d) Venus
   e) Jupiter

3. The fact that Mercury is very dense implies that it contains a(n) ___________.
   a) asthenosphere
   b) liquid interior
   c) magma reservoir
   d) iron core
   e) highly volatile surface

4. The densely cratered lunar highlands have been estimated to be about ___________.
   a) the same age as Earth
   b) 200 million years old
   c) much older than Earth
   d) 2.6 billion years old
   e) 360 million years old

5. The large dark regions on the Moon are ___________.
   a) highlands
   b) craters
   c) mountains
   d) maria
   e) rays

6. The most important agents currently modifying the Moon's surface are ___________.
   a) moonquakes
   b) volcanoes
   c) faults
   d) wind storms
   e) micrometeorites

7. Which planet has a mass greater than the combined mass of all the remaining planets, satellites, and asteroids?
   a) Venus
   b) Mars
   c) Jupiter
   d) Uranus
   e) Saturn

8. The oldest lunar features are ___________.
   a) highlands
   b) rayed craters
   c) ejecta
   d) maria
   e) regolith deposits

9. Galileo is credited with discovering ___________ of Jupiter's moons.
   a) four
   b) six
   c) seven
   d) nine
   e) eleven

10. Which one of the following is NOT a material commonly found in lunar regolith?
    a) glass beads
    b) clay
    c) lunar dust
    d) igneous rocks

11. The most obvious difference between the terrestrial and the Jovian planets is their ___________.
    a) color
    b) size
    c) composition
    d) length of day
    e) orbital velocity

12. Which planet experiences high surface temperatures as a consequence of its dense carbon dioxide atmosphere?
    a) Earth
    b) Mars
    c) Venus
    d) Mercury
    e) Uranus

13. Which one of Jupiter's moons is volcanically active?
    a) Callisto
    b) Europa
    c) Ganymede
    d) Io

14. Which features on Earth offer clear evidence that comets and asteroids have struck its surface?
    a) rayed craters
    b) terra
    c) astroblemes
    d) volcanoes
    e) giant faults

15. Which chemical is responsible for the greenish-blue color of Uranus and Neptune?
    a) hydrogen sulfide
    b) methane
    c) ammonia
    d) hydrogen chloride
    e) carbon dioxide

16. Which planet has a density less than that of water?
    a) Mercury
    b) Venus
    c) Jupiter
    d) Saturn
    e) Neptune

17. Which planet might best be described as a large, dirty iceball?
    a) Mercury
    b) Venus
    c) Mars
    d) Saturn
    e) Pluto

18. The Martian polar caps are made of __________, covered by a thin layer of frozen carbon dioxide.
    a) silicate minerals
    b) ammonia ice
    c) volcanic rocks
    d) water ice
    e) frozen methane

19. Although this planet is shrouded in thick clouds, radar mapping has revealed a varied topography consisting of subdued plains, highlands, and thousands of volcanic structures.
    a) Mercury
    b) Venus
    c) Mars
    d) Jupiter
    e) Uranus

20. Which one of the following is currently responsible for shaping the surface of Mars?
    a) marsquakes
    b) water
    c) wind
    d) tectonics
    e) volcanoes

21. Which moon is the only satellite in the solar system with a substantial atmosphere?
    a) Io
    b) Triton
    c) Miranda
    d) Titan
    e) Europa

22. The lowest surface temperature in the solar system (–200°C) occurs on __________.
    a) Titan
    b) Pluto
    c) Neptune
    d) Triton
    e) Callisto

23. The Moon's density is comparable to that of Earth's __________.
    a) core
    b) atmosphere
    c) crustal rocks
    d) asthenosphere
    e) lower mantle

24. Which planet's atmosphere contains the Great Dark Spot?
    a) Venus
    b) Jupiter
    c) Saturn
    d) Uranus
    e) Neptune

25. Which feature(s) on Mars have raised the question about the possibility of liquid water on the planet?
    a) mountain ranges with faults
    b) impact craters with sharp rims
    c) volcanic cones with craters

d) deep, long canyons
e) valleys with tributaries

26. The Jovian planets contain a large percentage of the gases ___________.

a) nitrogen and argon
b) hydrogen and oxygen
c) oxygen and nitrogen
d) helium and oxygen
e) hydrogen and helium

27. Which one of the following is NOT included in the solar system?

a) comets
b) galaxies
c) Sun
d) planets
e) asteroids

28. Second only to the Moon in brilliance in the night sky is ___________.

a) Mercury
b) Venus
c) Mars
d) Jupiter
e) Saturn

29. Which planet has the most eccentric orbit?

a) Mercury
b) Pluto
c) Earth
d) Uranus
e) Mars

30. Although the atmosphere of this planet is very thin, extensive dust storms with wind speeds in excess of 150 miles per hour do occur.

a) Mercury
b) Venus
c) Mars
d) Saturn
e) Uranus

31. Which planet's axis lies only 8 degrees from the plane of its orbit?

a) Mercury
b) Mars
c) Jupiter
d) Saturn
e) Uranus

32. Which planet completes one rotation in slightly less than 10 hours?

a) Mercury
b) Venus
c) Mars
d) Jupiter
e) Saturn

33. Relatively young lunar craters often exhibit ___________.

a) rays
b) maria
c) eroded rims
d) highlands
e) active volcanoes

34. Which one of the following planets does NOT have rings?

a) Mars
b) Jupiter
c) Saturn
d) Uranus
e) Neptune

35. Through the telescope, this planet appears as a reddish ball interrupted by some permanent dark regions that change intensity.

a) Mercury
b) Venus
c) Mars
d) Saturn
e) Uranus

36. Most asteroids lie between the orbits of ___________.

a) Mercury and Venus
b) Venus and Earth
c) Earth and Mars
d) Mars and Jupiter
e) Jupiter and Saturn

37. Most meteor showers are associated with the orbits of ___________.

a) satellites
b) comets
c) planets
d) asteroids
e) meteorites

True/false. For the following true/false questions, if a statement is not completely true, mark it false. For each false statement, change the **italicized** *word to correct the statement.*

1. _____ The orbital planes of seven planets lie within *three* degrees of the plane of the Sun's equator.
2. _____ The most prominent feature of *Pluto* is its ring system.
3. _____ Compared to other planet-satellite systems, Earth's Moon is unusually *large*.
4. _____ Most of the volcanoes on Venus are small *shield* cones.
5. _____ *Earth* has the greatest temperature extremes of any planet.
6. _____ Most asteroids orbit the Sun between the orbits of Mars and *Earth*.
7. _____ The Jovian planets have *high* densities and large masses.
8. _____ Bombardment of the Moon by *micrometeorites* gradually smooths the landscape.
9. _____ Maria basins are filled with very fluid basaltic *ejecta*.
10. _____ The Martian atmosphere is *50* percent as dense as that of Earth.
11. _____ Most of the lunar surface is made up of *maria*.
12. _____ Comets typically *gain* material with every pass by the Sun.
13. _____ Over 90 percent of the mass of the solar system is contained within the *planets*.
14. _____ The most striking feature on the face of *Mars* is the Great Red Spot.
15. _____ The escape velocity for Earth is *seven* miles per second.
16. _____ The formation of rings is believed be related to the *gravitational* force of a planet.
17. _____ The *surface* of Venus reaches temperatures of 475°C (900°F).
18. _____ The *retrograde* motion of Triton suggests that it formed independently of Neptune.
19. _____ In 1978 the moon Charon was discovered orbiting *Pluto*.
20. _____ The *Voyager 2* spacecraft has surveyed the greatest number of planets, including Neptune in 1989.
21. _____ On July 4, 1997, the *Mars Pathfinder* landed on the surface of Mars and deployed its wheeled companion, Sojourner.

Word choice. Complete each of the following statements by selecting the most appropriate response.

1. The tail of a comet always points [toward/away from] the Sun.
2. The Moon's density is considerably [more/less] than Earth's average density, and its gravity is approximately one-[quarter/sixth/eighth] of Earth's.
3. The processes that formed the very large valleys on Mars would be comparable to those that created [glacial/rift] valleys on Earth.
4. Because of its high temperature and small size, the planet [Mercury/Venus/Mars] was unable to retain an atmosphere during its formation.
5. Terrestrial bodies with no atmosphere [reflect/absorb] most of the sunlight that strikes them.
6. The process of plate tectonics [does/does not] appear to have contributed to the present Venusian topography.
7. Jupiter is covered with alternating bands of multicolored clouds that are [parallel/perpendicular] to its equator.
8. The planets orbit the Sun in the [same/different] direction(s).

9. A bright streak of light caused by a solid particle entering Earth's atmosphere is called a [meteor/meteorite].
10. The most prominent features of Mars visible through a telescope are the planet's [canals/polar caps].
11. The gradual smoothing of the lunar landscape is caused by [tectonic forces/micrometeorite impacts].
12. Most asteroids are about [1/10/100] kilometer(s) across.
13. The thickness of Saturn's ring system is approximately a few [hundred/thousand] meters.
14. On the Moon, greater crater density indicates that an area is [older/younger] than an area with lower crater density.
15. Compared to features on Earth, most Martian features are [old/young].
16. Earth and [Mercury/Venus/Mars] are often referred to as "twins."
17. Most comets are believed to orbit the Sun [within/beyond] the orbit of Pluto.
18. The most prominent feature of Venus visible through a telescope is the planet's [barren surface/cloud cover].
19. Currently Pluto is [closer to/farther from] the Sun than Neptune.
20. Mercury rotates [slowly/rapidly] and revolves [slowly/quickly].
21. In 2004 the *Mars Opportunity* rover discovered [iron-rich/salt-laden] sediments on Mars.

Written questions

1. Compare the physical characteristics of the terrestrial planets to those of the Jovian planets.

2. Distinguish between a meteoroid, meteor, and meteorite.

3. Briefly compare the two different types of lunar terrain.

For other interesting and pertinent
information, be sure to visit
the *Earth Science* companion website at

http://www.prenhall.com/tarbuck

23

Light, Astronomical Observations, and the Sun

Almost everything astronomers know about the universe beyond our solar system comes from the analysis of light and other forms of electromagnetic radiation from distant sources. Consequently, an understanding of the nature of electromagnetic radiation is basic to modern astronomy. This chapter begins with the study of light and how it is used to gather information concerning the state of matter, composition, temperature, and motion of stars and other celestial objects. After investigating light, the focus shifts to the tools and instruments astronomers use to gather and analyze electromagnetic radiation emitted by distant objects in the universe. In addition, the nearest source of light, the Sun, is also discussed in detail. By understanding the processes that operate on and within the Sun, astronomers can better grasp the nature of more distant celestial objects.

Learning Objectives

After reading, studying, and discussing this chapter, you should be able to:

Describe electromagnetic radiation and the two models used to explain its properties.

Visible light constitutes only a small part of an array of energy generally referred to as *electromagnetic radiation*. Light, a type of electromagnetic radiation, can be described in two ways (1) as waves and (2) as a stream of particles, called *photons*. The wavelengths of electromagnetic radiation vary from several kilometers for *radio waves* to less than a billionth of a centimeter for *gamma rays*. The shorter wavelengths correspond to more energetic photons.

List and describe the three types of light spectra and how they can be used to investigate the properties of a star.

Spectroscopy is the study of the properties of light that depend on wavelength. When a prism is used to disperse visible light into its component parts (wavelengths), one of three possible types of *spectra* (a *spectrum*, the singular form of spectra, is the light pattern produced by passing light through a prism) is produced. The three types of spectra are (1) *continuous spectrum*, (2) *dark-line (absorption) spectrum*, and (3) *bright-line (emission) spectrum*. The spectra of most stars are of the dark-line type. Spectroscopy can be used to determine (1) the composition of stars and other gaseous objects, (2) the temperature of a radiating body, and (3) the motion of an object. Motion (direction toward or away and velocity) is determined using the *Doppler effect*—the apparent change in the wavelength of radiation emitted by an object caused by the relative motions of the source and the observer.

Describe the two types of optical telescopes and list their component parts.

There are two types of optical telescopes: (1) the *refracting telescope*, which uses a *lens* to bend or refract light, and (2) the *reflecting telescope*, which uses a *concave mirror* to focus (gather) the light.

Describe how the properties of optical telescopes aid astronomers in their work.

Telescopes simply collect light. They become useful when the collected light is detected and analyzed. Historically, astronomers relied on their eyes as detectors. Then, photographic film was developed which was a revolutionary advancemnt. Presently, light is collected using a *charged coupled device* (CCD). A CCD camera produces a "digital image" akin to that of a digital camera.

Describe radio telescopes.

The detection of *radio waves* is accomplished by "big dishes" known as *radio telescopes*. A parabolic-shaped dish, often consisting of wire mesh, operates in a manner similar to the mirror of a reflecting telescope. Of great importance is a narrow band of radio waves that is able to penetrate Earth's atmosphere. Because this radiation is produced by neutral hydrogen, it has permitted us to map the galactic distribution of the material from which stars are made.

List and describe the four parts of the Sun.

The Sun is one of the 200 billion stars that make up the Milky Way Galaxy. The Sun can be divided into four parts: (1) the *solar interior*, (2) the *photosphere* (visible surface) and the two layers of its atmosphere, (3) the *chromosphere*, and (4) the *corona*. The *photosphere* radiates most of the light we see. Unlike most surfaces, it consists of a layer of incandescent gas less than 500 kilometers (300 miles) thick with a grainy texture consisting of numerous, relatively small, bright markings called *granules*. Just above the photosphere lies the *chromosphere*, a relatively thin layer of hot, incandescent gases a few thousand kilometers thick. At the edge of the uppermost portion of the solar atmosphere, called the *corona*, ionized gases escape the gravitational pull of the Sun and stream toward Earth at high speeds, producing the *solar wind*.

Describe several features found on the active Sun.

Numerous features have been identified on the active Sun. *Sunspots* are dark blemishes with a black center, the *umbra*, which is rimmed by a lighter region, the *penumbra*. The number of sunspots observable on the solar disk varies in an 11-year cycle. *Plages* are large "clouds" that appears as bright centers of solar activity often directly above sunspot clusters. *Prominences*, huge cloudlike structures best observed when they are on the edge, or limb, of the Sun, are produced by ionized chromospheric gases trapped by magnetic fields that extend from regions of intense solar activity. The most explosive events associated with sunspots are *solar flares*. Flares are brief outbursts that release enormous quantities of energy that appear as a sudden brightening of the region above sunspot clusters. During the event, radiation and fast-moving atomic particles are ejected, causing the solar wind to intensify. When the ejected particles reach Earth and disturb the ionosphere, radio communication is disrupted and the *auroras*, also called the Northern and Southern Lights, occur.

Describe the source of the Sun's energy.

The source of the Sun's energy is *nuclear fusion*. Deep in the solar interior, at a temperature of 15 million K, a nuclear reaction called the *proton-proton chain* converts four hydrogen nuclei (protons) into the nucleus of a helium atom. During the reaction some of the matter is converted to the energy of the Sun. A star the size of the Sun can exist in its present stable state for 10 billion years. Since the Sun is already 5 billion years old, it is a "middle-aged" star.

Key Terms

aurora
bright-line (emission) spectrum
chromatic aberration
chromosphere
continuous spectrum
corona
dark-line (absorption) spectrum
Doppler effect
electromagnetic radiation
granules
nuclear fusion
photon
photosphere
prominence
proton-proton chain
radiation pressure
radio interferometer
radio telescope
reflecting telescope
refracting telescope
solar flare
solar wind
spectroscope
spectroscopy
spicule
sunspot

Vocabulary Review

Choosing from the list of key terms, furnish the most appropriate response for the following statements.

1. The array of energy that consists of radiations of electromagnetic waves is referred to as __________.
2. The very tenuous, outermost portion of the solar atmosphere is called the __________.
3. A(n) __________ uses a lens to bend and concentrate the light from distant objects.
4. The __________ is an instrument for directly viewing the spectrum of a light source.
5. The visible surface region of the Sun that radiates energy to space is called the __________.
6. The source of the Sun's energy is __________.
7. A(n) __________ is the type of spectrum produced by an incandescent solid, liquid, or gas under high pressure.
8. The __________ concentrates light from distant objects by using a concave mirror.
9. A(n) __________ is a discrete amount (quantum) of electromagnetic energy that moves as a unit with the velocity of light.
10. A(n) __________ is designed to make observations using the radio wavelengths of electromagnetic radiation.
11. The apparent change in wavelength of radiation caused by the relative motions of the source and the observer is called the __________.
12. The streams of protons and electrons that "boil" from the Sun's corona constitute the __________.
13. Deep in the Sun's interior, a nuclear reaction called the __________ converts four hydrogen nuclei (protons) into the nucleus of a helium atom.
14. A sudden brightening of an area on the Sun that normally lasts an hour or so but releases enormous quantities of energy is called a(n) __________.
15. A(n) __________ is the type of spectrum produced by a hot (incandescent) gas under low pressure.
16. The Sun's __________ is a relatively thin layer of hot, incandescent gases a few thousand kilometers thick that lies just above the photosphere.
17. A solar __________ is a huge cloudlike concentration of material that appears as a bright arch-like structure above the solar surface.
18. __________ is the study of the properties of light that depend on wavelength.
19. A conspicuous dark blemish on the surface of the Sun is called a(n) __________.
20. A(n) __________ is the type of spectrum produced when white light is passed through a comparatively cool gas under low pressure.
21. __________ is the property of a lens whereby light of different colors is focused at different places.
22. The pressure ("push") on matter exerted by photons is called __________.
23. The numerous, relatively small, bright markings on the Sun's surface are called __________.
24. When two or more radio telescopes are wired together, the resulting network is called a(n) __________.

25. A narrow jet of rising material that extends upward from the Sun's chromosphere into the corona is called a(n) __________.

26. The __________ is a bright display of ever-changing light caused by solar radiation interacting with the upper atmosphere of Earth in the region of the poles.

Comprehensive Review

1. List and describe the two models used to explain the properties of light.

   1)

   2)

2. Other than the colors of visible light, list three types of the electromagnetic radiation. In what way(s) are they similar? In what way(s) are they different?

3. Briefly describe what happens to white light as it passes through a prism and emerges as the "colors of the rainbow."

4. List and describe the three types of spectra.

   1)

   2)

   3)

5. How do astronomers identify the elements that are present in the Sun and other stars?

6. What two aspects of stellar motion can be detected from a star's Doppler shift?

7. List and describe the two basic types of optical telescopes.

   1)

   2)

8. List and briefly describe the three properties of optical telescopes that astronomers are most interested in.

   1)

   2)

   3)

9. What are some advantages that a reflecting telescope has over a refracting telescope?

10. How do astronomers detect and study invisible infrared and ultraviolet radiation?

11. List two advantages that radio telescopes have over visible-light telescopes.

    1)

    2)

12. Using Figure 23.1, select the letter that identifies each of the following solar features.

    a) Corona: _____

    b) Core: _____

    c) Prominence: _____

    d) Spicules: _____

    e) Chromosphere: _____

    f) Sunspots: _____

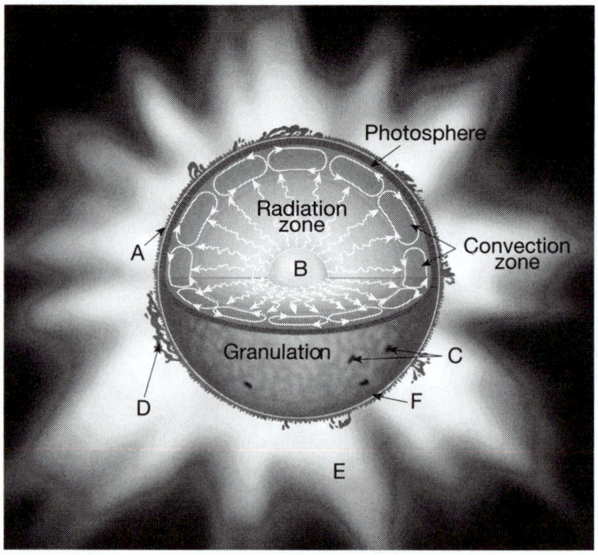

Figure 23.1

13. What is the effect on Earth's atmosphere of a strong solar flare?

14. Briefly describe how the Sun (as well as most other stars) produce their energy.

Practice Test

Multiple choice. Choose the best answer for the following multiple-choice questions.

1. Gamma rays, X-rays, visible light, and radio waves are all types of __________.
   a) electromagnetic radiation
   b) nuclear energy
   c) gravitational energy
   d) ultraviolet radiation
   e) chromatic aberration

2. The fact that light can exert a pressure ("push") on matter suggests it is made of particles called __________.
   a) electrons
   b) protons
   c) photons
   d) nuclei
   e) neutrons

3. The solar wind consists of streams of __________ that travel outward from the Sun at very high speeds.
   a) helium atoms
   b) photons
   c) protons and electrons
   d) radio waves
   e) protons and neutrons

4. The source of the Sun's energy is __________.
   a) magnetism
   b) nuclear fusion
   c) radiation pressure
   d) nuclear fission
   e) combustion

5. When several radio telescopes are wired together, the resulting network is called a radio __________.
   a) receiver
   b) interferometer
   c) refractor
   d) tuner
   e) exoresolver

6. The primary material from which stars are made is __________.
   a) hydrogen
   b) silicates
   c) helium
   d) oxygen
   e) nitrogen

7. The Sun should be able to exist in its present state for approximately __________ billion more years.
   a) 1
   b) 3
   c) 5
   d) 7
   e) 9

8. If the temperature of a radiating body is increased, the total amount of energy emitted __________.
   a) remains the same
   b) increases
   c) decreases

9. The numerous, relatively small, bright markings on the Sun's photosphere are referred to as __________.
   a) spicules
   b) plages
   c) sunspots
   d) granules
   e) prominences

10. Which one of the following stellar properties can NOT be determined from the analysis of the star's light?
    a) rate of movement
    b) composition
    c) temperature
    d) direction of movement
    e) surface features

11. Which layer of the Sun is considered its "surface"?
    a) corona
    b) spicule
    c) photosphere
    d) auroral
    e) chromosphere

12. The energy of a photon is related to its __________.
    a) size
    b) mass
    c) density
    d) speed
    e) wavelength

13. The largest earthbound telescopes cannot photograph lunar features less than __________ kilometer in size.
    a) 0.1
    b) 0.2
    c) 0.3
    d) 0.4
    e) 0.5

14. As the temperature of an object increases, a larger proportion of its energy is radiated at __________ wavelengths.
    a) longer
    b) the same
    c) shorter
    d) faster
    e) slower

15. Which color of light has the most energetic photons?
    a) red
    b) orange
    c) yellow
    d) blue
    e) violet

16. The spectra of most stars are __________.
   a) bright-line
   b) continuous
   c) dark-line
   d) emission
   e) invisible

17. The thin red rim seen around the Sun during a total solar eclipse is the __________.
   a) aurora
   b) chromosphere
   c) corona
   d) solar wind
   e) photosphere

18. As light passes through a lens, which color will undergo the greatest amount of bending?
   a) red
   b) orange
   c) yellow
   d) blue
   e) violet

19. The apparent change in the wavelength as a light-emitting object moves rapidly away from an observer is called __________.
   a) diffraction
   b) the Doppler effect
   c) relativity
   d) chromatic aberration
   e) refraction

20. __________ are huge cloudlike structures that appear as great arches when they are on the limb of the Sun.
   a) Spicules
   b) Prominences
   c) Umbras
   d) Flares
   e) Sunspots

True/false. For the following true/false questions, if a statement is not completely true, mark it false. For each false statement, change the **italicized** *word to correct the statement.*

1. _____ The outermost portion of the Sun's atmosphere is the *photosphere.*
2. _____ Compared to other stars, the Sun is best described as *middle-aged.*
3. _____ In a *reflecting* telescope, the light is focused in front of the objective.
4. _____ The energy-producing reaction in the Sun converts *hydrogen* nuclei into the nuclei of helium atoms.
5. _____ The internal temperature of the Sun is estimated to be approximately 15 million K.
6. _____ The average density of the Sun is nearly the density of *lead.*
7. _____ The study of the properties of light that depend on wavelength is called *optimography.*
8. _____ *Convection* is the process believed to be responsible for the transfer of energy in the uppermost part of the Sun's interior.
9. _____ On a night when the stars twinkle, seeing is *poor.*
10. _____ The pressure ("push") on matter exerted by photons is called *particle* pressure.
11. _____ Neutral hydrogen (hydrogen atoms that lack an electrical charge) emit *radio* radiation with a 21-centimeter (8-inch) wavelength.
12. _____ The diameter of the Sun is equal to 109 Earth diameters.
13. _____ To record *ultraviolet* and infrared radiation from a star, balloons, rockets, and satellites must transport cameras "above" the atmosphere.
14. _____ *Refracting* telescopes have the ability to "see" through interstellar dust clouds.
15. _____ All radiant energy travels through the vacuum of space in a straight line at the rate of 186,000 miles per *hour.*
16. _____ The variation in sunspot activity has a(n) *20-year* cycle.
17. _____ During nuclear fusion, some of the matter is converted to *energy.*

18. _____ Upward from the photosphere, the temperature of the Sun *decreases.*

Word choice. Complete each of the following statements by selecting the most appropriate response.

1. Short wavelengths of radiation have [more/less] energetic photons than radiation with longer wavelengths.
2. Different wavelengths of light will appear as different [intensities/colors] to the human eye.
3. Radio wavelengths are [longer/shorter] than visible wavelengths.
4. The [chromosphere/photosphere] is the layer of the Sun that radiates most of the sunlight we see.
5. If a light source is moving very rapidly away from an observer, its light will appear [bluer/redder] than it actually is.
6. Increasing the magnification of a telescope [increases/decreases] the brightness of an object.
7. Sunspots appear dark because they are [warmer/cooler] than the surrounding solar surface.
8. X-rays are a type of [nuclear/electromagnetic] radiation.
9. The absorption spectrum of the Sun's photosphere reveals that 90 percent of the surface atoms are [hydrogen/helium].
10. The largest optical telescopes are [refracting/reflecting] telescopes.
11. The greater the Doppler shift, the [greater/slower] the velocity of the object.
12. The Sun is a(n) [very large/average/very small] star.
13. All radiation travels through the vacuum of space in a [curved/straight] line at the rate of [300,000/186,000] kilometers per second.
14. [Red/Violet] is the color of visible light with the shortest wavelength, while [gamma/radio] radiation is the shortest wavelength of electromagnetic radiation.
15. Radio telescopes can detect [hotter/cooler] objects than those that are detectable with optical telescopes.
16. On a night when the stars shine steadily, seeing is [good/poor].
17. The [chromosphere/corona] is the lower portion of the solar atmosphere.
18. When passing through a lens, the longer wavelengths of light are refracted at a [larger/smaller] angle than the shorter wavelengths.

Written questions

1. Describe the source of the Sun's energy.

2. What are three properties of a star that can be determined by spectroscopic analysis of its light?

3. List and briefly describe each of the four parts of the Sun.

4. Compared to a refracting telescope, what are some advantages of a reflecting telescope?

For other interesting and pertinent information, be sure to visit the *Earth Science* companion website at

http://www.prenhall.com/tarbuck

24

Beyond Our Solar System

Other than the Sun, the nearest star to Earth, a star called Proxima Centauri, is roughly 100 million times further away than the Moon. Yet it along with the Sun are but two stars among about 100 billion stars that make up our home galaxy, the Milky Way. Extending beyond the Milky Way into the far reaches of the universe are hundreds of billions of other galaxies, each containing hundreds of billions of stars. What is the nature of the vast cosmos beyond our solar system? Are stars distributed evenly, or are they organized into distinct clusters? Are galaxies moving randomly in all directions throughout the universe? In consideration of these and other questions, this chapter examines the properties and evolution of stars, the types and motions of galaxies, as well as the very origin of the universe.

Learning Objectives

After reading, studying, and discussing this chapter, you should be able to:

Discuss the principle of parallax and explain how it is used to measure the distance to a star.

One method for determining the distance to a star is to use a measurement called *stellar parallax*, the extremely slight back-and-forth shifting in a nearby star's position due to the orbital motion of Earth. *The farther away a star is, the less its parallax.* A unit used to express stellar distance is the *light-year*, which is the distance light travels in one Earth year—about 9.5 trillion kilometers (5.8 trillion miles).

List and describe the major intrinsic properties of stars.

The intrinsic properties of stars include *brightness, color, temperature, mass*, and *size*. Three factors control the brightness of a star as seen from Earth: how big it is, how hot it is, and how far away it is. *Magnitude* is the measure of a star's brightness. *Apparent magnitude* is how bright a star appears when viewed from Earth. *Absolute magnitude* is the "true" brightness if a star were at a standard distance of about 32.6 light-years. The difference between the two magnitudes is directly related to a star's distance. Color is a manifestation of a star's temperature. Very hot stars (surface temperatures above 30,000 K) appear blue; red stars are much cooler (surface temperatures generally less than 3000 K). Stars with surface temperatures between 5000 and 6000 K appear yellow, like the Sun. The center of mass of orbiting *binary stars* (two stars revolving around a common center of mass under their mutual gravitational attraction) is used to determine the mass of the individual stars in a binary system.

Explain variable stars.

Variable stars fluctuate in brightness. Some, called *pulsating variables*, fluctuate regularly in brightness by expanding and contracting in size. When a star explosively brightens, it is called a *nova*. During the outburst, the outer layer of the star is ejected at high speed. After reaching maximum brightness in a few days, the nova slowly returns in a year or so to its original brightness.

Describe how a Hertzsprung-Russell diagram is constructed.

A *Hertzsprung-Russell diagram* is constructed by plotting the absolute magnitudes and temperatures of stars on a graph. A great deal about the sizes of stars can be learned from H-R diagrams. Stars located in the upper-right position of an H-R diagram are called *giants*, luminous stars of large radius. *Supergiants* are very large. Very small *white dwarf* stars are located in the lower-central portion of an H-R diagram. Ninety percent of all stars, called *main-sequence stars*, are in a band that runs from the upper-left corner to the lower-right corner of an H-R diagram.

Describe the different types of nebulae.

New stars are born out of enormous accumulations of dust and gases, called *nebula*, that are scattered between existing stars. A *bright nebula* glows because the matter is close to a very hot (blue) star. The two main types of bright nebulae are *emission nebulae* (which derive their visible light from the fluorescence of the ultraviolet light from a star in or near the nebula) and *reflection nebulae* (relatively dense dust clouds in interstellar space that are illuminated by reflecting the light of nearby stars). When a nebula is not close enough to a bright star to be illuminated, it is referred to as a *dark nebula*.

Discuss the life cycle of a star.

Stars are born when their nuclear furnaces are ignited by the unimaginable pressures and temperatures in collapsing nebulae. New stars not yet hot enough for nuclear fusion are called *protostars*. When collapse causes the core of a protostar to reach a temperature of at least 10 million K, the fusion of hydrogen nuclei into helium nuclei begins in a process called *hydrogen burning*. The opposing forces acting on a star are *gravity* trying to contract it and *gas pressure (thermal nuclear energy)* trying to expand it. When the two forces are balanced, the star becomes a stable *main-sequence star*. When the hydrogen in a star's core is consumed, its outer envelope expands enormously and a *red giant* star, hundreds to thousands of times larger than its main-sequence size, forms. When all the usable nuclear fuel in these giants is exhausted and gravity takes over, the stellar remnant collapses into a small dense body.

Describe the possible final states that a star may assume after it consumes its nuclear fuel and collapses.

The *final fate of a star is determined by its mass*. Stars with less than one half the mass of the Sun collapse into hot, dense *white dwarf* stars. Medium-mass stars (between 0.5 and 3.0 times the mass of the Sun) become red giants, collapse, and end up as white dwarf stars, often surrounded by expanding spherical clouds of glowing gas called *planetary nebulae*. Stars more than three times the mass of the Sun terminate in a brilliant explosion called a *supernova*. Supernovae events can produce small, extremely dense *neutron stars*, composed entirely of subatomic particles called neutrons; or even smaller and more dense *black holes*, objects that have such immense gravity that light cannot escape their surface.

Discuss the structure of the Milky Way Galaxy.

The *Milky Way Galaxy* is a large, disk-shaped *spiral galaxy* about 100,000 light-years wide and about 10,000 light-years thick at the center. There are three distinct *spiral arms* of stars, with some showing splintering. The Sun is positioned in one of these arms about two-thirds of the way from the galactic center, at a distance of about 30,000 light-years. Surrounding the galactic disk is a nearly spherical halo made of very tenuous gas and numerous *globular clusters* (nearly spherically shaped groups of densely packed stars).

List and describe the major types of galaxies.

The various types of galaxies include (1) *spiral galaxies*, which are typically disk-shaped with a somewhat greater concentration of stars near their centers, often containing arms of stars extending from their central nucleus; (2) *elliptical galaxies*, the most abundant type, which have an ellipsoidal shape that ranges to nearly spherical and lack spiral arms; and (3) *irregular galaxies*, which lack symmetry and account for only 10 percent of the known galaxies.

Describe the distribution of galaxies in the universe.

Galaxies are not randomly distributed throughout the universe. They are grouped in *galactic clusters*, some containing thousands of galaxies. Our own, called the *Local Group*, contains at least 28 galaxies.

Explain the Doppler effect and what it tells scientists about galactic motion in the universe.

By applying the *Doppler effect* (the apparent change in wavelength of radiation caused by the motions of the source and the observer) to the light of galaxies, galactic motion can be determined. Most galaxies have Doppler shifts toward the red end of the spectrum, indicating increasing distance. The amount of Doppler shift is dependent on the velocity at which the object is moving. Because the most distant galaxies have the greatest red shifts, Edwin Hubble concluded in the early 1900s that they were retreating from us with greater recessional velocities than more nearby galaxies. It was soon realized that an *expanding universe* can adequately account for the observed red shifts.

Describe the big bang theory of the origin of the universe.

The belief in the expanding universe led to the widely accepted *big bang theory* of the origin of the universe. According to this theory, the entire universe was at one time confined in a dense, hot, supermassive concentration. Almost 14 billion years ago a cataclysmic explosion hurled this material in all directions, creating all matter and space. Eventually the ejected masses of gas cooled and condensed, forming the stellar systems we now observe fleeing from their place of origin.

Key Terms

absolute magnitude
apparent magnitude
barred spiral galaxy
big bang
black hole
bright nebula
dark nebula
degenerate matter
elliptical galaxy
emission nebula
eruptive variables
galactic cluster
Hertzsprung-Russell (H-R) diagram
Hubble's law
hydrogen burning
interstellar dust
irregular galaxy
light-year
Local Group
magnitude
main-sequence stars
nebula
neutron star
nova
planetary nebula
protostar
pulsar
pulsating variables
red giant
reflection nebula
spiral galaxy
stellar parallax
supergiant
supernova
white dwarf

Vocabulary Review

Choosing from the list of key terms, furnish the most appropriate response for the following statements.

1. The measurement of a star's brightness is called its __________.
2. A(n) __________ is a type of galaxy that lacks symmetry.
3. A(n) __________ is a small, dense star that was once a low-mass or medium-mass star whose internal heat energy was able to keep these gaseous bodies from collapsing under their own weight.
4. A graph showing the relation between the true brightness and temperature of stars is called a(n) __________.
5. The fusion of groups of four hydrogen nuclei into a single helium nuclei is called __________.
6. __________ is a type of very dense matter that forms when electrons are displaced inward from their regular orbits around an atom's nucleus.
7. __________ is the extremely slight back-and-forth shifting in the position of a nearby star due to the orbital motion of Earth.
8. An interstellar cloud consisting of dust and gases is referred to as a(n) __________.
9. A(n) __________ is an exceptionally large star, such as Betelgeuse, with a radius about 800 times that of the Sun and high luminosity.
10. A(n) __________ is a gaseous interstellar mass that absorbs ultraviolet light from an embedded or nearby hot star and reradiates, or emits, this energy as visible light.
11. A collapsing cloud of gases and dust not yet hot enough to engage in nuclear fusion but destined to become a star is referred to as a(n) __________.
12. A star consisting of matter formed from the combination of electrons with protons that is smaller and more massive than a white dwarf is called a(n) __________.
13. A(n) __________ is a very large, cool, reddish-colored star of high luminosity.

14. Stars exceeding three solar masses have relatively short life spans and terminate in a brilliant explosion called a(n) __________.
15. When a dense cloud of interstellar material is not close enough to a bright star to be illuminated, it is referred to as a(n) __________.
16. Discovered in the early 1970s, a(n) __________ is a neutron star that radiates short bursts of radio energy.
17. __________ states that galaxies are receding from us at a speed that is proportional to their distance.
18. A star's brightness as it appears when viewed from Earth is referred to as its __________.
19. A type of galaxy that has an ellipsoidal shape that ranges to nearly spherical and lacks spiral arms is called a(n) __________.
20. A(n) __________ is a unit used to express stellar distance equal to about 9.5 trillion kilometers (5.8 trillion miles).
21. A(n) __________ is an interstellar cloud that glows because of its proximity to a very hot (blue) star.
22. An object even smaller and denser than a neutron star, with a surface gravity so immense that light cannot escape, is called a(n) __________.
23. An interstellar cloud of gases and dust that merely reflects the light of nearby stars is called a(n) __________.
24. The Milky Way is a type of galaxy called a(n) __________.
25. The brightness of a star, if it were viewed at a distance of 32.6 light-years, is called its __________.
26. A reflection nebula is thought to be composed of a rather dense cloud of large particles called __________.
27. On an H-R diagram, the "ordinary" stars that are located along a band that extends from the upper-left corner to the lower-right corner are called __________.
28. An expanding spherical cloud of gas, called a(n) __________, forms when a medium-mass star collapses from a red giant to a white dwarf star.
29. The theory that proposes that the universe originated as a single mass that subsequently exploded is referred to as the __________ theory.
30. Our own group of galaxies, called the __________, contains at least 28 galaxies.
31. Stars that fluctuate regularly in brightness by expanding and contracting in size are called __________.
32. A system of galaxies containing from several to thousands of member galaxies is called a(n) __________.
33. A(n) __________ is a star that explosively increases in brightness.

Comprehensive Review

1. Briefly describe the relation between a star's distance and its parallax.

2. What causes the difference between a star's apparent magnitude and its absolute magnitude?

3. List the three factors that control the apparent brightness of a star as seen from Earth.

   1)

   2)

   3)

4. What is the approximate surface temperature of a star with the following color?

   a) Red:

   b) Yellow:

   c) Blue:

5. Describe how binary stars are used to determine stellar mass.

6. List the two properties of stars that are used to construct an H-R diagram.

   1)

   2)

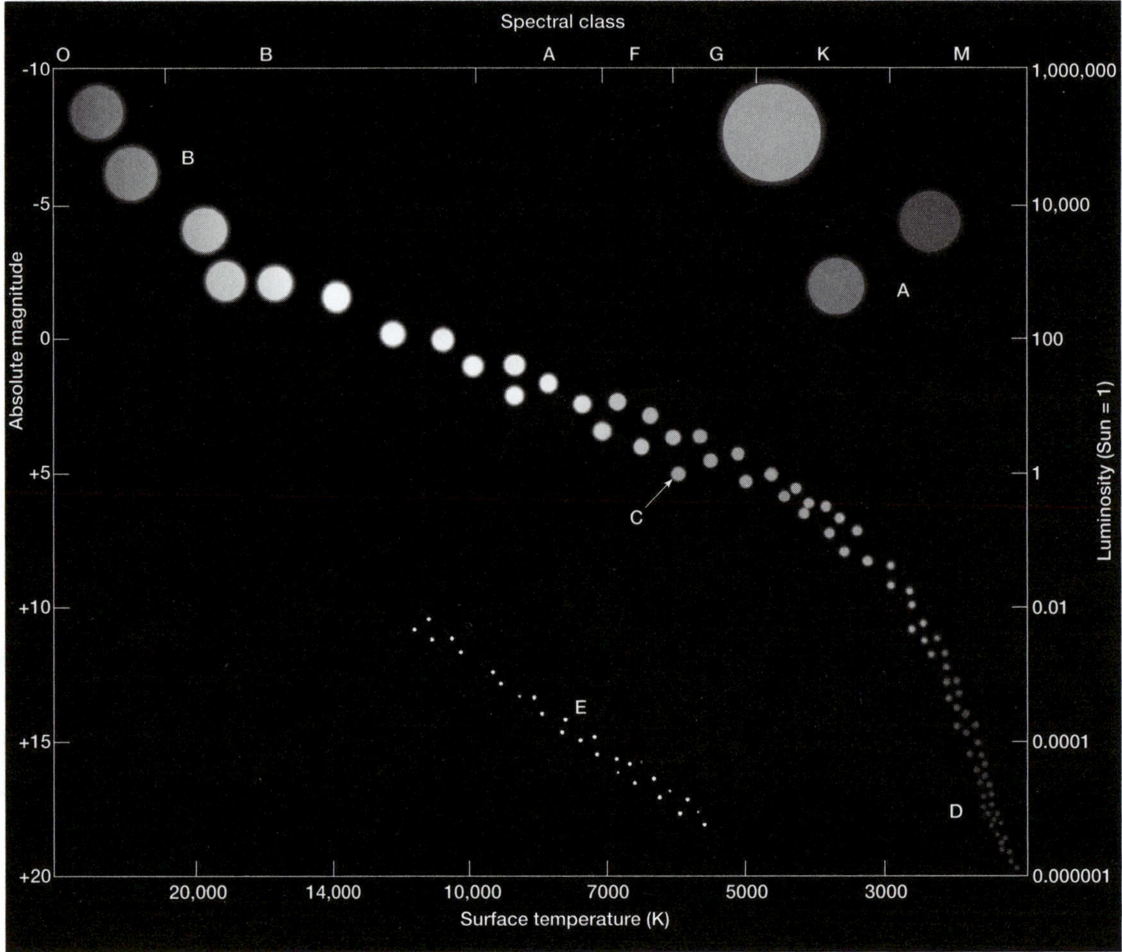

Figure 24.1

7. Using Figure 24.1, an idealized Hertzsprung-Russell diagram, select the letter that indicates the location of each of the following types of stars.

   a) White dwarfs: _____

   b) Blue stars: _____

   c) Red stars (low mass): _____

   d) Giants: _____

8. In Figure 24.1, which letter is most representative of our Sun? _____

9. In Figure 24.1, which stars are on the main sequence?

10. List and briefly describe the two main types of bright nebulae.

    1)

    2)

11. What circumstance produces a dark nebula?

12. Describe the process that astronomers refer to as hydrogen burning.

13. What will be the most likely final stage in the burnout and death of stars in each of the following mass categories?

    a) Low-mass stars:

    b) Medium-mass (sunlike) stars:

    c) Massive stars:

14. On Figure 24.2, beginning with dust and gases, draw a sequence of arrows that show in correct order the stages that a star about as massive as the Sun will follow in its evolution.

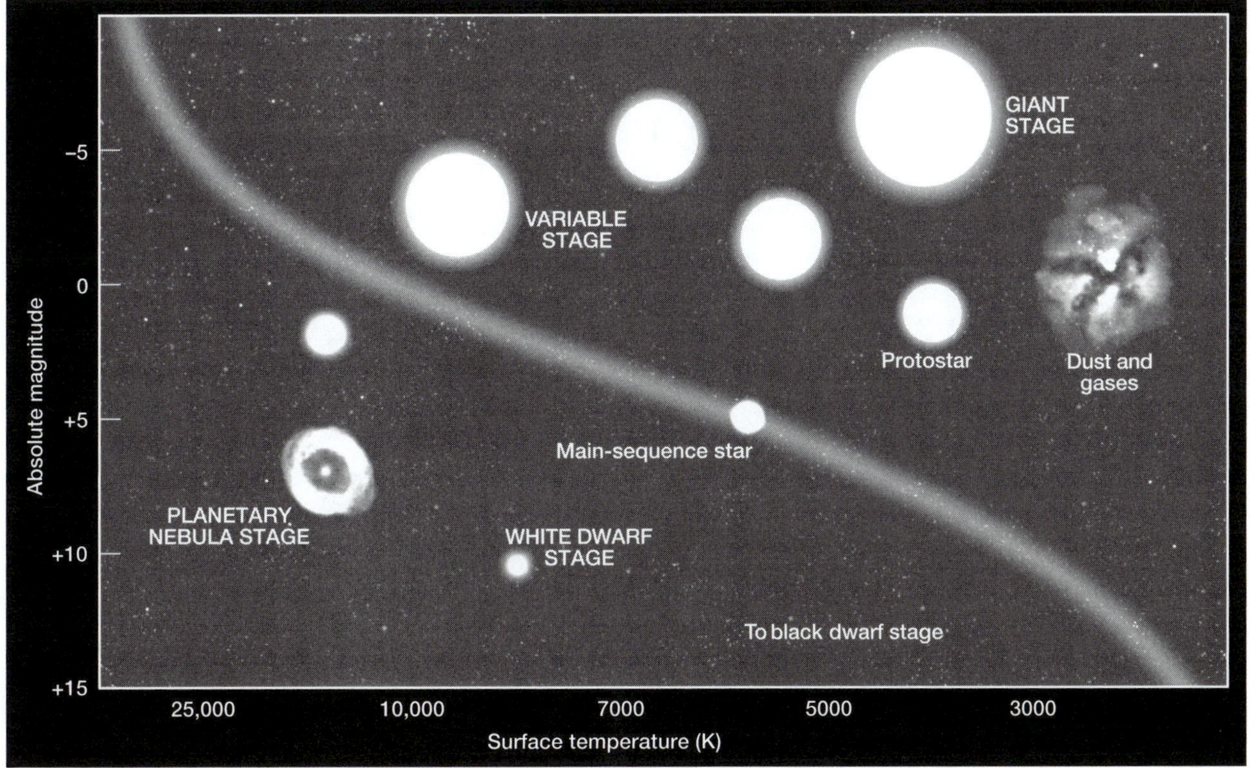

Figure 24.2

15. Why are radio telescopes rather than optical telescopes used to determine the gross structure of the Milky Way Galaxy?

16. Describe the size and structure of the Milky Way.

17. List and describe the three basic types of galaxies.

    1)

    2)

    3)

18. The larger galaxy shown in Figure 24.3 is of which type?

Figure 24.3

19. What is Hubble's law?

20. Briefly describe the big bang theory.

Practice Test

Multiple choice. Choose the best answer for the following multiple-choice questions.

1. The two properties of a star that are plotted on an H-R diagram are __________.
   a) size and distance
   b) brightness and temperature
   c) speed and distance
   d) size and temperature
   e) brightness and distance

2. Stars with surface temperatures between 5000 and 6000 K appear __________.
   a) red
   b) green
   c) violet
   d) yellow
   e) blue

3. A star with which one of the following magnitudes would appear the brightest?
   a) 15
   b) 10
   c) 5
   d) 1
   e) –5

4. Which force is most responsible for the formation of a star?
   a) magnetic
   b) gravity
   c) nuclear
   d) interstellar
   e) light pressure

5. The difference in the brightness of two stars having the same surface temperature is attributable to their relative __________.
   a) densities
   b) colors
   c) compositions
   d) ages
   e) sizes

6. In the cores of incredibly hot red giant stars, nuclear reactions convert helium to __________.
   a) carbon
   b) hydrogen
   c) oxygen
   d) argon
   e) nitrogen

7. Stellar parallax is used to determine a star's __________.
   a) motion
   b) brightness
   c) distance
   d) mass
   e) temperature

8. During the process referred to as hydrogen burning, hydrogen nuclei in the core of a star are fused together and become __________ nuclei.
   a) oxygen
   b) nitrogen
   c) helium
   d) carbon
   e) argon

9. Binary stars can be used to determine stellar __________.
   a) mass
   b) temperature
   c) brightness
   d) distance
   e) composition

10. One light-year is about __________.
    a) 6 trillion kilometers
    b) 9.5 trillion miles
    c) 8 trillion kilometers
    d) 5.8 trillion miles
    e) 11 trillion kilometers

11. Which one of the following stellar remnants has the greatest density?
    a) neutron star
    b) black hole
    c) white dwarf
    d) black dwarf

12. An interstellar accumulation of dust and gases is referred to as a __________.
    a) red giant
    b) nebula
    c) white dwarf
    d) galaxy
    e) neutron star

13. Which one of the following is NOT a type of galaxy?
    a) nebular
    b) irregular
    c) spiral
    d) elliptical

14. The most abundant gas in dark, cool, interstellar clouds in the neighborhood of the Milky Way is __________.
    a) hydrogen
    b) argon
    c) helium
    d) oxygen
    e) nitrogen

15. Very massive stars terminate in a brilliant explosion called a __________.
    a) red giant
    b) supernova
    c) protostar
    d) neutron star
    e) planetary nebula

16. According to the Big Bang theory, a cataclysmic explosion created the matter in the universe about __________ years ago.
    a) 600 million
    b) 2 billion
    c) 4.6 billion
    d) 12 billion
    e) 14 billion

17. Which one of the following is NOT a type of nebula?
    a) reflection
    b) emission
    c) spiral
    d) dark

18. The apparent change in the position of a celestial object due to the orbital motion of Earth is called __________.
    a) parallax
    b) recessional velocity
    c) magnitude
    d) brightness
    e) orbital shifting

19. Matter that is pulled into a black hole should become very hot and emit __________ before being engulfed.
    a) X-rays
    b) hydrogen nuclei
    c) atoms
    d) degenerate matter
    e) infrared radiation

20. Which one of the following stars would appear brightest in the night sky?
    a) Sirius
    b) Deneb
    c) Alpha Centauri
    d) Arcturus
    e) Betelgeuse

21. The Milky Way is a __________.
    a) constellation
    b) red giant
    c) spiral galaxy
    d) dark nebula
    e) galactic cluster

22. On an H-R diagram, main-sequence blue stars are massive and __________.
    a) cool
    b) white dwarfs
    c) hot
    d) novae
    e) supergiants

23. The measurement of a star's brightness is called __________.
    a) temperature
    b) magnitude
    c) parallax
    d) period
    e) pulsation

24. A very young, large, red object that is soon to become a star but is not yet hot enough for nuclear fusion is termed a ________.
    a) black hole
    b) supernova
    c) red giant
    d) protostar
    e) white dwarf

25. The apparent change in wavelength of radiation caused by the relative motions of the source and the observer is referred to as ________.
    a) Hubble's law
    b) the acceleration factor
    c) Newton's law of radiation
    d) relativity
    e) the Doppler effect

26. The density of a white dwarf is such that a spoonful of its matter would weigh ________.
    a) a kilogram
    b) 500 pounds
    c) 40 grams
    d) a ton
    e) several tons

27. The Sun is positioned about ________ of the way from the center of the galaxy.
    a) one-fourth
    b) one-third
    c) one-half
    d) two-thirds
    e) three-fourths

28. The brightness of a star, if it were at a distance of 32.6 light-years, is its ________.
    a) apparent magnitude
    b) parallax
    c) apparent brightness
    d) absolute magnitude
    e) mass-distance ratio

29. Which type of galaxy is composed mostly of young stars?
    a) nebulas
    b) elliptical
    c) irregular
    d) spiral
    e) nebular

30. The red shifts exhibited by galaxies can be explained by ________.
    a) calculation errors
    b) an expanding universe
    c) Kepler's first law
    d) expansion of the Milky Way
    e) a collapsing universe

31. On an H-R diagram, about 90 percent of the stars are ________.
    a) giants
    b) main-sequence stars
    c) white dwarfs
    d) super giants
    e) novae

32. The difference between a star's apparent and absolute magnitudes is related to ________.
    a) surface temperature
    b) color
    c) core temperature
    d) time
    e) distance

True/false. For the following true/false questions, if a statement is not completely true, mark it false. For each false statement, change the **italicized** *word to correct the statement.*

1. _____ A hot, massive blue star will age more *slowly* than a small (red) main-sequence star.
2. _____ The parallax of a far star is *less* than that of a near star.
3. _____ A stable main-sequence star is balanced between the force of gravity and internal *gas* pressure.
4. _____ Galaxies are receding from us at a speed that is proportional to their *size.*
5. _____ *Absolute* magnitude is the brightness of a star when viewed from Earth.
6. _____ Stars with equal surface *temperatures* radiate the same amount of energy per unit area.
7. _____ On an H-R diagram, the Sun is a *main-sequence* star.
8. _____ The larger the magnitude number, the *dimmer* the star.
9. _____ *Low-mass* stars evolve to become white dwarfs.
10. _____ On an H-R diagram, the hottest main-sequence stars are intrinsically the *brightest.*

11. _____ A star's color is primarily a manifestation of its *temperature*.
12. _____ Giant stars are *cooler* than white dwarfs.
13. _____ *Dark* nebulae glow because they are close to very hot stars.
14. _____ The light-year is a unit used to measure the *age* of a star.
15. _____ All stars eventually exhaust their nuclear fuel and collapse in response to *gravity*.
16. _____ A star with a surface temperature less than 3000 K will generally appear *red*.
17. _____ A rapidly rotating neutron star that radiates short pulses of radio energy is called a *pulsar*.
18. _____ The Milky Way is about 100,000 *kilometers* wide.
19. _____ *Spiral* galaxies are the most abundant type of galaxies.
20. _____ A first-magnitude star is about 100 times *dimmer* than a sixth-magnitude star.
21. _____ Large Doppler shifts indicate *low* velocities.
22. _____ Stars that orbit one another are called *neutron* stars.
23. _____ If there is enough *matter* in the universe, the outward expansion of the galaxies could eventually stop.

Word choice. Complete each of the following statements by selecting the most appropriate response.

1. There [are/are no] stars with negative magnitudes.
2. A [white dwarf/protostar] is the first stage in the evolution of a star.
3. As matter is pulled into a black hole, it emits a steady stream of [radio waves/X-rays].
4. On an H-R diagram, among main-sequence stars, hot stars are [more/less] massive than cooler stars.
5. The first naked-eye supernova in 383 years occurred in the southern sky in [1965/1973/1987].
6. Most of the stars in the Milky Way Galaxy are distributed in a [disk/sphere] with spiral arms.
7. Stellar parallax [is/is not] detectable with the unaided eye.
8. Stars with surface temperatures less than 3000 K appear [blue/yellow/red].
9. Neutron stars are believed to be produced by [nuclear fission/supernovae/degenerate matter].
10. The apparent magnitude of the Sun is [18.7/5.0/−26.7], while its absolute magnitude is [18.7/5.0/−26.7].
11. By determining the light period of a cepheid variable, its [color/size/absolute magnitude] can be determined.
12. The atoms that compose our Sun, as well as the rest of the solar system, formed during a [supernova/galactic collision] event trillions of kilometers away.
13. Planetary nebulae are the cast-off outer [planets/atmospheres] of collapsing red giants that formed from medium-mass stars.
14. The core of a red giant star is [expanding/contracting].
15. A star that is 5 units of magnitude brighter than another is [2.5/5.0/100] times brighter.
16. Concentrations of interstellar dust and gases are called [nebulae/novas].
17. If the sizes of the orbits of binary stars can be measured, a determination of their individual [masses/temperatures] can be made.
18. The term big bang describes the [beginning/end] of the [solar system/galaxy/universe].
19. A star's brightness as viewed from Earth is called its [absolute/apparent] brightness.
20. Neutron stars rotate very [slowly/quickly].

Written questions

1. Describe the movement through the H-R diagram that an average star like the Sun follows during its lifetime.

2. What is the final state for (1) a low-mass (red) main-sequence star, (2) a medium-mass (Sunlike) star, and (3) a very massive star?

3. Describe the big bang theory.

For other interesting and pertinent information, be sure to visit the *Earth Science* companion website at

http://www.prenhall.com/tarbuck

CHAPTER ONE

Vocabulary Review

1. core; mantle; crust
2. hypothesis
3. continental slope
4. geology
5. Meteorology
6. geologic time scale
7. theory
8. geosphere
9. hydrosphere
10. deep-ocean trench
11. oceanography
12. renewable resource
13. inner core
14. biosphere
15. continental shelf
16. nonrenewable resource
17. Astronomy
18. shield
19. atmosphere
20. oceanic (mid-ocean) ridge
21. nebular hypothesis
22. lithosphere
23. asthenosphere
24. system
25. abyssal plain
26. environment
27. positive-feedback mechanisms
28. plate tectonics
29. paradigm

Comprehensive Review

1. The sciences traditionally included in Earth science are geology, oceanography, meteorology, and astronomy.
2. Renewable resources, such as forest products and wind energy, can be replenished over relatively short time spans. By contrast, nonrenewable resources, such as oil and copper, form so slowly that significant deposits often take millions of years to accumulate. Therefore, practically speaking, the Earth contains fixed quantities of these materials.
3. (1) hydrosphere: a dynamic mass of water that is continually on the move from the oceans to the atmosphere, precipitating back to the land, and returning back to the ocean to begin the cycle again; (2) atmosphere: Earth's blanket of air; (3) solid Earth: which includes the core, mantle, and crust; (4) biosphere: which includes all life on Earth.
4. Earth science is the science that collectively seeks to understand Earth and its neighbors in space. It includes geology, oceanography, meteorology, and astronomy.
5. A scientific hypothesis is a tentative explanation, while a theory is a well-tested and widely accepted view that best explains certain observable facts.
6. Within about one second after the Big Bang, protons and neutrons appear and quickly become atoms of hydrogen and helium. During the first billion years, galaxies, including the Milky Way, form. Massive stars in the Milky Way die violently and produce complex atoms such as oxygen, carbon, and iron. It was from this debris, scattered during the death of a star or stars, that our solar system formed.
7. (1) collection of facts (data) through observation and measurement; (2) development of a working hypothesis to explain the facts; (3) construction of experiments to validate the hypothesis; (4) acceptance, modification, or rejection of the hypothesis on the basis of extensive testing
8. A (inner core); B (outer core); C (mantle); D (crust)
9. a) Physical geology studies the materials that compose Earth and seeks to understand the many processes that operate beneath and upon its surface. b) Historical geology seeks to understand the origin of Earth and the development of the planet through its 4.6-billion-year history.
10. On Earth, the hydrosphere, atmosphere, biosphere, and solid Earth act as a group of interrelated, interacting, or interdependent parts that form a complex whole that we call the Earth system.

11. The five layers of Earth based on physical properties include the (1) lithosphere, the outer, relatively cool, rigid shell of the Earth; (2) asthenosphere, a soft and comparatively weak layer beneath the lithosphere; (3) lower mantle, a rigid zone within Earth between the depths of 660 and 2900 kilometers; (4) outer core, a liquid shell 2260 kilometers thick; and (5) inner core, a solid sphere having a radius of about 1220 kilometers.
12. According to the theory of plate tectonics, Earth's rigid outer shell consists of numerous slabs called lithospheric plates, which are in continual motion. The interaction of these plates along their margins is responsible for earthquakes and mountain building, among other events. The movement of lithospheric plates is ultimately driven by the unequal distribution of heat within Earth.
13. The major features of deep-ocean basins include (1) incredibly flat regions called abyssal plains, (2) extremely deep, relatively narrow depressions called deep-ocean trenches, and (3) submerged volcanic structures called seamounts.
14. The most consistent set of ideas that explains the origin of the solar system is the nebular hypothesis. The hypothesis suggests that the solar system formed from an enormous nebular cloud consisting mostly of hydrogen and helium. As the cloud began to contract due to gravity and rotate, it formed a nebulous disk. Following the formation of the Sun, the inner planets Mercury, Venus, Earth, and Mars began forming from rocky and metallic matter. At the same time, the larger, outer planets began forming from ices of water, carbon dioxide, ammonia, and methane, as well as rocky and metallic debris.

Practice Test

Multiple choice

1. d
2. d
3. c
4. a
5. d
6. b
7. b
8. a
9. d
10. a
11. d
12. d
13. c
14. b
15. a
16. e
17. d
18. c
19. d
20. b
21. c
22. a
23. b

True/false

1. T
2. F (hydrosphere)
3. F (nonrenewable)
4. F (historical)
5. T
6. F (predictable)
7. T
8. T
9. F (environment)
10. T
11. F (oceanic ridge)
12. F (crust)
13. T
14. T
15. T
16. F (outer)
17. T

Word choice

1. dynamic
2. 15
3. mantle
4. nebular
5. 6; 30
6. slope
7. thinnest
8. negative
9. paradigm

Written questions

1. The four steps that scientists often use to conduct experiments and gain scientific knowledge are (1) collection of facts (data) through observation and measurement, (2) development of a working hypothesis to explain the facts, (3) construction of experiments to validate the hypothesis, and (4) acceptance, modification, or rejection of the hypothesis on the basis of extensive testing.
2. Earth's four "spheres" include the (1) hydrosphere, a dynamic mass of water that is continually on the move from the oceans to the atmosphere, precipitating back to the land and returning back to the ocean to begin the cycle again; (2) atmosphere, Earth's blanket of air; (3) solid Earth (or geosphere), which includes the core, mantle, and crust; and (4) biosphere, which includes all life on Earth.
3. Earth science is the science that collectively seeks to understand Earth and its neighbors in space. It includes geology, oceanography, meteorology, and astronomy.
4. The Earth system involves the complex and continuous interaction of the hydrosphere, atmosphere, biosphere, and geosphere and all their components.
5. Positive-feedback mechanisms enhance or drive change in the system, while negative-feedback mechanisms work to maintain the system as it is.
6. According to the theory of plate tectonics, Earth's rigid outer shell consists of numerous slabs called lithospheric plates, which are in continual motion. The interaction of these plates along their margins is responsible for earthquakes and mountain building, among other events. The movement of lithospheric plates is ultimately driven by the unequal distribution of heat within Earth.

ANSWER KEY

CHAPTER TWO

Vocabulary Review

1. neutron
2. atom
3. ore
4. mineral
5. ion
6. cleavage
7. proton
8. compound
9. silicate
10. reserve
11. electron
12. rock
13. nucleus
14. atomic number
15. element
16. mass number
17. energy level
18. covalent bond
19. isotope
20. tenacity
21. radioactive decay
22. luster
23. silicon-oxygen tetrahedron
24. ionic bond
25. mineral resource
26. streak
27. hardness
28. fracture
29. valence electrons
30. specific gravity
31. Mohs hardness scale
32. color
33. rock-forming minerals
34. periodic table

Comprehensive Review

1. (1) It must be naturally occurring. (2) It must be inorganic. (3) It must be solid. (4) It must possess an orderly crystalline structure. (5) It must have a well-defined chemical composition.
2. (1) protons: particles with positive electrical charges that are found in an atom's nucleus; (2) neutrons: particles with neutral electrical charges that are found in an atom's nucleus; (3) electrons: negative electrical charges surrounding the atom's nucleus in energy levels
3. Minerals are the same throughout, while most rocks are mixtures, or aggregates, of minerals, where each mineral retains its distinctive properties.
4. a) 6; b) 14
5. a) 17; b) 17; c) 18
6. The sodium atom loses one electron and becomes a positive ion. The chlorine atom gains one electron and becomes a negative ion. These oppositely charged ions attract one another, and the bond produces the compound sodium chloride.
7. Yes, the atom is an ion because it contains two more electrons (10) than protons (8). a) 13; b) 8
8. No, the atom contains an equal number of protons (17) and electrons (17) and is neutral.
9. A mineral is a naturally occurring inorganic solid that possesses an orderly crystalline structure.
10. An isotope will have a different mass number due to a different number of neutrons.
11. a) the appearance of light reflected from the surface of a mineral; b) the external expression of a mineral's orderly arrangement of atoms; c) the color of a mineral in its powdered form; d) a measure of the resistance of a mineral to abrasion or scratching; e) the tendency of a mineral to break along planes of weak bonding; f) the uneven, or irregular, breaking of a mineral; minerals that do not exhibit cleavage when broken are said to fracture; g) the weight of a mineral compared to the weight of an equal volume of water
12. Diagram A: three directions of cleavage that meet at 90-degree angles; Diagram B: three directions of cleavage that meet at approximately 70- and 120-degree angles
13. oxygen (symbol: O, 46.6%) and silicon (symbol: Si, 27.7%)
14. letter A (oxygen); letter B (silicon)
15. a) feldspars, quartz; b) micas (biotite, muscovite); c) amphibole group; d) pyroxene group; e) olivine
16. a) feldspar, quartz, and muscovite; b) calcite
17. A rock is an aggregate of minerals.
18. Most silicates crystallize from molten rock as it cools.
19. a) cinnabar; b) galena; c) sphalerite; d) hematite (see textbook Table 2.1)
20. As an isotope decays through a process called radioactivity, it actively radiates energy and particles. The decay often produces a different isotope of the same element, but with a different mass number.

21. No. Before aluminum can be extracted profitably, it must be concentrated to four times its average crustal percentage of 8.13 percent. The sample, with only 10 percent aluminum, lacks the necessary concentration.
22. Diagram A: amphibole group; Diagram B: pyroxene group
23. iron (Fe), magnesium (Mg), potassium (K), sodium (Na), and calcium (Ca)

Practice Test

Multiple choice

1. c
2. b
3. e
4. c
5. a
6. a
7. c
8. c
9. c
10. b
11. d
12. b
13. a
14. e
15. b
16. c
17. a
18. b
19. e
20. d
21. b
22. e
23. a
24. a
25. c

True/false

1. F (electrons)
2. T
3. F (four-thousand)
4. T
5. F (oxygen)
6. T
7. T
8. T
9. F (between)
10. T
11. F (8 percent)
12. F (carbonate)
13. F (Gypsum)
14. T
15. F (Habit)
16. T
17. F (nonmetallic)
18. T
19. F (compound)
20. T
21. F (electrons)
22. T
23. F (atom)
24. T
25. F (harder)
26. T

Word choice

1. atomic number; mass number
2. atom
3. mass; atomic; similar
4. luster
5. fracture
6. 8
7. silicon; tetrahedron
8. cleavage
9. calcite
10. halite; calcite
11. hematite; galena
12. covalent; ionic

Written questions

1. Protons and neutrons are found in the atom's nucleus and represent practically all of an atom's mass. Protons are positively charged, while neutrons are neutral. Electrons orbit the nucleus in regions called energy levels (or shells) and carry a negative charge.
2. Some mineral samples don't demonstrate visibly their common habit because their crystal growth is severely constrained because of space restrictions where and when the mineral forms.
3. The basic building block of all silicate minerals is the silicon-oxygen tetrahedron. It consists of four oxygen atoms surrounding a much smaller silicon atom.
4. For any material to be considered a mineral, it must be (1) naturally occurring, (2) inorganic, (3) solid, (4) possess an orderly crystalline structure, and (5) have a well-defined chemical composition.
5. A mineral resource is any useful mineral that can be recovered for use. Resources include already identified deposits from which minerals can be extracted profitably, called reserves.

ANSWER KEY

CHAPTER THREE

Vocabulary Review

1. rock cycle
2. vein deposit
3. Lithification
4. metamorphic rock
5. Magma
6. crystallization
7. regional metamorphism
8. sedimentary rock
9. lava
10. texture
11. foliated texture
12. extrusive (volcanic)
13. sediment
14. igneous rock
15. nonfoliated texture
16. fine-grained texture
17. hydrothermal solution
18. chemical sedimentary rock
19. porphyritic texture
20. strata (beds)
21. evaporite deposit
22. contact metamorphism
23. glassy texture
24. detrital sedimentary rock
25. coarse-grained texture
26. disseminated deposit
27. pegmatite
28. Fossils
29. felsic
30. ultramafic

Comprehensive Review

1. A (Crystallization); B (Igneous); C (Weathering, transportation, and deposition); D (Sediment); E (Lithification); F (Sedimentary); G (Metamorphism); H (Metamorphic); I (Melting)
2. The rock cycle allows us to view many of the interrelationships among different parts of the Earth system. It helps us understand how each type of rock (igneous, sedimentary, and metamorphic) is linked to the others by the processes that act upon and within the planet.
3. weathering, transportation, and deposition
4. running water, wind, waves, and glacial ice
5. a) Igneous rocks form as magma cools and solidifies either beneath or at Earth's surface. b) Sedimentary rocks are the lithified products of weathering. c) Metamorphic rocks can form from igneous, sedimentary, or other metamorphic rocks that are subjected to heat, pressure, and chemically active fluids.
6. The rate of cooling most influences the size of mineral crystals in igneous rock. Slow cooling results in the formation of large crystals, while rapid cooling often results in a mass of very small intergrown crystals.
7. Magma is molten rock beneath the surface. Lava, molten rock on the surface, is similar to magma except that most of the gaseous component has escaped.
8. (1) texture: the overall appearance of an igneous rock based on the size and arrangement of its interlocking crystals; (2) mineral composition: the mineral makeup of an igneous rock, which depends on the composition of the magma from which it originated
9. a) rapid cooling at the surface or as small masses within the upper crust (letter A.); b) very rapid cooling, perhaps when molten rock is ejected into the atmosphere (letter D.); c) slow cooling of large masses of magma far below the surface (letter B.); d) two different rates of cooling, resulting in two different crystal sizes (letter C.)
10. N.L. Bowen discovered that as magma cools in the laboratory, certain minerals crystallize first at very high temperatures. At successively lower temperatures, other minerals crystallize.
11. olivine; quartz
12. Yes, providing that they have different textures resulting from different rates of cooling.
13. Basaltic group: pyroxene, amphibole, plagioclase feldspar; Granitic group: potassium feldspar, muscovite mica, quartz
14. If the earlier formed minerals in a cooling magma are denser (heavier) than the liquid portion of the magma, they will settle to the bottom of the magma chamber. The process is called crystal settling.
15. compaction and cementation

16. a) Detrital sedimentary rocks are made of a wide variety of mineral and rock fragments; clay minerals and quartz dominate (example: sandstone). b) Chemical sedimentary rocks are derived from material that is carried in solution to lakes and seas and, when conditions are right, precipitates to form chemical sediments (example: limestone).
17. particle size
18. Angular fragments indicate that the particles were not transported very far from their source prior to deposition.
19. mineral composition
20. Limestone is the most abundant chemical sedimentary rock. Ninety percent of limestone is biochemical sediment. The rest precipitates directly from seawater.
21. swamp environment, peat, lignite, bituminous, anthracite
22. a) detrital; b) breccia; c) C
23. Fossils are traces or remains of prehistoric life. They are useful for interpreting past environments, as time indicators, and for matching up rocks from different places that are the same age.
24. a) Regional metamorphism occurs during mountain building when great quantities of rock are subjected to intense stress and deformation. b) Contact metamorphism occurs when rock is in contact with or near a mass of magma.
25. The three agents of metamorphism include heat, pressure, and chemically active fluids.
26. a) A foliated texture results from the alignment of mineral crystals with a preferred orientation, giving a metamorphic rock a layered or banded appearance (example: slate). b) The minerals in a nonfoliated metamorphic rock are generally equidimensional and not aligned (example: marble).
27. foliated (rock name: gneiss)
28. The ore deposits in both areas were generated from hot metal-rich fluids, called hydrothermal solutions, that were associated with cooling magma bodies.

Practice Test

Multiple choice

1. b
2. b
3. e
4. d
5. c
6. b
7. d
8. a
9. d
10. c
11. b
12. b
13. a
14. d
15. d
16. d
17. b
18. a
19. d
20. b
21. c
22. b
23. d
24. a
25. a

True/false

1. T
2. F (basalt)
3. T
4. T
5. T
6. T
7. F (Sedimentary)
8. F (metamorphic)
9. T
10. F (less)
11. T
12. T
13. F (crystallization)
14. T
15. F (Metamorphism)
16. F (lava)
17. T
18. T
19. F (sedimentary)
20. T
21. T
22. T
23. F (gypsum)
24. T
25. F (large)

Word choice

1. igneous
2. less; toward
3. silicate
4. size
5. deficient; not
6. slate
7. calcite
8. heat
9. biochemical
10. calcite
11. gneiss
12. can

Written questions

1. Magma, the material of igneous rocks, is produced when any rock is melted. Sedimentary rocks are formed from the weathered products of any preexisting rock—igneous, sedimentary, or metamorphic. Metamorphic rocks are created when any rock type undergoes metamorphism.

2. Bowen's reaction series illustrates the sequence in which minerals crystallize from magma. It can be used to help explain the mineral makeup of an igneous rock, in that minerals that crystallize at about the same time (temperature) are most often found together in the same igneous rock.

3. Quartz and clay minerals are the chief constituents of detrital sedimentary rocks. Clay minerals are the most abundant product of chemical weathering. Quartz is an abundant sediment because it is very resistant to chemical weathering.

4. Metamorphic processes cause many changes in existing rocks, including increased density, growth of larger crystals, foliation, and the transformation of low-temperature minerals into high-temperature minerals. Furthermore, the introduction of ions generates new minerals, some of which are economically important.

CHAPTER FOUR

Vocabulary Review

1. erosion
2. frost wedging
3. regolith
4. Leaching
5. exfoliation dome
6. Soil Taxonomy
7. slump
8. weathering
9. permafrost
10. Secondary enrichment
11. Soil
12. Solifluction
13. mechanical weathering
14. spheroidal weathering
15. rockslide
16. chemical weathering
17. Eluviation
18. lahar
19. mass wasting
20. Differential weathering
21. soil texture
22. sheeting
23. horizon
24. Creep
25. parent material
26. talus slope
27. solum
28. angle of repose
29. soil profile

Comprehensive Review

1. The three processes that are continually removing material from higher elevations and transporting them to lower elevations are (1) weathering, the disintegration and decomposition of rock at or near Earth's surface, (2) mass wasting, the transfer of rock material downslope under the influence of gravity, and (3) erosion, the incorporation and transportation of material by a mobile agent, usually water, wind, or ice.
2. (1) Mechanical weathering: the breaking of rock into smaller pieces, each retaining the characteristics of the original material. (2) Chemical weathering: those processes that alter the internal structure of a mineral by removing and/or adding elements.
3. The three natural processes that break rocks into smaller fragments (mechanically weather) include frost wedging, unloading, and biological activity.
4. Water becomes mildly acidic when carbon dioxide (CO_2) dissolves in rainwater and produces the weak acid, carbonic acid (H_2CO_3). Decaying organisms also produce acids.
5. The three factors that influence the rate of weathering are (1) the degree of mechanical weathering, which determines the amount of surface area exposed to chemical weathering, (2) mineral makeup—minerals that crystallize first from magma (Bowen's reaction series) are least resistant to chemical weathering, and (3) climate—the optimum environment for chemical weathering is a combination of warm temperatures and abundant moisture.
6. Originally angular blocks formed along cracks called joints. Chemical weathering by the water that entered the joints caused the corners and edges of blocks to become more rounded. The corners are attacked by the acidic water more readily because of their greater surface area, as compared to the edges and faces of the blocks. This process, called spheroidal weathering, gives the weathered rock a more rounded or spherical shape.
7. clay minerals, soluble salt (potassium bicarbonate), silica in solution
8. Regolith is the layer of rock and mineral fragments produced by weathering. Soil, on the other hand, is a combination of mineral and organic matter, water, and air—that portion of the regolith that supports the growth of plants.
9. The three particle-size classes for determining soil texture are (1) clay, (2) silt, and (3) sand.
10. Clay: 20 percent; Silt: 40 percent; Sand: 40 percent
11. silty loam
12. The four basic soil structures are (1) platy, (2) prismatic, (3) blocky, and (4) spheroidal.
13. The five controls of soil formation include (1) parent material, (2) time, (3) climate, (4) plants and animals, and (5) topography.
14. From top to bottom the horizons are: O, A, E, B, and C.
15. a) at location B; b) location A; c) location C

16. Strong winds and the expansion of agriculture brought about by mechanization were two other factors that contributed to the Dust Bowl of the 1930s.
17. Secondary enrichment takes place in one of two ways: (1) chemical weathering coupled with the downward-percolating water in soil removes undesired materials from decomposing rock, leaving the desired elements enriched in the upper zones; or (2) desirable elements near the surface are removed and carried to lower soil zones where they are redeposited and become more concentrated.
18. The combined effects of mass wasting and running water produce stream valleys.
19. Among the factors are (1) saturation of the material with water, (2) oversteepening of the slopes, (3) removal of anchoring vegetation, and (4) ground vibrations from earthquakes.
20. Diagram A: slump; Diagram B: rockslide; Diagram C: debris flow; Diagram D: earthflow

Practice Test

Multiple choice

1. b
2. b
3. e
4. b
5. d
6. b
7. b
8. d
9. b
10. a
11. b
12. a
13. c
14. a
15. d
16. d
17. d
18. d
19. c
20. d
21. b
22. d
23. b
24. e
25. c
26. d
27. a
28. a
29. a
30. b

True/false

1. T
2. F (subsoil)
3. T
4. F (first)
5. T
6. F (profile)
7. T
8. T
9. T
10. F (thinner)
11. T
12. T
13. T
14. F (horizon)
15. T
16. F (flow)
17. T
18. F (flowing water)
19. T

Word choice

1. more
2. immature
3. microorganisms
4. middle
5. topsoil
6. poor
7. bauxite
8. oxidized
9. increase
10. hardpan
11. weak; acid
12. greater
13. will not
14. leaching
15. falls
16. expands
17. curved
18. reduction; removed
19. creep

Written questions

1. The two types of weathering are mechanical weathering and chemical weathering. Mechanical weathering involves the breaking of rock into smaller pieces, each retaining the characteristics of the original material. Chemical weathering refers to those processes that alter the internal structure of a mineral by removing and/or adding elements. Advanced mechanical weathering aids chemical weathering by increasing the amount of surface area available for chemical attack.

2. Soil is a combination of mineral and organic matter, water, and air—that portion of the regolith that supports the growth of plants. Since the soil-forming processes operate from the surface downward, variations in composition, texture, structure, and color evolve at varying depths. These vertical differences divide the soil into zones or layers that soil scientists call horizons. Such a vertical section through all the soil horizons constitutes the soil profile.

3. Gravity is the controlling force of mass wasting. Other important factors include (1) saturation of the material with water, (2) oversteepening of the slopes, (3) removal of anchoring vegetation, and (4) ground vibrations from earthquakes.

ANSWER KEY

CHAPTER FIVE

Vocabulary Review

1. Infiltration
2. drainage basin
3. zone of saturation
4. porosity
5. Gradient
6. alluvium
7. capacity
8. groundwater
9. yazoo tributary
10. transpiration
11. meander
12. water table
13. discharge
14. artesian well
15. natural levee
16. permeability
17. hydrologic cycle
18. floodplain
19. competence
20. divide
21. aquitard
22. evapotranspiration
23. Base level
24. spring
25. aquifer
26. Laminar flow
27. suspended load
28. karst topography
29. braided stream
30. runoff

Comprehensive Review

1. mass wasting, weathering, and erosion by running water
2. a) C; b) B; c) A; d) D
3. a) the distance that water travels in a unit of time; b) the slope of a stream channel expressed as the vertical drop of a stream over a specified distance; c) the volume of water flowing past a certain point in a given unit of time
4. 4 meters/kilometer (40 meters divided by 10 kilometers)
5. a) decreases; b) increases; c) decreases; d) increases; e) increase
6. a) Ultimate base level is sea level, the lowest level to which stream erosion could lower the land. b) Temporary base levels include lakes and main streams, which act as a base level for the streams and tributaries entering them.
7. (1) as dissolved load; brought to a stream by groundwater; (2) as suspended load; sand, silt, and clay carried in the water; (3) as bed load; coarser particles moving along the bottom of a stream
8. During a flood, the increase in discharge of a stream results in a greater capacity, and the increase in velocity results in greater competence.
9. a) C; b) B; c) A; d) D
10. (1) A narrow valley is typically V-shaped because the primary work of the stream has been downcutting toward base level. The features of narrow valleys include rapids and waterfalls. (2) In a wide valley the stream's energy is directed from side to side, and downward erosion becomes less dominant. The features of wide valleys include meanders, floodplains, bars, cutoffs, and oxbow lakes.
11. radial pattern; where streams diverge from a central area like the spokes of a wheel
12. Engineering efforts to lessen the effect of floods include the construction of artificial levees, the building of flood-control dams, and river channelization.
13. floodplain, widespread meanders, natural levees, oxbow lakes, cutoffs, braided channels
14. a) D; b) B; c) A; d) C; e) E
15. The shape of the water table is usually a subdued replica of the surface, reaching its highest elevations beneath hills and decreasing in height toward valleys.
16. The source of heat for most hot springs and geysers is in cooling igneous rock beneath the surface.
17. A rock layer or sediment that transmits water freely because of its high porosity and high permeability is called an aquifer.
18. (1) land subsidence caused by groundwater withdrawal; (2) groundwater contamination
19. (1) slowly, over many years by dissolving the limestone below the soil; (2) suddenly, when the roof of a cavern collapses under its own weight
20. Diagram A: dendritic pattern; Diagram B: rectangular pattern

Practice Test

Multiple choice

1. c
2. c
3. d
4. c
5. b
6. b
7. a
8. d
9. b
10. d
11. e
12. a
13. b
14. c
15. b
16. c
17. e
18. e
19. d
20. b
21. c
22. a
23. b
24. d
25. a

True/false

1. T
2. F (Temporary)
3. T
4. T
5. T
6. F (wide)
7. T
8. F (decreases)
9. F (rectangular)
10. T
11. T
12. F (Impermeable)
13. T
14. T
15. T
16. F (suspension)
17. T
18. T
19. T
20. T
21. T
22. T
23. F (transpiration)
24. T
25. T
26. T
27. T
28. T
29. T
30. F (Running water)

Word choice

1. center
2. trellis
3. distributaries
4. water table
5. decreases
6. aquifer
7. decreases; increases
8. limestone; wet
9. gentle; light
10. discharge
11. raise
12. outside
13. carbonic
14. increases
15. porosity

Written questions

1. Evaporation, primarily from the ocean, transport via the atmosphere, and eventually precipitation back to the surface. If the water falls on the continents, much of it will find its way back to the atmosphere by evaporation and transpiration. Some water, however, will be absorbed by the surface (infiltration) and some will run off back to the ocean.
2. Base level, the lowest point to which a stream can erode its channel, is perhaps the most important variable controlling the work of streams. Any change in base level will cause a corresponding adjustment of stream activities. For example, when base level is raised, the stream gradient is reduced, lowering its velocity and hence its sediment-transporting ability. On the other hand, if base level is lowered, gradient and velocity would increase and the stream would downcut to establish balance with its new base level.
3. Porosity is the percentage of the total volume of rock or sediment that consists of pore spaces. Permeability is a measure of a material's ability to transmit water through interconnected pore spaces. It is possible for a material to have a high porosity but a low permeability if the pore spaces lack interconnections.
4. Generally, for karst topography to develop, the area must be humid and underlain by soluble rock, usually limestone.

ANSWER KEY

CHAPTER SIX

Vocabulary Review

1. ice sheet
2. ephemeral stream
3. glacial drift
4. drumlin
5. Pleistocene epoch
6. glacial striations
7. interior drainage
8. medial moraine
9. zone of wastage
10. glacial trough
11. cirque
12. barchan dune
13. zone of accumulation
14. alluvial fan
15. longitudinal dune
16. loess
17. till
18. outwash plain
19. blowout
20. cross beds
21. glacier
22. piedmont glacier
23. ablation
24. ice shelf

Comprehensive Review

1. (1) Valley, or alpine, glaciers are relatively small streams of ice that exist in mountain areas, where they usually follow valleys originally occupied by streams. (2) Ice sheets exist on a large scale and flow out in all directions from one or more centers, completely obscuring all but the very high areas of underlying terrain.
2. This upper part of the glacier consists of brittle ice that is subjected to tension when the ice flows over irregular terrain, with cracks called crevasses resulting.
3. a) Plucking occurs when a glacier flows over a fractured bedrock surface and the ice incorporates loose blocks of rock and carries them off. b) Abrasion occurs as the ice with its load of rock fragments moves along and acts as a giant rasp or file, grinding the surface below as well as the rocks within the ice.
4. a) D; b) E; c) A; d) C; e) B
5. Till is unsorted material deposited directly by the glacier. Stratified drift is characteristically sorted sediment laid down by glacial meltwater.
6. Two of the many hypotheses for the cause of glacial ages involve (1) plate tectonics and (2) variations in Earth's orbit.
7. B; a) C; b) D
8. a) forms at the terminus of a glacier; b) forms along the side of a valley glacier; c) a layer of till that is laid down as the glacier recedes
9. a) A; b) G; c) C; d) B
10. Periodic slides of sand down the slip face of a dune cause the slow migration of the dune in the direction of air movement.
11. The indirect effects of Ice Age glaciers include plant and animal migrations, changes in stream and river courses, adjustment of the crust by rebounding, climate changes, and the worldwide change in sea level that accompanied each advance and retreat of the ice sheets.
12. moraines, glacial striations, glacial troughs, hanging valleys, cirques, horns, arêtes, fiords, glacial drift, glacial erratics
13. The existence of several layers of glacial drift with well-developed zones of chemical weathering and soil formation, as well as the remains of plants that require warm temperatures, indicate several glacial advances separated by periods of warmer climates.
14. Wind is not capable of picking up and transporting coarse materials, and because it is not confined to channels, it can spread over large areas as well as high into the atmosphere.
15. Diagram A: barchan dunes; Diagram B: transverse dunes; Diagram C: barchanoid dunes; Diagram D: longitudinal dunes

Practice Test

Multiple choice

1. c
2. c
3. b
4. d
5. a
6. a
7. d
8. b
9. b
10. b
11. b
12. d
13. a
14. c
15. e
16. b
17. a
18. b
19. e
20. d
21. a
22. b
23. b
24. d
25. a

True/false

1. T
2. F (accumulation)
3. T
4. F (lateral)
5. T
6. F (50)
7. T
8. T
9. F (10)
10. F (30)
11. F (meters)
12. T
13. F (U-shaped)
14. F (drumlins)
15. T
16. T
17. T
18. F (horn)
19. T
20. F (Cirques)
21. T
22. T
23. T

Word choice

1. till
2. flour
3. mechanical
4. interior
5. Fiords
6. erratic
7. calving
8. sorted
9. few
10. subdue
11. deflation
12. is not
13. lack

Written questions

1. (1) moraines: layers or ridges of till; (2) drumlins: streamlined, asymmetrical hills composed of till; (3) eskers: ridges of sand and gravel deposited by streams flowing beneath the ice, near the glacier's terminus; (4) kames: steep-sided hills formed when glacial meltwater washes sediment into openings and depressions in the stagnant, wasting terminus of a glacier (answers may vary)

2. (1) glacial trough: a U-shaped glaciated valley; (2) horn: a pyramid-shaped peak; (3) cirque: a bowl-shaped depression at the head of a valley glacier where snow accumulation and ice formation occur (answers may vary)

3. (1) plant and animal migrations; (2) changes in stream and river courses; (3) worldwide change in sea level that accompanied each advance and retreat of the ice sheets (answers may vary)

4. Periodic slides of sand down the slip face of a dune cause the slow migration of the dune in the direction of air movement.

ANSWER KEY

CHAPTER SEVEN

Vocabulary Review

1. continental drift
2. plate
3. Seafloor spreading
4. convergent plate boundary
5. paleomagnetism
6. plate tectonics
7. normal polarity
8. divergent plate boundary
9. hot spot
10. Pangaea
11. subduction zone
12. rift (rift valley)
13. transform boundary
14. Curie point
15. reverse polarity
16. volcanic island arc
17. deep-ocean trench
18. oceanic ridge system

Comprehensive Review

1. (1) fit of the continents; (2) similar fossils on different landmasses; (3) similar rock types and structures on different landmasses; (4) similar ancient climates on different landmasses
2. One of the main objections to Wegener's continental-drift hypothesis was his inability to provide a mechanism that was capable of moving the continents across the globe.
3. a) C; b) A; c) B
4. The theory of plate tectonics holds that Earth's rigid outer shell consists of several rigid slabs called plates that are in continuous slow motion relative to each other. These plates interact in various ways and thereby produce earthquakes, volcanoes, mountains, and the crust itself.
5. (1) oceanic-continental convergence; (2) oceanic-oceanic convergence; (3) continental-continental convergence
6. a) Paleomagnetism is used to show that the continents have moved through time and, using the patterns of magnetic reversals found on the ocean floors, to explain seafloor spreading. b) Ocean drilling has shown that the ocean basins are geologically youthful and confirms seafloor spreading by the fact that the age of the deepest ocean sediment increases with increasing distance from the ridge.
7. a) D; b) B; c) C; d) F
8. According to the plate tectonics theory, the Himalayas formed as two plates, each with continental lithosphere, collided. As the continental masses collide, sediments along the margins are folded and deformed into mountains.
9. (1) slab push and slab pull hypothesis: as new ocean crust moves away from the ridge, it cools, increases in density, and eventually descends, pulling the trailing lithosphere along; (2) hot plumes hypothesis: hot plumes within the mantle reach the lithosphere and spread laterally, facilitating the plate motion away from the zone of upwelling
10. a) F; b) B; c) C; d) E; e) A; f) D; g) G
11. Using satellites, scientists are now able to confirm the fact that plates shift in relation to one another by directly measuring their relative motions using timed laser pulses that are bounced off the satellites.

Practice Test

Multiple choice

1. b
2. b
3. b
4. e
5. a
6. d
7. e
8. a
9. a
10. b
11. e
12. a
13. b
14. c
15. b
16. a
17. a
18. c
19. b
20. e
21. b
22. a
23. a
24. d
25. b
26. d

True/false

1. T
2. T
3. F (centimeters)
4. F (divergent)
5. T
6. F (mantle)
7. F (younger)
8. T
9. T
10. T
11. F (trenches)
12. T
13. T
14. F (mantle)
15. T
16. F (normal)
17. T
18. F (youngest)
19. T
20. T
21. T
22. T
23. F (thinnest)
24. T
25. T

Word choice

1. lithosphere
2. continental shelf
3. consumed
4. also
5. oceanic
6. rigid
7. 200
8. tropical
9. subduction
10. more than
11. created
12. South Pole
13. heat
14. thicker
15. cooler; more
16. subduction zones
17. geographic poles
18. rate
19. mantle plumes
20. increases
21. will not
22. parallel
23. shallow
24. rifts
25. volcanic arcs
26. faster
27. slab pull

Written questions

1. Hot spots are relatively stationary plumes of molten rock rising from Earth's mantle. According to the plate tectonics theory, as a plate moves over a hot spot, magma often penetrates the surface, thereby generating a volcanic structure. In the case of the Hawaiian Islands, as the Pacific plate moved over a hot spot, the associated igneous activity produced a chain of major volcanoes. Currently, because the rising plume of mantle material is located below it, the only active island in the chain is Hawaii. Kauai, the oldest island in the chain, formed approximately 5 million years ago when it was positioned over the hot spot and has now moved northwest as a result of plate motion.

2. (1) divergent plate boundaries—where the plates are moving apart; (2) convergent plate boundaries—where the plates are moving toward each other; (3) transform fault boundaries—where the plates are sliding past each other along faults

3. The lines of evidence used to support plate tectonics include (1) paleomagnetism (polar wandering and magnetic reversals), (2) earthquake patterns, (3) the age and distribution of ocean sediments, and (4) evidence from ocean-floor volcanoes that are, or were, located over hot spots.

ANSWER KEY

CHAPTER EIGHT

Vocabulary Review

1. earthquake
2. seismic sea wave (tsunami)
3. foreshock
4. focus
5. elastic rebound
6. surface wave
7. outer core
8. epicenter
9. asthenosphere
10. fault
11. crust
12. secondary (S) wave
13. seismogram
14. Mohorovicic discontinuity (Moho)
15. magnitude
16. inner core
17. primary (P) wave
18. lithosphere
19. Richter scale
20. shadow zone
21. mantle
22. seismograph
23. Moment magnitude

Comprehensive Review

1. (1) surface waves: travel around the outer layer of Earth; (2) primary (P) waves: travel through the body of Earth with a push-pull motion in the direction the wave is traveling; (3) secondary (S) waves: travel through Earth by "shaking" the rock particles at right angles to their direction of travel. P waves travel through all materials, whereas S waves are transmitted only through solids. Further, in all types of rock, P waves travel faster than S waves.
2. a) A; b) B
3. a) C; b) B; c) A
4. Primary (P) waves have a greater velocity than secondary (S) waves. P waves can travel through solids, liquids, and gases, while S waves are only capable of being transmitted through solids.
5. By determining the difference in arrival times between the first P wave and first S wave and using a travel-time graph, three different recording stations determine how far they each are from the epicenter. Next, for each station, a circle is drawn on a globe corresponding to the station's distance from the epicenter. The point where the three circles intersect is the epicenter of the quake.
6. (1) magnitude of the earthquake; (2) proximity to a populated area
7. surrounding the Pacific Ocean basin and along the mid-ocean ridge (answers may vary); Most earthquake epicenters are closely correlated with plate boundaries.
8. a) E (including D and part of C); b) G; c) A; d) B; e) D; f) F; g) C
9. a) Denser rocks like peridotite are thought to make up the mantle and provide the lava for oceanic eruptions. b) The molten outer core is thought to be mainly iron and nickel.

Practice Test

Multiple choice

1. e
2. c
3. e
4. b
5. a
6. a
7. c
8. d
9. b
10. e
11. e
12. a
13. b
14. c
15. d
16. b
17. c
18. b
19. a
20. d

True/false

1. T
2. F (aftershocks)
3. F (two)
4. T
5. T
6. T
7. F (vertical)
8. F (seismology)
9. T
10. F (can)
11. F (before)
12. F (Mohorovicic)
13. T
14. F (above)
15. T
16. T
17. T
18. T
19. T
20. F (less)
21. T
22. T

Word choice

1. S
2. epicenter
3. surface
4. rebound
5. Moho
6. seismic waves
7. faults
8. Richter
9. asthenosphere
10. seismographs; seismograms
11. on land
12. flexible
13. Inertia
14. lithosphere
15. stationary

Written questions

1. An earthquake is the vibration of Earth produced by the rapid release of energy, usually along a fault. This energy radiates in all directions from the earthquake's source, called the focus, in the form of waves.

2. P waves are "push-pull" waves that compress and expand the rock in the direction they are traveling; S waves, on the other hand, "shake" the materials at right angles to the direction of travel. P waves travel through all materials, whereas S waves are transmitted only through solids. Further, in all types of rock, P waves travel faster than S waves.

3. Oceanic crust is basaltic in composition, while the continental crust has an average composition similar to that of granite. The composition of the mantle is thought to be similar to the rock peridotite, which contains iron and magnesium-rich silicate minerals. Both the inner and outer cores are thought to be enriched in iron and nickel, with lesser amounts of other heavy elements.

ANSWER KEY

CHAPTER NINE

Vocabulary Review

1. pyroclastic material
2. caldera
3. aa flow
4. volcanic neck
5. dike
6. vent
7. cinder cone
8. batholith
9. shield volcano
10. flood basalt
11. pahoehoe flow
12. hot spot
13. composite cone
14. volcano
15. laccolith
16. fissure eruption
17. sill
18. nuée ardente
19. partial melting
20. viscosity
21. lahar
22. geothermal gradient
23. Plutons
24. lava tube

Comprehensive Review

1. (1) the magma's composition; (2) the magma's temperature; (3) the amount of dissolved gases in the magma
2. a) B (Mauna Loa): Shield volcanoes have recurring eruptions over a long span of time and are composed of large amounts of basaltic lava. b) A (Paricutin): The eruption history of a cinder cone is short and produces a relatively small cone composed of loose pyroclastic material. c) C (Fujiyama): A composite cone may extrude viscous lava for a long period, and then suddenly the eruptive style changes and the volcano violently ejects pyroclastic material. Composite cones represent the most violent type of volcanic activity.
3. water vapor, carbon dioxide, nitrogen, sulfur gases
4. a) the igneous body cuts across existing sedimentary beds; b) the igneous body is parallel to existing sedimentary beds
5. The viscosity of magma is affected by its temperature, silica content (the more silica, the greater the viscosity), and quantity of dissolved gases.
6. a) C; b) E; c) F; d) D
7. a) One source of heat to melt rock is from friction in a subduction zone. Secondly, the gradual increase in temperature with increasing depth can melt subducting oceanic crust. Finally, a hot magma body could form at depth and migrate upward to the base of the crust. b) Reducing confining pressure and/or the introduction of volatiles lowers a rock's melting temperature sufficiently to trigger melting. c) Partial melting produces most, if not all, magma. A consequence of partial melting is the production of a magma with a higher silica content than the parent rock.
8. When compared to rhyolitic/felsic magma, basaltic magma contains less silica, is less viscous, and in a near-surface environment basaltic rocks melt at a higher temperature.
9. (1) Along the oceanic ridge system, as the rigid lithosphere pulls apart, pressure is lessened and the partial melting of the underlying mantle rocks produces large quantities of basaltic magma. (2) Adjacent to ocean trenches, magma is generated by the partial melting of descending slabs of oceanic crust. The melt slowly rises upward toward the surface because it is less dense than the surrounding rock. (3) Intraplate volcanism may be the result of rising plumes of hot mantle material that produce hot spots, volcanic regions a few hundred kilometers across.

Practice Test

Multiple choice

1. b	6. e	11. b	16. e	21. d
2. b	7. a	12. b	17. a	22. b
3. b	8. e	13. a	18. a	23. c
4. c	9. d	14. e	19. b	24. a
5. b	10. c	15. a	20. d	25. a

True/false

1. F (convergent)
2. T
3. T
4. T
5. F (cinder)
6. T
7. F (less)
8. T
9. F (crater)
10. T
11. T
12. F (increase)
13. F (silica)
14. F (composite)
15. T
16. T
17. T
18. F (sills)
19. T
20. T

Word choice

1. Decreasing
2. summit; magma chamber
3. sulfur
4. high; large
5. higher
6. oceanic ridge system
7. upper mantle rock; basaltic
8. gases; upper
9. increase; temperature
10. Increasing
11. cinder
12. decrease; pressure
13. basaltic
14. convergent plate; water
15. reduced

Written questions

1. Magma and lava both refer to the material from which igneous rocks form. However, magma is molten rock below Earth's surface, including dissolved gases and crystals, while lava refers to molten rock that reaches Earth's surface.

2. (1) Shield cones are among the largest on Earth. These gently sloping domes are associated with relatively quiet eruptions of fluid basaltic lava. (2) Cinder cones are composed almost exclusively of pyroclastic material, are steep-sided, and are the smallest of the volcanoes. (3) Composite cones, as the name suggests, are composed of alternating layers of lava and pyroclastic debris. Their slopes are steeper than those of a shield volcano but more gentle than those of a cinder cone.

3. (1) Dikes are tabular, discordant plutons. (2) Sills are tabular, concordant plutons. (3) Batholiths, by far the largest of all intrusive features, are massive, discordant plutons (answers may vary).

ANSWER KEY

CHAPTER TEN

Vocabulary Review

1. strike-slip fault
2. horst
3. Isostasy
4. fold
5. basin
6. Orogenesis
7. fault
8. thrust fault
9. graben
10. passive continental margin
11. anticline
12. syncline
13. strike-slip faults
14. Monoclines
15. dip-slip fault
16. Deformation
17. normal fault
18. isostatic adjustment
19. joint
20. accretionary wedge
21. terrane
22. dome
23. reverse fault

Comprehensive Review

1. The presence of fossilized shells of aquatic organisms at high elevations is evidence that the sedimentary rocks were once below sea level.
2. The two reasons that the crust beneath the oceans is at lower elevation than the continental crust are (1) the crust beneath the oceans is thinner than that of the continents, and (2) oceanic rocks have a greater density than continental rocks.
3. Isostasy refers to the fact that Earth's less dense crust is "floating" on top of the denser, and more easily deformed rocks of the mantle.
4. Changes in rock resulting from elastic deformation are reversible; that is, the rock will return to nearly its original size and shape when the stress is removed.
5. a) An anticline is most commonly formed by the upfolding, or arching, of rock layers (letter A); b) Synclines, usually found in association with anticlines, are downfolds, or troughs (letter B).
6. Diagram A: dome; Diagram B: basin
7. a) hanging wall moves up relative to the footwall (diagram B); b) dominant displacement is along the trend, or strike, of the fault (diagram D); c) hanging wall moves down relative to the footwall (diagram A); d) hanging wall moves up relative to the footwall at a low horizontal angle (diagram C)
8. a) B; b) A
9. normal faults
10. reverse faults and thrust faults
11. a) Where oceanic and continental crusts converge, the first stage in the development of a mountain belt is the formation of a subduction zone. Deposition of sediments occurs along the continental margin. Convergence of the continental block and the subducting oceanic plate leads to deformation and metamorphism of the continental margin. A volcanic island arc forms. Sediment derived from the land as well as that scraped from the subducting plate becomes plastered against the landward side of the trench, forming an accretionary wedge consisting of folded, faulted, and metamorphosed sediments and volcanic debris. b) Where continental crusts converge, the continental lithosphere is too buoyant to undergo any appreciable subduction, and a collision between continental fragments eventually results. The result of the collision is the formation of a mountain range.
12. a) C; b) A; c) E; d) D; e) B
13. Any two of the following: Himalaya Mountains, Ural Mountains, Alps, and Appalachians.
14. Because of the great length and complexity of the San Andreas Fault, it is more appropriately referred to as a "fault system." This major fault system consists primarily of the San Andreas Fault and several major branches, including the Hayward and Calaveras faults of central California.

Practice Test

Multiple choice

1. d
2. c
3. e
4. b
5. a
6. a
7. c
8. d
9. b
10. e
11. e
12. a
13. e
14. d
15. c
16. c
17. d
18. a
19. b
20. c
21. c
22. a
23. c
24. b
25. c

True/false

1. F (raises)
2. T
3. T
4. F (Faults)
5. T
6. T
7. T
8. F (compressional)
9. T
10. T
11. F (youngest)
12. F (elastically)
13. F (vertical)
14. T
15. F (folds)
16. T
17. T
18. T
19. F (folded)
20. F (synclines)
21. F (downward)
22. T
23. T
24. F (convergent)
25. T
26. T
27. F (California)

Word choice

1. ductile
2. tensional
3. 29
4. isostasy
5. dip-slip
6. compressional
7. isostatic
8. tensional
9. subducting
10. synclines
11. margins
12. divergence
13. less
14. expose
15. igneous
16. is

Written questions

1. In time, extensive erosion often levels the hills on the surface, leaving only the upfolded rocks below.

2. An accretionary wedge is a chaotic accumulation of metamorphic rocks, sedimentary rocks derived from the land as well as sediment scraped from a subducting plate, and occasional pieces of ocean crust that is plastered against the landward side of a trench.

3. Most of Earth's major mountain systems have formed along convergent plate boundaries. Convergence can occur between one oceanic and one continental plate, between two oceanic plates, or between continental plates. In the first two, subduction zones often develop and the sediments and volcanics are squeezed, faulted, and deformed into mountain ranges as the plates converge. When continents collide, no subduction zone forms, and the collision between the continental fragments deforms the continental material into mountains.

ANSWER KEY

CHAPTER ELEVEN

Vocabulary Review

1. uniformitarianism
2. Cenozoic era
3. numerical date
4. correlation
5. half-life
6. catastrophism
7. relative dating
8. Precambrian
9. original horizontality
10. law of superposition
11. unconformity
12. fossil
13. radioactivity
14. period
15. inclusions
16. Paleozoic era
17. index fossil
18. principle of fossil succession
19. radiometric dating
20. eon
21. paleontology

Comprehensive Review

1. Catastrophism states that Earth's landscape was developed by great catastrophes over a short span of time. On the other hand, uniformitarianism states that the physical, chemical, and biological laws that operate today have also operated in the geologic past. Uniformitarianism implies that the forces and processes that we observe today have been at work for a very long time.
2. a) Large coal swamps flourished in North America before the extinction of dinosaurs. b) Dinosaurs became extinct about 66 million years ago.
3. a) In an undeformed sequence of sedimentary rocks, each bed is older than the one above it and younger than the one below. b) Most layers of sediment are deposited in a horizontal position. c) Fossil organisms succeed each other in a definite and determinable order, and therefore any time period can be recognized by its fossil content.
4. a) A; b) C; c) B
5. (1) rapid burial; (2) possession of hard parts
6. a) 1; b) 4
7. a) B; b) A
8. An atom's atomic number is the number of protons in the nucleus, while the mass number is determined by adding together the number of protons and neutrons in the nucleus.
9. (1) emission of alpha particles (2 protons and 2 neutrons) from the nucleus; (2) emission of beta particles, or electrons (derived from neutrons), from the nucleus; (3) electrons are captured by the nucleus and combine with protons to form additional neutrons
10. Because the rates of decay for many isotopes have been precisely measured and do not vary under the physical conditions that exist in Earth's outer layers.
11. Carbon-14 is continuously produced in the upper atmosphere as a consequence of cosmic-ray bombardment. Cosmic rays shatter the nuclei of gas atoms, releasing neutrons. These neutrons are absorbed by nitrogen atoms, which in turn emit a proton and become the element carbon-14.
12. The primary problem in assigning numerical dates to units of time of the geologic time scale is the fact that not all rocks can be dated radiometrically.
13. a) younger (principle of cross-cutting relationships); b) younger (law of superposition); c) younger; d) older; e) older
14. 100,000 years (two half-lives, each 50,000 years)

Practice Test

Multiple choice

1. a
2. b
3. d
4. e
5. e
6. c
7. c
8. a
9. a
10. d
11. b
12. b
13. c
14. b
15. c
16. b
17. b
18. a
19. a
20. e
21. b
22. d
23. d
24. e
25. b

True/false

1. F (relative)
2. F (long)
3. T
4. T
5. F (hard)
6. T
7. F (cannot)
8. F (protons)
9. T
10. F (younger)
11. F (increases)
12. T
13. T
14. T
15. T
16. T
17. T
18. F (uniformitarianism)
19. T
20. T
21. T

Word choice

1. index
2. eons
3. neutrons
4. James Hutton
5. organic
6. different from
7. daughter product
8. conformable
9. wide; short
10. do
11. relative
12. fine
13. daughter
14. older
15. radon
16. older
17. three
18. relative
19. folding and erosion
20. Sedimentary

Written questions

1. Radiometric dating uses the radioactive decay of particular isotopes to arrive at a numerical date, which pinpoints the time in history when something took place; for example, the extinction of the dinosaurs about 65 million years ago. Relative dating involves placing rocks or events in their proper sequence using various laws and principles. Relative dating cannot tell us how long ago something took place, only that it followed one event and preceded another.

2. (1) law of superposition; (2) principle of original horizontality; (3) principle of cross-cutting relationships

3. Fossils help researchers understand past environmental conditions, are useful as time indicators, and play a key role in correlating rocks of similar ages that are from different places.

ANSWER KEY

CHAPTER TWELVE

Vocabulary Review

1. stromatolites
2. Protoplanets
3. prokaryotes
4. supernova
5. solar nebula
6. planetesimals
7. outgassing
8. banded iron formations
9. Supercontinents
10. Eukaryotes

Comprehensive Review

1. The source of the gases that formed Earth's original atmosphere was volcanic outgassing. The gases were probably water vapor, carbon dioxide, nitrogen, and several trace gases—very similar to those gases released in volcanic emissions today.
2. From green plants releasing oxygen during photosynthesis.
3. a) Cenozoic; b) Precambrian; c) Mesozoic; d) Paleozoic; e) Paleozoic; f) Mesozoic; g) Paleozoic; h) Cenozoic; i) Paleozoic; j) Mesozoic
4. In the late Archean, between 3-2.5 billion years ago, an episode of major crustal growth occurred that created the initial nucleus of North America. During this time span, the accretion of numerous island arcs and other crustal fragments generated several large crustal provinces, including the Superior and Hearne/Rae cratons. About 1.9 billion years ago these crustal provinces collided to produce the Trans-Hudson mountain belt, which produced the North American craton. Several large and numerous small crustal fragments were later added to the craton including the Blue Ridge and Piedmont provinces of the Appalachians. Finally, several other terranes were added to the western margin of North America during the Mesozoic and Cenozoic eras to generate the mountainous North American Cordillera.
5. The beginning of the Paleozoic era is marked by the appearance of the first life forms with hard parts—shells, scales, bones, or teeth.
6. The two hypotheses most often cited for the extinction of the dinosaurs are (1) an impact of a large asteroid or comet with Earth and (2) extensive volcanism. Both could produce a dust-laden atmosphere that would reduce the amount of sunlight that penetrated Earth's surface, causing delicate food chains to collapse.
7. The stable continental margin of eastern North America was the site of abundant sedimentation. In the West, the Rocky Mountains were eroding and a great wedge of sediment was building the Great Plains, crustal movements created the Basin and Range Province, the Rockies were re-elevated, volcanic activity was common, and the Coast Ranges and Sierra Nevada formed.
8. The development and specialization of mammals took four principal directions; (1) increase in size, (2) increase in brain capacity, (3) specialization of teeth, and (4) specialization of limbs.
9. a) Cenozoic era: (65 million years ago through today), the "age of mammals," the "age of flowering plants," humans evolve, large mammals become extinct; b) Mesozoic era: (248 to 65 million years ago), "age of dinosaurs," breakup of Pangaea, widespread igneous activity in the west, gymnosperms dominant, first birds, mammals evolve, mass extinction at the close; c) Paleozoic era: (540 to 248 million years ago), first life forms with hard parts, trilobites dominant and then become extinct, first fishes, first land plants, large coal swamps, amphibians, first reptiles, Pangaea forms, mass extinction at the close.

Practice Test

Multiple choice

1. d
2. e
3. b
4. e
5. d
6. e
7. a
8. c
9. b
10. c
11. a
12. d
13. b
14. d
15. a
16. d
17. a
18. d
19. a
20. c
21. a
22. b
23. b
24. b
25. b
26. a
27. a

True/false

1. T
2. F (bacteria)
3. T
4. T
5. T
6. F (drier)
7. F (epoch)
8. F (reptiles)
9. T
10. T
11. T
12. T
13. F (Paleozoic)
14. T
15. F (Paleozoic)
16. T
17. T
18. F (Cenozoic)
19. F (stromatolites)
20. T

Word choice

1. life forms
2. iron
3. Paleozoic
4. molten rock
5. angiosperms
6. near
7. Precambrian
8. Tertiary
9. before
10. Paleozoic
11. Cambrian
12. drier
13. increases
14. above
15. western

Written questions

1. The two hypotheses most often cited for the extinction of the dinosaurs are (1) an impact of a large asteroid or comet with Earth and (2) extensive volcanism. Both could produce a dust-laden atmosphere that would reduce the amount of sunlight that penetrated Earth's surface, causing delicate food chains to collapse.

2. So little is known about the Precambrian because the early rock record has been obscured by Earth processes such as plate tectonics, erosion, and deposition.

3. Unlike amphibians, reptiles have shell-covered eggs that can be laid on land. The watery fluid within the reptilian egg closely resembles seawater in chemical composition. Therefore, reptiles eliminated the need for a water-dwelling stage (like the tadpole stage in frogs).

4. Mammals (1) have their young born live, (2) maintain a steady body temperature—that is, they are "warm blooded," (3) have insulating body hair, and (4) have a more efficient heart and lungs.

ANSWER KEY

CHAPTER THIRTEEN

Vocabulary Review

1. Oceanography
2. Bathymetry
3. deep-ocean trench
4. continental shelf
5. turbidity current
6. abyssal plain
7. oceanic (mid-ocean) ridge
8. terrigenous sediment
9. seamount
10. hydrogenous sediment
11. biogenous sediment
12. continental rise
13. guyot
14. rift valley
15. continental slope
16. echo sounder
17. turbidite
18. manganese nodule
19. Gas hydrate
20. volcanic island arc
21. oceanic plateau

Comprehensive Review

1. In the Northern Hemisphere nearly 61 percent of the surface is water. About 81 percent of the Southern Hemisphere is water. The Northern Hemisphere has about twice as much land (39 percent compared to 19 percent of the surface) as the Southern Hemisphere.
2. The first device to measure water depth is an instrument, called an echo sounder, that bounces sound-waves off the ocean floor. Since World War II, sidescan sonar and high-resolution multibeam instruments have also been developed.
3. a) D; b) C; c) G; d) F; e) B
4. Abyssal plains are incredibly flat features found adjacent to continental slopes on ocean basin floors in all oceans. They consist of thick accumulations of sediment transported by turbidity currents far out to sea. (letter E)
5. Submarine canyons that are not seaward extensions of river valleys have probably been excavated by turbidity currents as these sand- and mud-laden currents repeatedly sweep downslope.
6. (1) warm waters with an average annual temperature about 24°C (75°F); (2) clear sunlit water; (3) a depth of water no greater than 45 meters (150 feet)
7. Coral islands, called atolls, form on the flanks of sinking volcanic islands from corals and other organisms that continue to build the reef complex upward.
8. (1) Terrigenous sediment consists primarily of mineral grains that were weathered from continental rocks and transported to the ocean. (2) Biogenous sediment consists of shells and skeletons of marine animals and plants. (3) Hydrogenous sediment consists of minerals that crystallize directly from seawater through various chemical reactions.
9. a) C; b) F; c) B; d) A; e) E
10. Oil, natural gas, gas hydrates, sand and gravel, evaporative salts, manganese nodules (any three)

Practice Test

Multiple choice

1. c
2. a
3. b
4. a
5. a
6. c
7. c
8. c
9. b
10. c
11. b
12. e
13. e
14. b
15. b
16. b
17. d
18. b
19. c
20. c

True/false

1. T
2. F (Atlantic)
3. T
4. T
5. F (slope)
6. T
7. F (mud)
8. T
9. T
10. T
11. T
12. T
13. F (slow)

Word choice

1. Mariana
2. continental
3. Pacific
4. larger
5. subsidence
6. lower
7. deep-sea fans
8. Pacific
9. Biogenous
10. passive
11. rift
12. mantle
13. more
14. living organisms
15. mud
16. biogenous
17. divergent

Written questions

1. The ocean basin floor lies between the continental margin and the mid-oceanic ridge system. The features of the ocean basin floor include deep-ocean trenches (the deepest parts of the ocean, where moving crustal plates descend back into the mantle), abyssal plains (the most level places on Earth, consisting of thick accumulations of sediments that were deposited atop the low, rough portions of the ocean floor), and seamounts and guyots (isolated volcanic peaks on the ocean floor that originate near the mid-ocean ridge or in association with volcanic hot spots).

2. (1) continental shelf; (2) continental slope; (3) continental rise

3. Oceanic ridges are spreading centers; that is, they are divergent plate boundaries where magma from Earth's interior rises and forms new oceanic crust. Ocean trenches form where lithospheric plates plunge into the mantle. Thus, these long, narrow, linear depressions are associated with convergent plate boundaries and the destruction of crustal material.

ANSWER KEY

CHAPTER FOURTEEN

Vocabulary Review

1. benthos
2. salinity
3. photic zone
4. Plankton
5. intertidal zone
6. primary productivity
7. food chain
8. pycnocline
9. Nekton
10. neritic zone
11. pelagic zone
12. Density
13. aphotic zone
14. oceanic zone
15. phytoplankton
16. thermocline
17. food web
18. photosynthesis
19. zooplankton
20. abyssal zone
21. trophic level
22. biomass
23. euphotic

Comprehensive Review

1. Sodium chloride and magnesium chloride are the two most abundant salts in seawater. The two primary sources for most dissolved substances in the ocean are (1) chemical weathering of rocks on the continents, and (2) from outgassing through volcanic eruptions.
2. Increase seawater salinity: evaporation and the formation of sea ice. Decrease seawater salinity: precipitation, runoff from land, icebergs melting, and sea ice melting (any two).
3. Near the equator salinity is low due to precipitation; in the tropics higher salinities occur because of high evaporation rates; salinity decreases toward the poles (see textbook Figure 14.3).
4. Seawater density is influenced by two main factors: *salinity* and *temperature.* Temperature has the greatest influence on surface seawater density because variations in surface seawater temperature are greater than salinity variations.
5. In low latitudes, both temperature and density decrease rapidly below the surface zone. The layer of rapidly changing temperature is called the *thermocline,* while the layer of changing density is referred to as the *pycnocline.* At high latitudes, both temperature and density remain relatively uniform with increasing depth (see textbook Figures 14.4 and 14.6).
6. The generally recognized three-layer structure in most parts of the open ocean includes (1) the *surface mixed zone,* with nearly uniform temperatures, (2) the *transition zone,* where temperature falls abruptly with depth, and (3) the *deep zone,* where sunlight never reaches and water temperatures are just a few degrees above freezing. In high latitudes, the three-layer structure of ocean layering does not exist, because the water column is both isothermal and isopycnal.
7. Organisms that inhabit the water column are classified as either (1) *plankton,* those that drift with ocean currents, (2) *nekton,* including all animals capable of moving independently of the ocean currents, and (3) *benthos,* organisms living on or in the ocean bottom.
8. a) E; b) B; c) F; d) A; e) D; f) C
9. The two factors that influence a region's photosynthetic productivity are (1) *availability of nutrients,* and (2) the *amount of solar radiation.*
10. High-latitude surface waters typically have high nutrient concentrations. However, the availability of solar energy limits photosynthetic productivity. By contrast, productivity is low in tropical regions of the open ocean because a permanent thermocline produces a stratification of water masses that prevents mixing between surface waters and nutrient-rich deeper waters.
11. A food chain is a sequence of organisms through which energy is passed, starting with an organism that is the primary producer. In a food web, top carnivores feed on a number of different animals, each of which has its own simple or complex feeding relationships. Animals in a food web are more likely to survive because they have alternate foods to eat should one of their food sources diminish or even disappear.

Practice Test

Multiple choice

1. c	5. d	9. a	13. a	17. c
2. d	6. b	10. a	14. e	18. b
3. c	7. d	11. d	15. a	19. b
4. c	8. b	12. a	16. a	20. e

True/false

1. T	6. T	11. T
2. T	7. F (inefficient)	12. T
3. T	8. F (lower)	13. T
4. T	9. T	14. T
5. F (intertidal)	10. T	15. F (lowers)

Word choice

1. Evaporation	6. euphotic	11. abyssal
2. Precipitation	7. cold	12. 300
3. resistant	8. mixed	13. trophic
4. high	9. Temperature	14. water
5. thousand	10. less	15. phytoplankton

Written questions

1. Salinity refers to the total amount of solid material dissolved in water, generally expressed in parts-per-thousand (‰). Salinity variations in the open ocean normally range from 33‰ to 38‰ and average about 35‰. Near the equator salinity is low due to precipitation; in the tropics higher salinities occur because of high evaporation rates; salinity decreases toward the poles (see textbook Figure 14.3).

2. The generally recognized three-layer structure in most parts of the open ocean includes (1) the *surface mixed zone,* with nearly uniform temperatures, (2) the *transition zone,* where temperature falls abruptly with depth, and (3) the *deep zone,* where sunlight never reaches and water temperatures are just a few degrees above freezing. In high latitudes, the three-layer structure of ocean layering does not exist, because the water column is both isothermal and isopycnal.

3. The factors used to divide the ocean into distinct marine-life zones include (1) *availability of sunlight,* (2) *distance from the shore,* and (3) *water depth*. Based on availability of sunlight, the zones include the *photic zone* and *aphotic zone*. Classified according to distance from the shore, the zones are the *intertidal zone, neritic zone,* and *oceanic zone*. Based on water depth, the zones include the *pelagic zone, benthic zone,* and *abyssal zone*.

ANSWER KEY

CHAPTER FIFTEEN

Vocabulary Review

1. spit
2. Beach drift
3. gyre
4. wave height
5. thermohaline circulation
6. surf
7. Coriolis effect
8. Beach nourishment
9. wavelength
10. barrier island
11. tidal current
12. estuary
13. upwelling
14. tombolo
15. wave period
16. sea stack
17. emergent coast
18. baymouth bar
19. tidal flat
20. wave refraction
21. nip current
22. longshore current
23. wave-cut platform
24. submergent coast
25. beach
26. groin
27. beach face
28. spring tide
29. tidal delta
30. backshore

Comprehensive Review

1. The Coriolis effect, the deflective force of Earth's rotation on all free-moving objects, causes the movement of ocean waters to be deflected to the right, forming clockwise gyres, in the Northern Hemisphere and to the left, forming counterclockwise gyres, in the Southern Hemisphere.
2. a) C; b) F; c) D; d) H; e) G; f) A; g) B; h) E
3. In addition to influencing temperatures of adjacent land areas, cold currents also transform some tropical deserts into relatively cool, damp places that are often shrouded in fog.
4. (1) the shape of the coastline; (2) the configuration of the ocean basin (also water depth and Coriolis effect)
5. a) B; b) A
6. a) crest; b) trough; c) wavelength; d) wave height
7. (1) wind speed; (2) length of time the wind has blown; (3) the distance the wind has traveled across the open water, called fetch
8. groins, breakwaters, seawalls (any two)
9. (1) temperature; (2) salinity
10. (1) Arctic waters; (2) Antarctic waters
11. Wave refraction in a bay causes waves to diverge and expend less energy. As a consequence of wave refraction, sediment often accumulates and forms sheltered sandy beaches.
12. Sediment is transported along a coast by (1) beach drift, which transports sediment in a zigzag pattern along the beach and (2) by longshore currents, where turbulent water moves sediment in the surf zone parallel to the shore.
13. a) F; b) B; c) C; d) A; e) E; f) D
14. (1) as spits that were subsequently severed from the mainland by wave erosion; (2) as former sand dune ridges that originated along the shore during the last glacial period, when sea level was lower
15. (1) the topography and composition of the land; (2) prevailing winds and weather patterns; (3) the configuration of the coastline and nearshore areas (answers may vary)
16. a) Emergent coasts develop either because an area experiences uplift or as a result of a drop in sea level. Feature: elevated wave-cut platform (answer may vary); b) Submergent coasts are created when sea level rises or the land adjacent to the sea subsides. Feature: estuary (answer may vary)
17. a) Semidiurnal tidal patterns have two high and two low tides each tidal day, with a relatively small difference in the high and low water heights. b) Diurnal tidal patterns have a single high- and low-water height each tidal day. c) Mixed tidal patterns have two high and two low waters each day and are characterized by a large inequality in high-water heights, low-water heights, or both.

Practice Test

Multiple choice

1. b
2. b
3. d
4. a
5. e
6. b
7. b
8. b
9. d
10. b
11. b
12. c
13. b
14. c
15. d
16. a
17. c
18. d
19. c
20. a
21. a
22. a
23. c
24. c
25. b
26. a
27. c

True/false

1. T
2. F (more)
3. F (Fetch)
4. T
5. T
6. T
7. T
8. T
9. F (arch)
10. T
11. T
12. T
13. F (right)
14. T
15. F (surf)
16. T

Word choice

1. circular
2. increase
3. waves
4. clockwise; counterclock-wise
5. rotation
6. emergent
7. rising
8. heat
9. uplift
10. polar
11. surf
12. two
13. wind
14. gain
15. narrow
16. Gulf Stream
17. beach
18. current
19. lose
20. Barrier
21. submergent
22. more
23. Fetch
24. refraction
25. winds
26. berm

Written questions

1. Coastal upwelling occurs where winds are blowing equatorward and parallel to the coast. Due to the Coriolis effect, the surface water moves away from the shore area and is replaced by deeper, colder water with greater concentrations of dissolved nutrients. The nutrient-enriched waters from below promote the growth of plankton, which in turn supports extensive populations of fish.

2. Deep-ocean circulation is called thermohaline circulation. It occurs when water at the surface is made colder and/or more salty and its density increases. The cold, dense water sinks toward the ocean bottom and flows away from its source, which is generally near the poles.

3. Tides result from the gravitational force exerted on Earth by the Moon and, to a lesser extent, the Sun. The gravitational force causes water bulges on both the side of Earth toward the Moon and the side away from it. These tidal bulges remain in place while Earth rotates "through" them, producing periods of high and low water at any one location along a coast.

4. In deep-water waves, each water particle moves in a nearly circular path to a depth equal to one half the wavelength. As a wave approaches the shore, the movement of water particles at the base of the wave are interfered with by the bottom. As the speed and length of the wave diminish, wave height increases. Eventually, water particles at the top of the wave pitch forward and the wave collapses, or breaks, and forms surf.

CHAPTER SIXTEEN

Vocabulary Review

1. troposphere
2. Weather
3. Rotation
4. thermosphere
5. Climate
6. heat
7. Tropic of Cancer
8. element (of weather and climate)
9. Convection
10. inclination of the axis
11. autumnal equinox
12. environmental lapse rate
13. Tropic of Capricorn
14. circle of illumination
15. mesosphere
16. radiation
17. revolution
18. Conduction
19. stratosphere
20. infrared
21. winter solstice
22. aerosols
23. albedo
24. greenhouse effect
25. summer solstice
26. Reflection
27. isotherm

Comprehensive Review

1. Weather is the state of the atmosphere at a particular place over a short period of time. Climate, on the other hand, is a generalization of the weather conditions of a place over a long period of time.
2. a) C; b) F; c) E; d) G; e) D; f) A
3. (1) rotation: the spinning of Earth about its axis; (2) revolution: the movement of Earth in its orbit around the Sun
4. (1) air temperature; (2) humidity; (3) type and amount of cloudiness; (4) type and amount of precipitation; (5) air pressure; (6) the speed and direction of the wind
5. (1) Solid particles act as surfaces upon which water vapor may condense. (2) Aerosols may absorb or reflect incoming solar radiation.
6. (1) When the Sun is high in the sky, the solar rays are most concentrated. (2) The angle of the Sun determines the amount of atmosphere the rays must penetrate.
7. a) spring equinox, March 21–22, equator; b) summer solstice, June 21–22, $23\frac{1}{2}$°N (Tropic of Cancer); c) autumnal equinox, September 22–23, equator; d) winter solstice, December 21–22, $23\frac{1}{2}$°S (Tropic of Capricorn)
8. nitrogen (78 percent) and oxygen (21 percent)
9. Ozone (O_3) has three oxygen atoms per molecule, while oxygen (O_2) has two atoms per molecule. Ozone in the atmosphere absorbs the potentially harmful ultraviolet (UV) radiation from the Sun.
10. a) 20; b) 50; c) 22
11. Most of the energy that is absorbed at Earth's surface is reradiated skyward in the form of infrared radiation. This terrestrial radiation is readily absorbed by water vapor and carbon dioxide in the atmosphere, thereby heating the atmosphere from the ground up.
12. carbon dioxide and water vapor
13. (1) Conduction is the transfer of heat through matter by molecular activity. (2) Convection is the transfer of heat by mass movement within a substance. (3) Radiation is the transfer of energy (heat) through space by electromagnetic waves. It is the only mechanism of heat transfer that can transmit heat through the relative emptiness of space.
14. a) The daily mean temperature is determined by adding the maximum and minimum temperatures and then dividing by two. b) The annual mean temperature is an average of the 12 monthly means. c) The annual temperature range is computed by finding the difference between the highest and lowest monthly means.

15. a) Land heats more rapidly and to higher temperatures than water. Land also cools more rapidly and to lower temperatures than water. b) Due to the mechanism by which the atmosphere is heated, temperature generally decreases with an increase in altitude. c) A windward coastal location will often experience cool summers and mild winters when compared to an inland station at the same latitude. Leeward coastal sites will tend to have a more continental temperature pattern because the winds do not carry the ocean's influence onshore. d) Many clouds have a high albedo and reflect a significant portion of the radiation that strikes them back to space; therefore, less radiation reaches the surface and daytime temperatures will be lower than if the clouds were not present.
16. January and July are selected most often for analysis because, for most locations, they represent the temperature extremes.

Practice Test

Multiple choice

1. a
2. d
3. a
4. b
5. d
6. b
7. c
8. a
9. b
10. b
11. a
12. c
13. e
14. b
15. c
16. c
17. b
18. a
19. c
20. d
21. b
22. b
23. a
24. b
25. d
26. e
27. e
28. c
29. b
30. c

True/false

1. T
2. F (Weather)
3. F (lower)
4. T
5. T
6. F (conduction)
7. T
8. T
9. T
10. F (temperature)
11. T
12. F (troposphere)
13. F (illumination)
14. F (winter)
15. F (gradual)
16. T
17. T
18. F (decreased)
19. F (fast)
20. T
21. T
22. T
23. F (smaller)
24. T
25. T

Word choice

1. varies
2. Cancer
3. decreases
4. insignificant
5. more
6. 3.5
7. smaller
8. heat energy
9. greater
10. decreases
11. 23.5
12. latent
13. ultraviolet
14. equator
15. climate
16. arid
17. temperature
18. land
19. is not
20. shorter
21. least
22. radiation
23. decrease
24. Earth
25. convection
26. greater

Written questions

1. CFCs, chlorofluorocarbons, are a group of chemicals that were once commonly used as propellants for aerosol sprays and in the production of certain plastics and refrigerants. When CFCs reach the stratosphere, chlorine breaks up some of the ozone molecules. The net effect is depletion of the ozone layer and a reduction in the layer's ability to absorb harmful ultraviolet (UV) radiation.

2. The amount of solar energy reaching places on Earth's surface varies with the seasons because Earth's orientation to the Sun continually changes as it travels along its orbit. This changing orientation is due to the facts that (1) Earth's axis is inclined and (2) the axis remains pointed in the same direction (toward the North Star) as Earth journeys around the Sun.

3. The atmosphere allows most of the short-wave solar energy to pass through and be absorbed at Earth's surface. Earth reradiates this energy in the form of longer wavelength terrestrial radiation, which is absorbed by carbon dioxide and water vapor in the lower atmosphere. This mechanism, referred to as the greenhouse effect, explains the general drop in temperature with increasing altitude experienced in the troposphere.

ANSWER KEY

CHAPTER SEVENTEEN

Vocabulary Review

1. humidity
2. condensation
3. Relative humidity
4. Orographic lifting
5. vapor pressure
6. adiabatic temperature change
7. Latent heat
8. cloud
9. sleet
10. saturation
11. wet adiabatic rate
12. Sublimation
13. stratus
14. mixing ratio
15. Frontal wedging
16. condensation nuclei
17. melting
18. rain
19. dew-point temperature
20. Rime
21. deposition
22. hail
23. dry adiabatic rate
24. hygroscopic nuclei
25. evaporation
26. Radiation fog
27. fog
28. rainshadow desert
29. cirrus
30. convergence
31. calorie
32. supercooled
33. hygrometer

Comprehensive Review

1. a) D; b) C; c) F; d) B; e) A; f) E
2. (1) adding moisture to the air; (2) lowering the air temperature
3. a) 25 percent (5 grams per kilogram/20 grams per kilogram); b) 60 percent (3 grams per kilogram/5 grams per kilogram)
4. a) 15°C(59°F); b) 14°F(–10°C)
5. The psychrometer consists of two thermometers mounted side by side: the dry-bulb, which gives the present temperature, and the wet-bulb, which has a moist, thin muslin wick tied around the end. The lower the air's relative humidity, the more evaporation (and hence cooling) there will be from the wet-bulb thermometer. Thus, the greater the difference between the wet- and dry-bulb temperatures, the lower the relative humidity.
6. 42 percent
7. Once the air is cooled sufficiently to reach the dew point and condensation occurs, latent heat of condensation stored in the water vapor will be liberated, thereby reducing the rate at which the air cools.
8. Stable air resists vertical movement because the environmental lapse rate is less than the wet adiabatic rate. Unstable air wants to rise because it is less dense than the surrounding air. Unstable conditions prevail when the environmental lapse rate is greater than the dry adiabatic rate.
9. a) Orographic lifting occurs when elevated terrains, such as mountain barriers, act as barriers to flowing air. b) Frontal wedging occurs when cool air acts as a barrier over which warmer, less dense air rises. c) Convergence occurs whenever air masses flow together. Because the air must go somewhere, it moves upward. d) Unequal heating causes pockets of air to rise.
10. (1) the form of the cloud; (2) the height of the cloud
11. a) stratus; b) cirrus; c) altocumulus
12. a) B; b) A
13. a) Steam fog, a type of evaporation fog, occurs when cool air moves over warm water and moisture evaporated from the water surface condenses. b) Frontal fog occurs when warm air is lifted over colder air by frontal wedging. If the resulting clouds yield rain, and the cold air below is near the dew point, enough rain will evaporate to produce fog.
14. (1) The Bergeron process occurs when ice crystals in a cloud collect available water vapor, eventually growing large enough to fall as snowflakes. When the surface temperature is about 4°C(39°F) or higher, snowflakes usually melt before they reach the ground and continue their descent as rain. (2) The collision-coalescence process occurs when large droplets form on large hygroscopic condensation nuclei and,

because the bigger droplets fall faster in a cloud, they collide and join with smaller water droplets. After many collisions the droplets are large enough to fall to the ground as rain.

15. a) Sleet, a wintertime phenomenon, forms when raindrops freeze as they fall through colder air that lies below warmer air. b) Hail is produced in a large cumulonimbus cloud with violent updrafts and an abundance of supercooled water. As they fall through the cloud, ice pellets grow by collecting supercooled water droplets. The process continues until the hailstone encounters a downdraft or grows too heavy to remain suspended by the thunderstorm's updraft.

Practice Test

Multiple choice

1. d
2. b
3. b
4. a
5. b
6. c
7. b
8. e
9. c
10. b
11. a
12. c
13. b
14. a
15. c
16. b
17. d
18. e
19. b
20. d
21. a
22. d
23. a
24. a
25. a
26. b
27. d
28. c
29. c
30. a

True/false

1. T
2. F (calorie)
3. F (height)
4. T
5. F (cool)
6. T
7. T
8. T
9. F (wet-bulb)
10. T
11. T
12. F (wet)
13. F (decrease)
14. T
15. F (vaporization)
16. T
17. F (fog)
18. T

Word choice

1. releases
2. collision-coalescence
3. altitude
4. relative
5. less
6. Sleet
7. More
8. unstable
9. increases
10. compresses; warms
11. absolute instability
12. absorbs
13. Stratus
14. ice crystals
15. decreasing; increasing
16. lower; latent; condensation
17. Bergeron
18. lower
19. convergence; upward
20. increases

Written questions

1. Clouds are classified on the basis of their form (cirrus, cumulus, and stratus) and height (high, middle, low, and clouds of vertical development).

2. As air rises, it expands because air pressure decreases with an increase in altitude. Expanding air cools because it pushes on (does work on) the surrounding air and expends energy.

3. Stable air resists vertical movement because the environmental lapse rate is less than the wet adiabatic rate. Unstable air wants to rise because it is less dense than the surrounding air. Unstable conditions prevail when the environmental lapse rate is greater than the dry adiabatic rate.

4. (answers will vary) Depending on the place and season, the process will be either orographic lifting, frontal wedging, or convergence. In either case, when air rises, it will expand and cool adiabatically. If sufficient cooling takes place, the dew point may be reached and condensation may result in the formation of clouds and perhaps precipitation.

CHAPTER EIGHTEEN

Vocabulary Review

1. polar front
2. isobar
3. monsoon
4. Coriolis effect
5. wind
6. Barometric tendency
7. aneroid barometer
8. prevailing wind
9. wind vane
10. cyclone
11. Convergence
12. land breeze
13. equatorial low
14. chinook
15. sea breeze
16. cup anemometer
17. pressure gradient
18. jet stream
19. anticyclone
20. subpolar low
21. polar easterlies
22. barograph
23. Divergence
24. subtropical high
25. westerlies
26. trade winds
27. Southern Oscillation
28. El Niño

Comprehensive Review

1. a) The mercury barometer is an instrument used to measure air pressure. It consists of a glass tube, closed at one end, filled with mercury. The tube is inverted in a reservoir (pan) of mercury. If the air pressure decreases, the height of the column of mercury in the tube falls; if pressure increases, the column rises. b) The wind vane is used to determine wind direction by pointing into the wind. c) The aneroid barometer measures air pressure using partially evacuated metal chambers that are compressed as air pressure increases and expand as pressure decreases. d) The cup anemometer measures wind speed using several cups mounted on a shaft that is rotated by the moving wind. e) A barograph is an aneroid barometer that continuously records pressure changes.
2. (1) pressure gradient force: the amount of pressure change over a given distance; (2) Coriolis effect is the deflective force of Earth's rotation on all free-moving objects. Deflection is to the right in the Northern Hemisphere. (3) friction with Earth's surface: slows down air movement within the first few kilometers of Earth's surface and, as a consequence, alters wind direction
3. (The diagrams in Figure 18.1 should show the following pressures and surface air movements.) Northern Hemisphere high: highest pressure in center, air moves out (divergence), clockwise. Southern Hemisphere high: highest pressure in center, air moves out (divergence), counterclockwise. Northern Hemisphere low: lowest pressure in center, air moves inward (convergence), counterclockwise. Southern Hemisphere low: lowest pressure in center, air moves inward (convergence), clockwise.
4. a) inward (convergent) and counterclockwise; b) outward (divergent) and counterclockwise
5. a) B; b) A; c) B; d) A; e) B
6. In a high-pressure system (anticyclone), outflow near the surface is accompanied by convergence aloft and general subsidence of the air column. Because descending air is compressed and warmed, cloud formation and precipitation are unlikely, and "fair" weather can usually be expected.
7. A: subpolar low; B: subtropical high; C: equatorial low; D: subtropical high; E: subpolar low; F: polar easterlies; G: westerlies; H: trade winds; I: trade winds; J: westerlies; K: polar easterlies
8. a) subpolar low; b) subtropical high; c) equatorial low; d) polar high
9. a) warm, dry winds often found on the leeward sides of mountains, especially the eastern slopes of the Rockies; b) cooler air over the water (higher pressure) moves toward the warmer land (lower pressure); c) a light wind blowing into the city from the surrounding countryside.
10. a) northeast; b) west; c) south

Practice Test

Multiple choice

1. d	6. c	11. c	16. e	21. b
2. a	7. a	12. a	17. d	22. b
3. a	8. d	13. e	18. b	23. a
4. c	9. d	14. d	19. c	24. b
5. b	10. c	15. e	20. d	25. e

True/false

1. T	7. T	13. F (subtropical)
2. F (faster)	8. F (cyclone)	14. T
3. T	9. F (dry)	15. F (convergence)
4. F (right)	10. T	16. F (colder)
5. T	11. T	17. T
6. T	12. F (greater)	18. F (Pacific)

Word choice

1. higher; lower	7. west-to-east	13. higher
2. rising	8. left	14. lower
3. lower	9. west-to-east	15. lower
4. weakest	10. low	16. less
5. maintains	11. does not	17. compression
6. spacing	12. westerlies	18. strongest

Written questions

1. The Coriolis effect, the deflective force of Earth's rotation on all free-moving objects, causes air to be deflected to the right of its path of motion in the Northern Hemisphere and to the left in the Southern Hemisphere.

2. A drop in barometric pressure indicates the approach of a low-pressure system, called a cyclone. The cyclone is associated with converging surface winds and ascending air; hence, adiabatic cooling, cloud formation, and precipitation often occur. Conversely, an anticyclone, with its high air pressure, is associated with subsidence of the air column and divergence at the surface. Because descending air is compressed and warmed, cloud formation and precipitation are unlikely in an anticyclone, and "fair" weather can usually be expected.

3. In a Northern Hemisphere cyclone (low pressure), surface winds are inward (convergence) and counterclockwise. As a consequence of surface convergence, there is a slow, net upward movement of air near the center of the cyclone. Furthermore, the surface convergence found in a cyclone is balanced by divergence of the air aloft.

CHAPTER NINETEEN

Vocabulary Review

1. air mass
2. hurricane
3. front
4. tropical storm
5. source region
6. continental (c) air mass
7. warm front
8. middle-latitude cyclone
9. eye wall
10. tropical (T) air mass
11. tropical depression
12. eye
13. air-mass weather
14. cold front
15. tornado
16. maritime (m) air mass
17. tornado watch
18. Doppler radar
19. thunderstorm
20. storm surge
21. tornado warning
22. polar (P) air mass
23. occluded front
24. lake-effect snow
25. stationary front
26. Occlusion

Comprehensive Review

1. Air masses are classified according to their source region using two criteria; (1) the nature of the surface (land or water) where the air mass originates is identified using a lowercase letter such as c (continental) or m (maritime), and (2) the latitude where the air mass originates is identified using an upper case letter such as P (polar) or T (tropical). Therefore, a cP air mass would have a continental polar source region.
2. A: maritime polar (mP)—cool, moist; B: continental polar (cP)—cold and dry in winter; F: continental tropical (cT)—hot and dry in summer; G: maritime tropical (mT)—warm, moist
3. continental polar (cP) from central Canada and maritime tropical (mT) from the Gulf of Mexico
4. (See textbook Figures 19.6 and 19.7) (1) A warm front occurs where the surface position of a front moves so that warm air occupies territory formerly covered by cooler air. The boundary separating the cooler air from the warmer air above has a small slope (about 1:200). (2) A cold front occurs when cold air actively advances into a region occupied by warmer air. On the average, the boundary separating the cold and warm air is about twice as steep as the slope along a warm front.
5. diagram: A; Because the cold front has caught up to the warm front, forming an occluded front.
6. a) warm front; b) cold front; c) maritime tropical (mT); d) continental polar (cP); e) southwest to south; f) northwest to north; g) thunderstorms and precipitation with the passage of the cold front, followed by cooler, drier conditions as the cP air moves over the area, change in wind direction from southwest to northwest after the cold front passes
7. a) surface: inward (convergent), counterclockwise (Northern Hemisphere), rising near center; aloft: divergence; b) surface: outward (divergent), clockwise (Northern Hemisphere), subsiding near center; aloft: convergence
8. (1) rate of movement; (2) steepness of slope
9. a) All thunderstorms require warm, moist air, which when lifted will release sufficient latent heat to provide the buoyancy necessary to maintain its upward flight. b) Tornadoes form in association with severe thunderstorms spawned along the cold front of a middle-latitude cyclone. These violent windstorms that take the form of a rapidly rotating column of air that extends downward from a cumulonimbus cloud are products of the interaction between strong updrafts in the thunderstorm and winds in the troposphere.
10. (1) when they move over waters that cannot supply warm, moist, tropical air; (2) when they move onto land; (3) when they reach a location where the large-scale flow aloft is unfavorable
11. (1) wind damage; (2) storm surge damage; (3) inland freshwater flooding

Practice Test

Multiple choice

1. b
2. b
3. b
4. a
5. b
6. a
7. d
8. b
9. d
10. d
11. b
12. e
13. c
14. b
15. a
16. b
17. b
18. a
19. d
20. b
21. c
22. c
23. e
24. d
25. d
26. b
27. e
28. a

True/false

1. T
2. F (one thousand)
3. T
4. T
5. F (more)
6. F (eastward)
7. F (Maritime)
8. T
9. T
10. T
11. T
12. F (cold)
13. F (twice)
14. T
15. T
16. F (central)
17. T
18. T
19. T
20. T
21. T
22. T

Word choice

1. cyclone
2. warm; cooler
3. hurricane
4. triangles
5. Pacific
6. dry
7. cumulonimbus
8. clear
9. low
10. occluded
11. counterclockwise
12. descends; heats; compression
13. leeward
14. spring; cold
15. parallel

Written questions

1. A continental polar (cP) air mass that originates in central Canada will most likely be cold and dry in the winter and cool and dry in the summer. A maritime tropical (mT) air mass with its source region in the Gulf of Mexico will be warm and moist.

2. As the warm front approaches, winds would blow from the east or southeast and air pressure would drop steadily. Cirrus clouds would be sighted first, followed by progressively lower clouds and perhaps nimbostratus clouds. Gentle precipitation could occur as the nimbostratus clouds moved overhead. As the warm front passed, temperatures would rise, precipitation would cease, and winds would shift to the south or southwest. Further, the sky clears and the pressure tendency steadies.

   Later, with the approach of the cold front, cumulonimbus clouds fill much of the sky and bring the likelihood of heavy precipitation and a possibility of hail and tornado activity. The passage of the cold front is accompanied by a drop in temperature, clearing sky, a wind shift to the northwest or north, and rising air pressure. Fair weather can probably be expected for the next few days.

3. A tornado watch alerts the public to the fact that conditions are right for the formation of tornadoes, whereas a tornado warning is issued when a tornado has actually been sighted in an area or is indicated by radar.

CHAPTER TWENTY

Vocabulary Review

1. climate system
2. humid subtropical climate
3. arid (or desert) climate
4. Köppen classification
5. ice cap climate
6. subarctic climate
7. marine west coast climate
8. tropical rain forest
9. rainshadow desert
10. humid continental climate
11. tropical wet and dry climate
12. tundra climate
13. semiarid (or steppe) climate
14. dry-summer subtropical climate
15. highland climate
16. climate-feedback mechanism

Comprehensive Review

1. The five parts of Earth's climate system are the atmosphere, hydrosphere, solid Earth, biosphere, and cryosphere.
2. The two most important elements in a climatic description are temperature and moisture. They are important because they have the greatest influence on people and their activities and also have an important impact on the distribution of such phenomena as vegetation and soils.
3. (1) Humid tropical (A): winterless climates; all months above 18°C. (2) Dry (B) climates: climates where evaporation exceeds precipitation; there is a constant water deficiency. (3) Humid middle-latitude climates with mild winters (C climates): mild winters; average temperature of the coldest month is below 18°C but above –3°C. (4) Humid middle-latitude climates with severe winters (D climates): severe winters; the average temperature of the coldest month is below –3°C and the warmest monthly mean exceeds 10°C. (5) Polar (E) climates: summerless climates; the average of the warmest month is below 10°C.
4. a) July, 26.4°C; b) January, –0.1°C; c) 26.5°C (–0.1°C to 26.4°C); d) June; e) late spring to early summer; f) Cfa, humid subtropical (average temperature of the coldest month is under 18°C and above –3°C, criteria for w and s cannot be met, and warmest month is over 22°C, with at least four months over 10°C)
5. The three temperature/precipitation features that characterize the wet tropics are (1) temperatures usually average 25°C or more each month, (2) the total precipitation for the year is high, often exceeding 200 centimeters, and (3) all months are usually wet. The most continuous expanses of the wet tropics lie near the equator in South America, Africa, and Indonesia.
6. The difference is primarily a matter of degree. The semiarid is a marginal more humid variant of the arid.
7. Low-latitude deserts and steppes coincide with the subtropical high-pressure belts. Here the air is subsiding, compressed, and warmed, causing clear skies, a maximum of sunshine, and drought conditions.
8. The climate is found in the south, southeastern, and eastern United States.
9. D climates are land-controlled climates, the result of broad continents in the middle latitudes. D climates are absent in the Southern Hemisphere where the middle-latitude zone is dominated by the oceans.
10. a) dry climates; b) polar climates; c) wet tropical climates; d) marine west coast climates; e) dry climates (low latitude); f) subarctic climate; g) wet tropical climates
11. In addition to having the lowest elevation in the Western Hemisphere, Death Valley is a desert. Clear skies allow a maximum of sunshine to reach the dry, barren surface. Because no energy is used to evaporate moisture, all of the energy is available to heat the ground. Also subsiding and warming air is common to the area.
12. (see textbook "Students Sometimes Ask") a) warmer b) more; c) slower
13. Water vapor and carbon dioxide are largely responsible for the greenhouse effect of the atmosphere due to their ability to absorb the longer-wavelength outgoing terrestrial radiation and alter temperatures in the lower atmosphere.

14. Humans have contributed to the buildup of atmospheric carbon dioxide by (1) the combustion of fossil fuels, and (2) by the clearing of forests.
15. Three potential weather changes that could occur are (1) warmer temperatures, (2) a higher frequency and greater intensity of hurricanes, and (3) shifts in the paths of large-scale cyclonic storms, which in turn would affect the distribution of precipitation.

Practice Test

Multiple choice

1. b	6. d	11. b	16. a	21. c
2. d	7. b	12. c	17. a	22. c
3. c	8. a	13. c	18. e	23. a
4. b	9. b	14. c	19. d	24. e
5. b	10. b	15. d	20. d	

True/false

1. F (five)	6. T	11. T
2. F (windward)	7. F (evaporation)	12. T
3. F (low)	8. T	13. T
4. F (precipitation)	9. T	14. F (industrialization)
5. T	10. F (below)	15. T

Word choice

1. five	6. subtropical high	11. windward
2. B	7. coniferous	12. 0.036
3. rise	8. 200	13. precipitation
4. temperature; precipitation	9. subsiding	14. water vapor
5. Europe	10. 0.6	15. will not

Written questions

1. In North America, climates east of the 100 degree west meridian vary north to south. A Cfa (humid subtropical warm, summer) climate located along the Gulf Coast and southeastern United States gives way to D (humid middle-latitude) climates in the central and northern portions of the continent. Polar (tundra) climate is located in the extreme northern reaches. West of the 100 degrees west meridian are middle-latitude steppes and deserts, along with highland climates in the mountains. The west coast has an area of dry-summer subtropical (Mediterranean) climate from about latitude 35 to 40 degree north, with a marine west coast climate extending northward along the coast into Alaska.

2. Wet tropical climates are restricted to elevations below 1000 meters because of the general decrease in temperature with altitude. At higher altitudes the temperature criterion for an A climate is not met.

3. The two types of climate-feedback mechanisms are (1) positive-feedback mechanisms, which reinforce the initial change, and (2) negative-feedback mechanisms, which produce results that are the opposite of the initial change and tend to offset it.

ANSWER KEY

CHAPTER TWENTY-ONE

Vocabulary Review

1. retrograde motion
2. sidereal day
3. heliocentric
4. sidereal month
5. geocentric
6. Rotation
7. astronomical unit (AU)
8. solar eclipse
9. declination
10. perturbation
11. perihelion
12. ecliptic
13. celestial sphere
14. constellation
15. plane of the ecliptic
16. Ptolemaic system
17. Revolution
18. phases of the Moon
19. right ascension
20. synodic month
21. lunar eclipse
22. equatorial system
23. aphelion
24. axial precession
25. mean solar day

Comprehensive Review

1. The geocentric model held by the early Greeks proposed that Earth was a sphere that stayed motionless at the center of the universe. Orbiting Earth were the Moon, Sun, and the known planets, Mercury through Jupiter. Beyond the planets was a transparent, hollow sphere, called the celestial sphere, on which the stars traveled daily around Earth.
2. Retrograde motion is the apparent periodic westward drift of a planet in the sky that results from a combination of the motion of Earth and the planet's own motion around the Sun. In Figure 20.1, retrograde motion is occurring between points 3 and 4. To explain retrograde motion, the geocentric model of Ptolemy had the planets moving in small circles, called epicycles, as they revolved around Earth along large circles, called deferents.
3. Early Greeks rejected the idea of a rotating Earth because Earth exhibited no sense of motion and seemed too large to be movable.
4. a) Galileo Galilei; b) Nicolaus Copernicus; c) Tycho Brahe; d) Johannes Kepler; e) Nicolaus Copernicus; f) Johannes Kepler; g) Galileo Galilei; h) Galileo Galilei; i) Sir Isaac Newton; j) Aristarchus; k) Hipparchus; l) Eratosthenes
5. Stellar parallax is the apparent shift in the position of a nearby star, when observed from extreme points in Earth's orbit six months apart, with respect to the more distant stars.
6. (1) The path of each planet around the Sun is an ellipse, with the Sun at one focus. (2) Each planet revolves so that an imaginary line connecting it to the Sun sweeps over equal areas in equal intervals of time. (3) The orbital periods of the planets and their distances to the Sun are proportional.
7. (1) four Moons orbiting Jupiter; (2) the planets are circular disks rather than just points of light; (3) Venus has phases just like the Moon; (4) the Moon's surface is not a smooth glass sphere
8. a) November through February; b) June to July; c) gravity keeps the planet from going in a straight line out to space; d) Using Kepler's third law, a planet's orbital period squared is equal to its mean solar distance cubed ($p^2 = d^3$), the planet's orbital period (p) would be 8 Earth-years (the square root of 4 cubed, or 64).
9. The equatorial system divides the celestial sphere into a coordinate system very similar to the latitude-longitude system used for locations on Earth's surface. In the sky, declination, like latitude, is the angular distance north or south of celestial equator and right ascension is the angular distance measured eastward along the celestial equator from the position of the vernal equinox.
10. a) C; b) A; c) F; d) D
11. 1. rotation; 2. revolution; 3. precession; 4. movement with the entire solar system in the direction of the star Vega at 20 kilometers per second; 5. revolution around the galaxy, a trip that requires 230 million years to traverse; 6. movement as a part of the galaxy through the universe toward the Great Galaxy of Andromeda
12. The synodic cycle is the movement of the Moon through its phases, which takes $29^1/_2$ days. The sidereal cycle, which takes only $27^1/_3$ days, is the true period of the Moon's revolution around Earth. The reason for the difference of nearly two between the cycles is that as the Moon orbits Earth, the Earth-Moon system also moves in an orbit about the Sun (see textbook Figure 21.26).
13. a) 5; b) 7; c) 2; d) 1; e) 6

Practice Test

Multiple choice

1. b
2. c
3. c
4. a
5. c
6. e
7. b
8. b
9. a
10. e
11. b
12. c
13. d
14. c
15. d
16. b
17. e
18. d
19. a
20. c
21. e
22. d
23. d
24. b
25. d
26. b
27. c
28. a
29. a
30. a

True/false

1. F (elliptical)
2. F (seven)
3. F (Hipparchus)
4. F (smaller)
5. F (synodic)
6. T
7. F (longer)
8. T
9. T
10. F (perturbation)
11. T
12. T
13. T
14. F (day)
15. T
16. F (geocentric)
17. F (same)
18. T
19. F (degrees)
20. F (Johannes Kepler)

Word choice

1. sidereal
2. eastward
3. solar
4. astronomical
5. constant; varies
6. longer
7. Gregorian
8. geocentric
9. reflected sunlight
10. inertia
11. lunar
12. perihelion
13. circular; Earth
14. Aristarchus
15. equal to
16. are not
17. axis
18. new-Moon
19. Galileo
20. pendulum

Written questions

1. The apparent westward drift of planets, called retrograde motion, results from the combination of the motion of Earth and the planet's own motion around the Sun. Periodically Earth overtakes a planet in its orbit, which makes it appear that for a period of time the planet moves backward in its orbit. (See textbook Figure 21.6.)

2. For an eclipse to occur, the Moon must be passing through the plane of the ecliptic during the new- or full-Moon phase. Because the Moon's orbit is inclined about 5 degrees to the plane that contains Earth and the Sun, the conditions are normally met only twice a year, and the usual number of eclipses is four.

3. Four of the six motions Earth continuously experiences are (1) rotation, (2) revolution, (3) axial precession, and (4) movement with the entire solar system in the direction of the star Vega (answers may vary).

4. To explain the observable motions of the planets, the geocentric model of Ptolemy had the planets revolving around Earth along large circles called deferents, while simultaneously orbiting on small circles called epicycles.

CHAPTER TWENTY-TWO

Vocabulary Review

1. coma
2. terrestrial planet
3. asteroid
4. maria
5. meteoroid
6. Kuiper belt
7. Jovian planet
8. Oört cloud
9. Lunar regolith
10. meteor
11. escape velocity
12. planetesimals
13. comet
14. meteorite
15. meteor shower
16. iron meteorite
17. cryovolcanism
18. terrae

Comprehensive Review

1. The solar system consists of those celestial objects that are under the direct gravitational influence of the Sun. These objects include the Sun, the eight planets and their satellites, asteroids, comets, meteoroids, and the dwarf planets.
2. a) hydrogen and helium; b) principally silicate minerals and metallic iron; c) ammonia, methane, carbon dioxide, and water
3. a) The terrestrial planets (Mercury, Venus, Earth, and Mars) are all similar to Earth in their physical characteristics. b) The Jovian planets (Jupiter, Saturn, Uranus, and Neptune) are all Jupiter-like in their physical characteristics.
4. a) The Jovian planets are all large compared to the terrestrial planets. b) Because the Jovian planets contain a large percentage of gases, their densities are less than those of the terrestrial planets. c) The Jovian planets have shorter periods of rotation. d) The Jovian planets are all more massive than the terrestrial planets. e) The Jovian planets have more moons. f) The periods of revolution of the Jovian planets are longer than those of the terrestrial planets. g) The terrestrial planets are mostly rocky and metallic material, with minor amounts of gases. The Jovian planets, on the other hand, contain a large percentage of gases (hydrogen and helium), with varying amounts of ices (mostly water, ammonia, and methane).
5. Because of the greater surface gravities of the Jovian planets, it is more difficult for gases to evaporate from them.
6. a) A; b) B; c) D; d) C; e) C
7. Because the gravitational attraction at the lunar surface is only one-sixth of that experienced on Earth's surface.
8. Lunar maria basins are enormous impact craters that were flooded with layer upon layer of very fluid basaltic lava.
9. The most widely accepted model for the origin of the Moon is that during the formative period of the solar system a Mars-size body impacted Earth. The ejected debris entered an orbit around Earth and coalesced to form the Moon.
10. a) 3; b) 1; c) 5; d) 4; e) 2
11. a) Mars; b) Venus; c) Jupiter; d) Saturn; e) Mercury; f) Saturn; g) Venus; h) Uranus; i) Jupiter; j) Neptune; k) Saturn; l) Mars; m) Jupiter; n) Venus; o) Neptune; p) Mercury; q) Pluto
12. The most probable origin of the Martian moons is that they are asteroids that have been captured by the gravity of Mars.
13. a) Jupiter; b) Mars; c) Jupiter; d) Neptune; e) Uranus; f) Pluto; g) Saturn
14. Planetary ring particles are thought to be debris ejected from moons that coexist with the rings.
15. (1) they might have formed from the breakup of a planet; (2) perhaps several large bodies once coexisted in close proximity and their collisions produced numerous smaller objects
16. a) B; b) A; c) C; d) D
17. (1) radiation pressure: pushes dust particles away from the coma; (2) solar wind: responsible for moving the ionized gases
18. A meteor is a brilliant streak of light produced when a meteoroid enters Earth's atmosphere and burns up. A meteorite is any portion of a meteoroid that survives its traverse through Earth's atmosphere and strikes the surface.

19. The three most common types of meteorites classified by their composition are (1) irons—mostly iron with 5–10 percent nickel; (2) stony—silicate minerals with inclusions of other minerals; and (3) stony-irons—mixtures.

Practice Test

Multiple choice

1. b
2. e
3. d
4. a
5. d
6. e
7. c
8. a
9. a
10. b
11. b
12. c
13. d
14. c
15. b
16. d
17. e
18. d
19. b
20. c
21. d
22. d
23. c
24. e
25. e
26. e
27. b
28. b
29. b
30. c
31. e
32. d
33. a
34. a
35. c
36. d
37. b

True/false

1. T
2. F (Saturn)
3. T
4. T
5. F (Mercury)
6. F (Jupiter)
7. F (low)
8. T
9. F (lava)
10. F (one)
11. F (highlands)
12. F (lose)
13. F (Sun)
14. F (Jupiter)
15. T
16. T
17. T
18. T
19. T
20. T
21. T

Word choice

1. away from
2. less; sixth
3. rift
4. Mercury
5. absorb
6. does not
7. parallel
8. same
9. meteor
10. polar caps
11. micrometeorite impacts
12. 1
13. hundred
14. older
15. old
16. Venus
17. beyond
18. cloud cover
19. farther from
20. slowly; quickly
21. salt-laden

Written questions

1. Compared to the Jovian planets, the terrestrial planets are small, less massive, more dense, have longer rotational periods and shorter periods of revolution. The terrestrial planets are also warmer and consist of a greater percentage of rocky material and less gases and ices.

2. Meteoroid is the name given to any solid particle that travels through interplanetary space. When a meteoroid enters Earth's atmosphere it vaporizes with a brilliant flash of light called a meteor. On occasions when meteoroids reach Earth's surface, the remains are termed meteorites.

3. The two different types of lunar terrain are (1) maria—the dark, smooth, lowland areas resembling seas on Earth that formed when asteroids punctured the surface, letting basaltic lava "bleed" out; and (2) highlands—the densely cratered areas that make up most of the lunar surface and contain remnants of the oldest lunar crust.

ANSWER KEY

CHAPTER TWENTY-THREE

Vocabulary Review

1. electromagnetic radiation
2. corona
3. refracting telescope
4. spectroscope
5. photosphere
6. nuclear fusion
7. continuous spectrum
8. reflecting telescope
9. photon
10. radio telescope
11. Doppler effect
12. solar wind
13. proton-proton chain
14. solar flare
15. bright-line (emission) spectrum
16. chromosphere
17. prominence
18. Spectroscopy
19. sunspot
20. dark-line (absorption) spectrum
21. Chromatic aberration
22. radiation pressure
23. granules
24. radio interferometer
25. spicule
26. aurora

Comprehensive Review

1. The properties of light are explained by two models; (1) by acting as a stream of particles, called photons, and (2) by behaving like waves.
2. Three other types of electromagnetic radiation would be gamma rays, ultraviolet light, and radio waves (answers may vary). Each is similar in that their properties can be explained using the particle and wave models. They are different in that their wavelengths are different; hence, they have different energies. The shortest wavelength, gamma rays, have the greatest energy.
3. As white light passes through a prism, the color with the shortest wavelength, violet, is bent more than blue, which is bent more than green, and so forth. Consequently, as the light emerges from the prism it will be separated into its component parts, the "colors of the rainbow." (see textbook Figure 23.2)
4. (1) Continuous spectra, uninterrupted bands of color, are produced by an incandescent solid, liquid, or gas under high pressure. (2) Dark-line (absorption) spectra, which appear as continuous spectra with dark lines running through them, are produced when white light is passed through a comparatively cool gas under low pressure. The spectra of most stars of this type. (3) Bright-line (emission) spectra are produced by a hot (incandescent) gas under low pressure. (see textbook Figure 23.3)
5. Each element or compound that is in gaseous form in a star produces a unique set of spectral lines which act as "fingerprints."
6. The two aspects of stellar motion that can be determined from a star's Doppler shift are (1) the direction, toward or away, and (2) the rate, or velocity, of the relative movement.
7. The two basic types of optical telescopes are (1) refracting telescopes, which use a lens as their objective to bend or refract light to an area called the focus, and (2) reflecting telescopes, which use a mirror to gather and focus the light.
8. (1) Light-gathering power is the ability of a telescope to collect light from a distant object. The larger the lens (or mirror) of a telescope, the greater its light-gathering ability. (2) Resolving power refers to the ability of a telescope to separate close objects. The greater the resolving power, the sharper the image and the finer the detail. (3) Magnifying power is the ability to make images larger. It is changed by changing the eyepiece. Magnification is limited by atmospheric conditions and the resolving power of the telescope.
9. Reflecting telescopes do not have chromatic aberration (color distortion), the glass does not have to be of optical quality, and large mirrors can be made because they can be supported from behind.
10. Invisible infrared and ultraviolet radiation is detected and studied by using photographic film that is sensitive to these wavelengths. Since most of this radiation cannot penetrate our atmosphere, balloons, rockets, and satellites must transport the cameras "above" the atmosphere to record it.

11. Radio telescopes are (1) less affected by weather and (2) can "see" through interstellar dust clouds that obscure our view at visible wavelengths. (answers may vary)
12. a) E; b) B; c) D; d) A; e) F; f) C
13. During a strong solar flare, fast-moving atomic particles are ejected from the Sun, causing the solar wind to intensify noticeably. In about a day, the ejected particles reach Earth and disturb the ionosphere, affecting long distance communication and producing the auroras, the Northern and Southern Lights.
14. The source of the Sun's energy is nuclear fusion. The nuclear reaction, called the proton-proton reaction, converts four hydrogen nuclei into the nucleus of a helium atom, releasing energy during the process by converting some of the matter to energy.

Practice Test

Multiple choice

1. a
2. c
3. c
4. b
5. b
6. a
7. c
8. b
9. d
10. e
11. c
12. e
13. e
14. c
15. e
16. c
17. b
18. e
19. b
20. b

True/false

1. F (corona)
2. T
3. T
4. T
5. T
6. F (water)
7. F (spectroscopy)
8. T
9. T
10. F (radiation)
11. T
12. T
13. T
14. F (Radio)
15. F (second)
16. F (eleven)
17. T
18. F (increases)

Word choice

1. more
2. colors
3. longer
4. photosphere
5. redder
6. decreases
7. cooler
8. electromagnetic
9. hydrogen
10. reflecting
11. greater
12. average
13. straight; 300,000
14. Violet; gamma
15. cooler
16. good
17. chromosphere
18. smaller

Written questions

1. The source of the Sun's energy is nuclear fusion. The nuclear reaction, called the proton-proton reaction, converts four hydrogen nuclei into the nucleus of a helium atom, releasing energy as some of the matter is converted to energy.

2. Spectroscopic analysis of the light from a star can determine the star's (1) composition, (2) temperature, and (3) motion (direction and rate), among other properties.

3. The four parts of the Sun include: (1) the solar interior, where, in the deep interior, nuclear fusion produces the star's energy; (2) photosphere, which is considered to be the Sun's surface and radiates most of the sunlight we see; and the two layers of the solar atmosphere: (3) the chromosphere, the lowermost portion of the solar atmosphere, is a relatively thin layer of hot, incandescent gases; and (4) the corona, the very tenuous outermost portion of the solar atmosphere.

4. A reflecting telescope does not have chromatic aberration (color distortion), it is usually less expensive than a comparable sized refractor because the glass does not have to be of optical quality, and large mirrors can be made because they are opaque and can be supported from behind.

ANSWER KEY

CHAPTER TWENTY-FOUR

Vocabulary Review

1. magnitude
2. irregular galaxy
3. white dwarf
4. Hertzsprung-Russell (H-R) diagram
5. hydrogen burning
6. Degenerate matter
7. Stellar parallax
8. nebula
9. supergiant
10. emission nebula
11. protostar
12. neutron star
13. red giant
14. supernova
15. dark nebula
16. pulsar
17. Hubble's law
18. apparent magnitude
19. elliptical galaxy
20. light-year
21. bright nebula
22. black hole
23. reflection nebula
24. spiral galaxy
25. absolute magnitude
26. interstellar dust
27. main-sequence stars
28. planetary nebula
29. big bang
30. Local Group
31. pulsating variables
32. galactic cluster
33. nova

Comprehensive Review

1. The nearest stars have the largest parallax angles, while those of distant stars are too slight to measure.
2. The difference between a star's apparent magnitude and its absolute magnitude is directly related to its distance.
3. (1) how big the star is; (2) how hot the star is; (3) how far away the star is
4. a) less than 3000 K; b) between 5000 and 6000 K; c) above 30,000 K
5. Binary stars orbit each other around a common point called the center of mass. If one star is more massive than the other, the center of mass will be located closer to the more massive star. Thus, by determining the sizes of their orbits, a determination of each star's mass can be made.
6. (1) luminosity (brightness); (2) temperature
7. a) E; b) B; c) D; d) A
8. letter C
9. B through C through D represents the main sequence
10. (1) Emission nebulae absorb ultraviolet radiation emitted by an embedded or nearby hot star and reradiate, or emit, this energy as visible light. (2) Reflection nebulae merely reflect the light of nearby stars.
11. Dark nebulae are produced when a dense cloud of interstellar material is not close enough to a bright star to be illuminated.
12. Hydrogen burning occurs when groups of four hydrogen nuclei are fused together into single helium nuclei in the hot (at least 10 million K) cores of stars.
13. a) white dwarfs; b) giants followed by white dwarfs, often with planetary nebulae; c) supernovae followed by either a neutron star or black hole.
14. (line will go from dust and gases to protostar, to main-sequence star, to giant stage, to variable stage, to planetary-nebula stage, to white-dwarf stage, to black-dwarf stage).
15. The large quantities of interstellar matter that lie in our line of sight block a lot of visible light.
16. The Milky Way is a rather large spiral galaxy with at least three distinct spiral arms whose disk is about 100,000 light-years wide and about 10,000 light-years thick at the nucleus.
17. (1) Spiral galaxies are typically disk-shaped with a somewhat greater concentration of stars near their centers. Arms are often seen extending from the central nucleus, giving the galaxy the appearance of a fireworks pinwheel. (2) Elliptical galaxies, the most abundant group, are generally smaller than spiral galaxies, have an ellipsoidal shape that ranges to nearly spherical, and lack spiral arms. (3) Irregular galaxies, about 10 percent of the known galaxies, lack symmetry.
18. spiral galaxy
19. Hubble's law states that galaxies are receding from us at a speed that is proportional to their distance.

20. According to the Big Bang theory, the entire universe was at one time confined to a dense, hot, supermassive ball. Almost 14 billion years ago, a cataclysmic explosion marking the inception of the universe occurred. Eventually the material that was hurled in all directions from the big bang cooled and condensed, forming the stellar systems we now observe fleeing from their birthplace.

Practice Test

Multiple choice

1. b
2. d
3. e
4. b
5. e
6. a
7. c
8. c
9. a
10. d
11. b
12. b
13. a
14. a
15. b
16. e
17. c
18. a
19. a
20. a
21. c
22. c
23. b
24. d
25. e
26. e
27. d
28. d
29. c
30. b
31. b
32. e

True/false

1. F (rapidly)
2. T
3. T
4. F (distance)
5. F (Apparent)
6. T
7. T
8. T
9. T
10. T
11. T
12. T
13. F (Bright)
14. F (distance)
15. T
16. T
17. T
18. F (light-years)
19. F (Elliptical)
20. F (brighter)
21. F (high)
22. F (binary)
23. T

Word choice

1. are
2. protostar
3. X-rays
4. more
5. 1987
6. disk
7. is not
8. red
9. supernovae
10. –26.7; 5
11. absolute magnitude
12. supernova
13. atmospheres
14. contracting
15. 100
16. nebulae
17. masses
18. beginning; universe
19. apparent
20. quickly

Written questions

1. Following the protostar phase, the star will become a main-sequence star. After about 10 billion years (90 percent of its life) the star will have depleted most of the hydrogen fuel in its core. The star then becomes a giant. After about a billion years the giant will collapse into an Earth-size body of great density called a white dwarf. Eventually the white dwarf becomes cooler and dimmer as it continually radiates its remaining thermal energy into space.

2. (1) A low-mass star never becomes a red giant. It will remain a main-sequence star until it consumes its fuel, collapses, and becomes a white dwarf. (2) A medium-mass star becomes a red giant. When its hydrogen and helium fuel is exhausted, the giant collapses and becomes a white dwarf. During the collapse, a medium-mass star may cast off its outer atmosphere and produce a planetary nebula. (3) A massive star terminates in a brilliant explosion called a supernova. The two possible products of a supernova event are a neutron star or a black hole.

3. According to the Big Bang theory, the entire universe was at one time confined to a dense, hot, supermassive ball. About 14 billion years ago a cataclysmic explosion marking the inception of the universe occurred. Eventually the material that was hurled in all directions from the big bang cooled and condensed, forming the stellar systems we now observe fleeing from their birthplace. One of the main lines of evidence in support of the Big Bang theory is the fact that galaxies are moving away from each other as indicated by the red shifts in their spectrums.

Photo Credits

Chapter 3

Page 28: Figure 3.2 A/ E.J. Tarbuck
Page 28: Figure 3.2 B/ E.J. Tarbuck
Page 28: Figure 3.2 C/ E.J. Tarbuck
Page 28: Figure 3.2 D/American Geological Institute AGI
Page 30: Figure 3.3B/E.J. Tarbuck
Page 30: Figure 3.3C/E.J. Tarbuck
Page 31: Figure 3.4/E.J. Tarbuck

Chapter 4

Page 40: Figure 4.1/E.J. Tarbuck

Chapter 11

Page 121: Figure 11.3A/E.J. Tarbuck
Page 121: Figure 11.3B/E.J. Tarbuck

Chapter 17

Page 188: Figure 17.2A/E.J. Tarbuck
Page 188: Figure 17.2B/E.J. Tarbuck

Chapter 22

Page 242: Figure 22.1/Photo ©UC Regents/Lick Observatory
Page 244: Figure 22.2/Peoria Astronomical Society, photograph by Eric Clifton and Craig Neaveill

Chapter 24

Page 268: Figure 24.3/U.S. Naval Observatory